HIGH-VOLTAGE DIRECT-CURRENT TRANSMISSION

T0335516

HIGH-VOLTAGE DIRECT-CURRENT TRANSMISSION
CONVERTERS, SYSTEMS AND DC GRIDS

Dragan Jovcic and Khaled Ahmed

School of Engineering, University of Aberdeen, UK

This edition first published 2015
© 2015 John Wiley & Sons, Ltd

Registered Office
John Wiley & Sons, Ltd, The Atrium, Southern Gate, Chichester, West Sussex, PO19 8SQ, United Kingdom

For details of our global editorial offices, for customer services and for information about how to apply for permission to reuse the copyright material in this book please see our website at www.wiley.com.

Library of Congress Cataloging-in-Publication Data

Jovcic, Dragan.
High-voltage direct-current transmission : converters, systems and DC grids / Dragan Jovcic, Khaled Ahmed,
School of Engineering, University of Aberdeen, Scotland, UK.
    pages   cm
  Includes bibliographical references and index.
  ISBN 978-1-118-84666-7 (cloth)
1. Electric power distribution–Direct current.   2. Electric power distribution–High tension.   3. Electric current converters.
I. Ahmed, Khaled.   II. Title.
  TK3111.J68 2015
  621.319′12–dc23
                                                2015011296

A catalogue record for this book is available from the British Library.

Set in 9.5/11.5pt Times by SPi Global, Pondicherry, India

1   2015

Contents

Preface

At the time of writing, there are over 170 high-voltage direct-current (HVDC) links installed worldwide. The largest installations operate at ±800 kV DC voltage and the highest DC current ratings are over 4500 A. Although alternating current was the predominant method for transmitting electrical energy in the twentieth century, HVDC was demonstrated to be the best solution for many specific application areas and the number of installations per year has been constantly increasing at the beginning of twenty-first century. Despite significant converter-station costs, HVDC is techno-economically preferred in general applications for:

- long-distance, large-scale power transfer;
- subsea and long-distance cable-power transmission;
- interconnecting asynchronous AC systems or systems with different frequencies;
- controllable power transfer between different nodes in an electricity market or markets;
- AC grid-stability support, ancillary service provision and resilience to blackouts;
- connecting isolated systems like offshore wind farms or oil platforms.

DC transmission technology was used in many instances in very early power systems but modern HVDC transmission begins with the 1954 Sweden–Gotland installation. This system and all the other HVDCs commissioned until the mid-1970s were based on mercury arc valves. A significant technical advance came with the introduction of solid-state valves (thyristors), although they only support the line-commutated converter (LCC) concept. In the first decade of the twenty-first century there has been very rapid development of fundamentally new technologies and an increasing demand for HVDC technology. The introduction of voltage-source converters (VSCs) requires new valves, which use insulated-gate bipolar transistors (IGBTs) and also new protection and control approaches. The modular multilevel converters have eventually emerged as the most cost effective VSC converter concept, which practically eliminates filtering needs with HVDC and removes voltage limits with VSC valves.

In the second decade of the twenty-first century it has become apparent that DC transmission grids are a technically feasible and viable solution to large-scale energy challenges. The primary application drivers come from initiatives like the North Sea DC grid, Medtech, Desertec, the European overlay super grid and Atlantic Wind. It is accepted that the DC transmission grids must have levels of reliability and technical performance that are similar to or better than an AC transmission system. This level of performance, security and reliability is technically feasible, although, in many aspects, DC grids will be substantially different from traditional AC systems. The development of DC grids brings significant technical advances in HVDC technologies, in particular related to DC circuit breakers (CBs), DC/DC converters and DC protection systems, and substantial further research and development are anticipated.

Nowadays, HVDC and DC grids are associated with green energy, as facilitators of large-scale renewable energy plants. This helps with public acceptance and image, and facilitates further investments in large public projects. HVDC is perceived as the technology that avoids pylons by using long underground cables, further strengthening arguments for future funding decisions.

The timing of this book is therefore in step with an increased interest in HVDC and a projected significant increase in its use.

The book is organized in three parts in order to study all three major HVDC concepts – line commutated HVDC, VSC HVDC and DC grids current research developments. Each part will review theoretical concepts and analyse aspects of technology, interaction with AC grids, modelling, control, faults and protection, with particular emphasis on practical implementation aspects and on reported operational issues.

The technical field of HVDC transmission and DC grids straddles three major traditional electrical engineering disciplines:

- *Power transmission engineering.* The impact of HVDC systems on the connecting AC transmission systems and the national grid is of primary importance. The influence of AC systems on HVDC is also of significance in terms of technical performance, stability, protection and power transfer security in general. Harmonic interaction will be studied in some depth.
- *Power electronics.* Each HVDC link involves at least two AC/DC converters whereas DC grids will have many more, including semiconductor DC CBs and DC/DC converters. These converters have features that are similar to those of traditional low-power converters but many other unique requirements exist to develop valves and converter assemblies capable of sustaining up to 800 kV and perhaps over 4500 A. The protection of valves and converters is very important and is a defining power electronics feature in HVDC.
- *Control engineering.* Modelling and simulation of HVDC is essential for design and operation and several different modelling approaches exist, depending on the model application. In particular, because of the high costs of HVDC testing and the consequences of any design issues, model accuracy and simulation speed play crucial role in the system design. The control systems for HVDC have evolved into very complex technologies, which are always multivariable, nonlinear and with multiple control layers.

The above three technical disciplines will be employed in this book in order to analyse all essential technical aspects of HVDC and DC grids which is aimed to facilitate learning by researchers and engineers who are interested in this field.

The material in this book includes contributions from many HVDC researchers and engineers and it is developed from research projects funded by several research councils and private firms. More importantly, the studies are inspired by and build on previous work by numerous great HVDC engineers.

The authors are particularly grateful to ALSTOM Grid, UK, for providing their comprehensive report, *HVDC: Connecting to the Future,* as well as to SIEMENS, Germany and ABB, Sweden, for their HVDC photographs. We are also indebted to the researchers at the University of Aberdeen Power Systems Group and, in particular, to Dr Weixing Lin, Dr Ali Jamshidifar, Dr Masood Hajian, Dr Huibin Zhang and Dr Lu Zhang for their contributions.

We would like to give our special thanks to SSE, Scotland, and in particular to Andrew Robertson, for their support for the HVDC course at University of Aberdeen, which provided important material for this book.

The authors are also grateful to the following organizations, which supported related research studies at the University of Aberdeen:

- Engineering and Physical Sciences Research Council (EPSRC) UK;
- European Research Council (ERC), FP 7 Ideas Programme;
- Réseau de Transport d'Électricité, (RTE), France.

Dragan Jovcic and Khaled Ahmed

Part I

HVDC with Current Source Converters

1

Introduction to Line-Commutated HVDC

1.1 HVDC Applications

Thyristor-based high-voltage direct-current (HVDC) transmission has been used in over 150 point-to-point installations worldwide. In each case it has proven to be technologically and/or economically superior to AC transmission. Typical HVDC applications can be grouped as follows:

- *Submarine power transmission.* The AC cables have large capacitance and for cables over 40–70 km the reactive power circulation is unacceptable. This distance can be extended somewhat with reactive power compensation. For larger distances, HVDC is more economical. A good example is the 580 km, 700 MW, ±450 kV NorNed HVDC between Norway and the Netherlands.
- *Long-distance overhead lines.* Long AC lines require variable reactive power compensation. Typically 600–800 km is the breakeven distance and, for larger distances, HVDC is more economical. A good example is the 1360 km, 3.1 GW, ±500 kV Pacific DC intertie along the west coast of the United States.
- *Interconnecting two AC networks of different frequencies.* A good example is the 500 MW, ±79 kV back-to-back Melo HVDC between Uruguay and Brazil. The Uruguay system operates at 50 Hz whereas Brazil's national grid runs at 60 Hz.
- *Interconnecting two unsynchronized AC grids.* If phase difference between two AC systems is large, they cannot be directly connected. A typical example is the 150 MW, ±42 kV McNeill back-to-back HVDC link between Alberta and Saskatchewan interconnecting asynchronous eastern and western American systems.
- *Controllable power exchange between two AC networks (for trading).* The AC power flow is determined by the line impedances and it cannot therefore be controlled directly in each line. In complex AC networks it is common to observe loop power flow or even overloading or underutilization of some AC lines. Many HVDC systems participate directly in trading power and one typical example is the 200 MW, ±57 kV Highgate HVDC between Quebec and Vermont.

There are other less common applications of LCC (line-commutated converter) HVDC technology, including the 300 MW Levis De-Icer HVDC project. Here, one standard HVDC converter station – a

High-Voltage Direct-Current Transmission: Converters, Systems and DC Grids, First Edition.
Dragan Jovcic and Khaled Ahmed.
© 2015 John Wiley & Sons, Ltd. Published 2015 by John Wiley & Sons, Ltd.

converter from a Static Var Compensator (SVC) – is used to provide a very high DC current of up to 7920 A (feeding essentially a DC short circuit) to enable heating of remote Canadian overhead lines in order to prevent ice buildup.

An important argument for selecting HVDC instead of an AC for a new transmission line is the contribution to the short-circuit level. High-voltage direct current is able to limit the fault current and therefore it will not require the upgrading of substation equipment.

Figure 1.1 shows a comparison of costs for DC and AC transmission lines. In the case of HVDC the initial capital investment is much higher because of the converter costs. As the transmission distance increases, the benefits of DC offset the capital investment and at certain distance the total cost of an HVDC system is same as an AC line. The breakeven distance is in the range of 40–70 km for submarine cables and in the range of 600–800 km for overhead lines. Figure 1.2 shows an aerial view of the

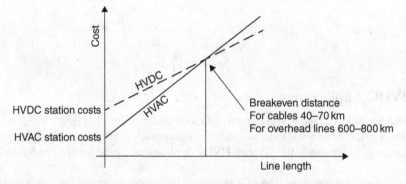

Figure 1.1   HVDC and HVAC transmission cost comparison.

Figure 1.2   Terminal station of Moyle HVDC interconnector (Bipole 2 × 250 MW, ±250 kV, with light triggered thyristors, commissioned in 2001). Reproduced with permission of Siemens.

terminal station of the 500 MW Moyle HVDC link. This HVDC enables a controllable bidirectional power exchange between Scotland and Northern Ireland.

1.2   Line-Commutated HVDC Components

Figure 1.3 shows a typical LCC HVDC schematic interconnecting AC systems 1 and 2. It consists of two terminals and a DC line between them. Each terminal (converter station) includes converters, transformers, filters, reactive power equipment, control station and a range of other components. There are two DC lines in this figure while one line is at ground potential.

As shown in Figure 1.3, the major components of an HVDC system include:

- *Converters.* They typically include one or more six-pulse thyristor (Graetz) bridges. Each bridge consists of six thyristor valves, which in turn contain hundreds of individual thyristors. With large systems, bridges are connected in series in 12-pulse or 24-pulse configuration. The 12-pulse converters can be connected into poles or bipoles.
- *Converter transformers.* These are a special converter transformer type, which is somewhat more expensive than typical AC transformers of the same rating. The converter transformers are designed to operate with high harmonic currents and they are designed to withstand AC and DC voltage stress. In most cases converter transformers will have tap changers, which enable optimization of HVDC operation.
- *Smoothing reactors on DC side.* Typical inductance for large HVDC systems is 0.1–0.5 H, which is determined considering DC fault responses, commutation failure and dynamic stability. The reactors are of air-core, natural air-cooling type and costs are modest.
- *Reactive power compensation.* The converters typically require reactive power of around 60% of the converter power rating. A large portion of this reactive power is supplied with filter banks and the remaining part with capacitor banks. Reactive power demand varies with DC power level, so the capacitors are arranged in switchable banks.
- *Filters.* A typical 12-pulse thyristor terminal will require 11th, 13th, 23rd and 25th filters on the AC side. A high-pass filter is frequently included. In some cases third harmonic filters are required. Some HVDC systems with overhead lines also employ DC-side filters.

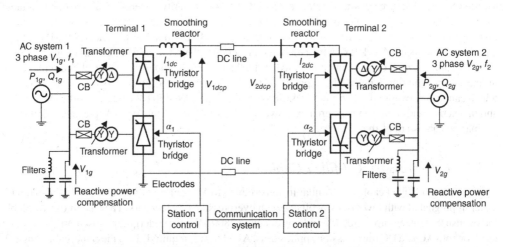

Figure 1.3   Typical HVDC schematic (12-pulse monopole with metallic return).

- *Electrodes.* Some old HVDC systems normally operate with sea/ground return but most grid operators no longer allow permanent ground currents for environmental reasons. Electrodes demand ongoing maintenance costs. Many new bipolar systems are allowed to operate with ground return at half power for a short time (10–20 minutes) in case of loss of an HVDC pole. This implies that electrodes are designed for full current, but carry no current in normal operation.
- *Control and communication system.* Each terminal will have a control system consisting of several hierarchical layers. A dedicated communication link between terminals is needed but speed is not critical. An HVDC link can operate in the event of a loss of a communication link.

1.3 DC Cables and Overhead Lines

1.3.1 Introduction

Line-commutated converter HVDC has been implemented using overhead lines and underground/subsea DC cables. Overhead lines are vulnerable to lightning strikes, which are essentially DC faults. Nevertheless DC faults only cause transient disturbances and they are readily managed by LCC HVDC. On the other hand, with voltage source converter (VSC) HVDC, as will be discussed later, DC faults cause much more serious disturbances.

The most common cable technologies that have been developed so far include:

- mass-impregnated (MI) cables;
- low-pressure oil-filled (LPOF) cables;
- extruded cross-linked polyethylene (XLPE) cables.

The above cable types have same conductors and their construction is similar but the insulation material is substantially different. The cable voltage rating depends on the capability of the insulation (dielectric) material, and there are two main types of dielectrics, namely lapped and extruded.

1.3.2 Mass-impregnated (MI) Cables

Since 1895, MI cables have been used in power transmission. In MI cables, the dielectric is lapped paper insulation, which is impregnated with high-viscosity fluid. For bulk power transmission, mass impregnated cables still prove to be the most suitable solution because of their capacity to work up to 500 kV DC. These cables also tolerate fast DC voltage polarity reversal, making them suitable for LCC HVDC. The MI cables have a long record of field operation at voltages of 500 kV and transmission capacity of over 800 MW (1.6 kA) for monopole HVDC but 600 kV and 1000 MW ratings have been announced. An HVDC with a bipolar connection is therefore able to transmit up to 2000 MW with MI cables. These cables can be installed at depths to 1000 m under the sea level and with nearly unlimited transmission length. The capacity of this system is limited by the conductor temperature, which can reduce overload capabilities. The 580 km-long 700 MW, 450 kV cable link between Norway and the Netherlands represents the greatest power and length for this cable type. At present over 90% of submarine cables are of the MI type.

1.3.3 Low-pressure Oil-filled Cables

Low-pressure oil-filled cables are similar in construction to MI cables but the cables are insulated with paper impregnated with low viscosity oil under an overpressure of a few bars. The technology available today ensures voltages up to 500 kV and powers up to 2800 MW for underground installation. It can be used for both AC and DC transmission applications. As oil flow is required along the cable, cable length is limited to around 80 km. The risk of oil leakage must be taken into account for environmental reasons.

Table 1.1 DC cables types for underground and submarine application.

Type	Mass impregnated	Oil filled	XLPE
Conductor	Cu/Al	Cu/Al	Cu/Al
Insulation	Paper and mass	Paper and fluid	Cross-linked PE
Voltage (kV)	600	500	320 *(525 kV is available)*
Capacity per cable (MW)	1000	2800	1000
Converter type	LCC or VSC	LCC or VSC	VSC or unidirectional LCC
Distance	Unlimited	Limited because of oil	Unlimited

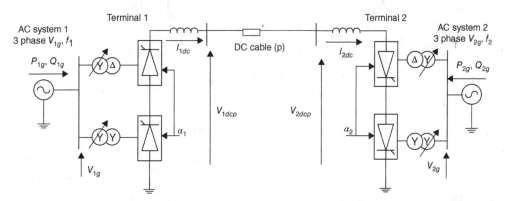

Figure 1.4 Twelve-pulse monopolar HVDC with ground return.

1.3.4 Extruded Cross-linked Polyethylene (XLPE) Cables

Extruded cross-linked polyethylene cables cannot withstand fast polarity reversal and they are not normally used with LCC HVDC (unless it is a unidirectional system). They will be discussed further with VSC HVDC.

The above three types of cables are used for both underground and submarine cables and their basic properties are shown in Table 1.1. The difference between the underground and submarine cables is in the conductor material and the armour layer. Armour strengthening is used in submarine cables to withstand the axial mechanical tension during laying and operation.

Cables with copper conductors are used for submarine applications whereas aluminium conductors are generally preferred for underground. Copper has high electrical conductivity and mechanical properties. It is also simpler to implement strong joints using copper. However it is heavy and more expensive and for these reasons it is used when the mechanical properties are mandatory, as in submarine cables. Aluminium has low conductivity and low mechanical properties. Splicing is more difficult. It is lighter and less expensive than copper.

1.4 LCC HVDC Topologies

High-voltage direct-current systems are divided into transmission systems and back-to-back HVDC. High-voltage direct-current transmission can be bipolar or monopolar.

Monopolar HVDC is typically used for smaller systems and the topology is shown in Figure 1.4. Typically, positive DC voltage is adopted because of less corona issues. The return current can run through ground or a dedicated cable can be employed. If a return cable is used (metallic return) it will

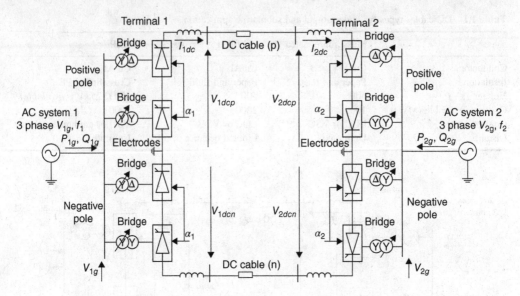

Figure 1.5   Bipolar HVDC (12-pulse) with ground return.

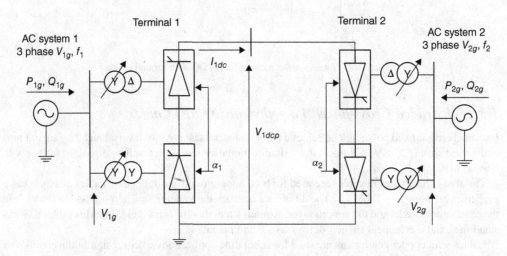

Figure 1.6   Back-to-back HVDC topology.

be at ground potential with low insulation level (typically around 10 kV) and costs are therefore lower than positive-pole DC cable. A 12-pulse topology is shown with two six-pulse converters in series.

Figure 1.5 shows a bipolar HVDC. Bipolar HVDC has two independent poles and it can operate at half power if one DC cable or pole is out of service. Normally the poles are balanced and there is no ground current but ground return would be used if one pole is out of service. In modern grid codes, ground current would not be allowed because of environmental concerns. In some national standards ground currents are allowed only for short periods of time in emergency situations (e.g. secondary reserve startup for 10–20 minutes). Instead of ground return a third cable or DC cable from the faulted pole can sometimes be used.

Figure 1.6 shows a back-to-back HVDC, which is frequently monopolar. In this topology both converter terminals are located in a single station and DC cables are very short. The main purpose of back-to-back HVDC is to provide controllable power transfer between two asynchronous AC systems or AC systems with different frequency. As DC cables are very short and therefore transmission losses are low, back-to-back HVDC are designed at low voltage (as high current as possible) in order to reduce costs (costs are proportional to insulation level). The smoothing reactors are very small or not required because there is a low probability of DC line faults. Back-to-back HVDC allows for operation with variable DC voltage and this facilitates some limited reactive power control capability.

1.5 Losses in LCC HVDC Systems

The losses in HVDC systems will include converter station losses and DC cable losses. Figure 1.7 shows the main components of typical HVDC station losses. The total LCC HVDC station losses will depend on the size of HVDC station, the voltage level, configuration and typically may amount to 0.5–1% of the power transfer.

At partial loading the percentage losses will generally increase. Figure 1.8 shows the load dependence of major loss components. As an example, magnetizing current in converter transformers will be constant irrespective of loading and at 10% loading the transformer losses are 20%.

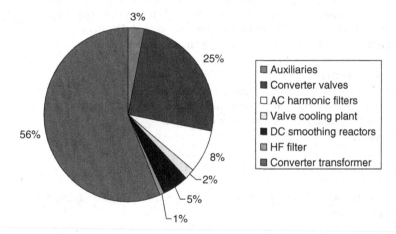

Figure 1.7 Breakdown of typical LCC HVDC station losses at 1 pu power.

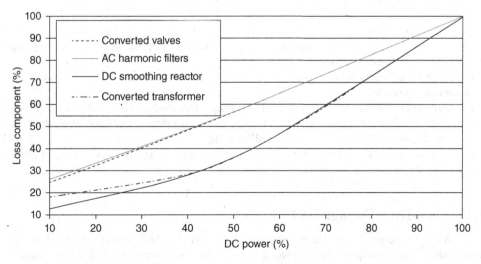

Figure 1.8 Variation of HVDC station losses with the DC power, shown relative to 1 pu losses.

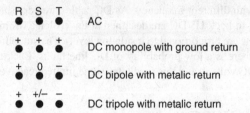

Figure 1.9 Options for conversion of three-phase AC lines into DC.

1.6   Conversion of AC Lines to DC

There have been many studies worldwide on converting existing AC lines into DC. This mainly results from the desire to increase AC line capacity or to remove stability constraints. These issues usually require costly line upgrades/reconductoring, series compensation or installing a device from the flexible AC transmission systems (FACTS) family. In such cases, conversion to HVDC can usually offer the highest capacity increase and a range of other benefits. Typically towers and conductors will not be changed but insulators may need to be upgraded to operate with DC lines.

The main advantages of converting existing AC line to HVDC are:

* an increase in capacity;
* fewer corona issues and a generally higher operating voltage;
* better control of active and reactive power and other system-level benefits;
* better stability limits and active stabilization of the grid;
* lower transmission losses.

Some of the disadvantages of conversion to HVDC include:

* more pollution is attracted to insulators energized with DC – insulator upgrade is recommended;
* converter station costs.

Figure 1.9 shows some common options for converting a single-circuit three-phase AC transmission into DC which include:

* The first option employs all three conductors for a single DC pole while the ground is used for return. This method will significantly increase current carrying capacity but ground return will not be allowed in many modern systems.
* The second option adopts DC bipole with metallic return. The neutral conductor can be used for monopolar operation.
* The third option is based on the tripole HVDC concept. This method uses the third conductor alternatively as a positive or negative pole, which exploits the long thermal constants of conductors. The capacity increase of around 37% is achieved (over bipole configuration) using lines and the RMS values of current in the conductors (over 10 minutes) are equal to the conductor rating. An additional bidirectional converter is required.

1.7   Ultra-High Voltage HVDC

The standard DC voltage for HVDC is 500 kV and the Itaipu 3150 MW, ±600 kV HVDC has used the highest DC voltage for a long period. However the emerging requirements for bulk power transmission over long distances of 5–10 GW in Asia, Africa and South America in late 1990s have resulted in the progressive development of UHVDC (ultra-high voltage direct current).

Xiangjiaba–Shanghai 6400 MW, ±800 kV UHVDC, implemented in 2010, was the first commercial UHVDC, and four other ±800 kV systems have been implemented in 2011–2013, while studies are underway for 1100 kV DC voltages. The progress towards UHVDC has demanded a lot of research and development effort and the main challenges are summarized below:

- improving insulation, in particular in polluted areas;
- transformer development, including bushings;
- developing ultra-high voltage (UHV) test centres.

It is important to appreciate that all the equipment, including auxiliaries that connect to DC lines, must be changed to UHV. In practice this translates to longer units – bushings, arresters, VT (voltage transducers), CT (current transducers), and so forth – with more series-connected basic elements. Frequently, the main challenge is the need for mechanical strength in the face of increased forces from seismic requirements, wind and other factors.

The use of new insulating materials and corona shields becomes a standard method of increasing insulation levels, although developing UHV insulators and bushings remains challenging.

The UHV valve design is not considered to be a significant obstacle.

2

Thyristors

2.1 Operating Characteristics

The thyristor is an essential component in high-voltage direct-current (HVDC) valves and it is still one of the most common devices used in power-switching applications in all industries. This is attributed to its high power ratings, robustness and high efficiency. Single devices have up to 8500 V, 4500 A capability, they are built on single wafers of up to 150 mm in diameter and have been in existence since the 1950s.

The thyristor is a four-layer, three-terminal device as shown in Figure 2.1. The three connections are A-anode, K-cathode and G-gate. When gate current is applied, the layer between J2 and J3 becomes negative (N) and the thyristor becomes a PN device similar to a diode, also shown in Figure 2.1. Functionally, it is similar to a diode but the start of conduction can be delayed using the gate circuit.

A thyristor can be considered as a controllable diode, as shown in operating curves in Figure 2.2. With no gate current $i_g = 0$ it behaves like an open circuit (OFF state) both in forward and reverse directions. A forward voltage across the device (A positive with respect to K) results in junctions J1 and J3 being forward biased, whereas J2 is reverse biased, and therefore only a small leakage current flows. If V_{AK} is increased to a critical limit, the device switches suddenly to a conducting state as the result of breakdown or breakover of J2. If a gate current i_g is applied then the magnitude of V_{AK} needed for breakover is dramatically reduced and the device behaves like a diode. The level of i_g required is small compared to the main power current. The current I_l is called the latching current, which is the anode current required to ensure thyristor switches to the ON state. Once the anode current reaches I_l, the gate current can be removed. The gate current is therefore a short pulse of 10–50 μs. Theoretically, gate pulse is required once per half cycle but, in practice, gate pulses are sent multiple times per half cycle to ensure firing under all operating conditions.

Once the device is conducting, i_g can be reduced and the device remains in the ON state. When the device is in conduction, its state is determined solely by the anode current. If the anode current I_A falls below some critical value, the holding current I_h (typically few milliamperes), the device switches off reverting to the blocking OFF state.

High-Voltage Direct-Current Transmission: Converters, Systems and DC Grids, First Edition.
Dragan Jovcic and Khaled Ahmed.
© 2015 John Wiley & Sons, Ltd. Published 2015 by John Wiley & Sons, Ltd.

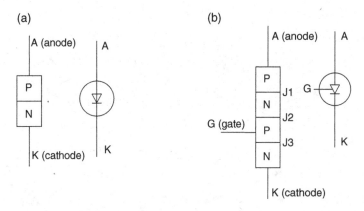

Figure 2.1 Structure and symbol for (a) diode and (b) thyristor.

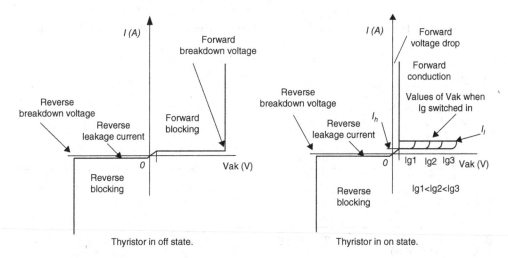

Figure 2.2 Thyristor operating curves.

If a reverse voltage is applied across the device, (*negative V_{AK}*), J1 and J3 become reverse biased, only J2 is forward biased and therefore only a small leakage current flows. If negative V_{AK} is increased sufficiently, then eventually avalanche breakdown occurs across J1 and J3 resulting in damage to the device unless steps are taken to limit the current. The reverse breakdown may not be destructive. The forward and reverse blocking capability are similar for a given thyristor and they have good temperature stability for typical operating temperatures below 125 °C. However, forward-blocking capability deteriorates very fast with temperatures above 125 °C.

Figure 2.3 illustrates the design of high-power press-pack thyristors.

2.2 Switching Characteristic

A typical switching characteristic for an operating cycle of a thyristor is shown in Figure 2.4. The top graph shows the gating circuit current and the bottom graph shows the anode current and V_{AK} voltage. If a device is forward biased (V_{AK} positive) and a gate-current pulse is applied, the device switches on.

Figure 2.3 High-power thyristors of press-pack design. Reproduced with permission from ABB.

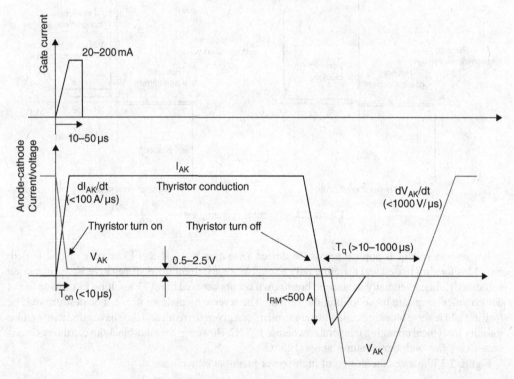

Figure 2.4 Thyristor switching characteristic.

Once a thyristor is in conduction, the gate has no control over the device. The device conducts even if the gate pulse is now turned off. There is a delay while the device switches on, which is termed the *on time*, t_{on}. During the time t_{on}, which is in the order of few microseconds, the voltage across thyristor reduces and the current increases. The rate of current rise at turn on should be limited (to around

100 A/μs), in order to allow current spreading across the entire PN junction surface. If the current rise is too fast the thyristor can be destroyed because of local thermal melting. For a large thyristor it may take around 1 ms for current to spread across whole surface area, in which interval the conduction loss is high.

In the conduction state, a typical voltage drop across a large thyristor is 1.5–2.5 V.

The device turns off when the anode current reduces to zero, which is driven by the external circuit in which thyristor is connected. There are two possible turn-off conditions:

- The current can fall to zero naturally, as would be the case in some resonant converters or with pulse power applications. The current can also naturally fall to zero in discontinuous converter mode, as for example in case of HVDC converter operation with very low DC current. Thyristor turn off during normal conduction interval is not desirable and this is prevented by sending repeated gate pulses.
- The thyristor current can fall to zero if another thyristor in the converter is fired and consequently the load current commutates to the other thyristor. This is a common commutation with HVDC converters.

On turning off, a thyristor is reverse biased in the converter circuit and it can immediately withstand full reverse blocking voltage. However, a thyristor cannot immediately withstand forward blocking voltage. After the current falls to zero it is necessary to keep the device reverse biased for a short period of time in order to allow full recombination of charge carriers on the PN junction. After this period, the thyristor is able to gain forward-blocking capability, as required in the next cycle. The minimum reverse bias time after current falls to zero is called the extinction time t_q. The extinction time is typically 10–50 μs for small thyristors but for those used with HVDC it is 300–1000 μs. If this condition is violated (a forward voltage is reapplied immediately after I_A goes to zero) the device will switch to ON state even without a gate pulse. This unwanted turn on can be destructive for a thyristor. A special firing logic will normally intentionally fire thyristors if such conditions are detected.

The rate of forward-blocking voltage increase should also be limited (typically to around 1000 V/μs) to prevent unwanted triggering. The PN junction behaves as a capacitor and therefore a sufficiently large dV/dt will generate anode current ($i = C(dV/dt)$), which can cause latching.

Figure 2.5 illustrates thyristor operation in the simplest AC/DC converter with an inductive-resistive load. In this single-phase, half-wave converter thyristor can be fired only in positive half cycles giving

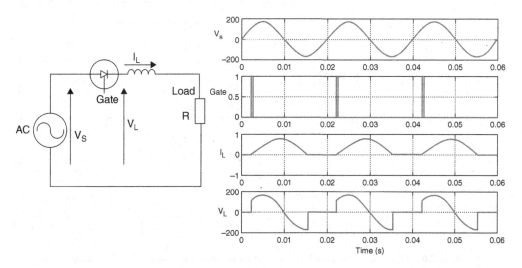

Figure 2.5   Thyristor in a single-phase half-wave converter. Firing angle is 40°.

a crude DC voltage consisting of positive and negative segments. The operation with firing angle of around 40° is shown in this figure. It should be observed that current lags voltage and therefore the thyristor conducts for periods while it is forward biased but also for some interval while it is reverse biased. The thyristor is turned off when anode current naturally falls to zero.

Example 2.1
Study a single-phase half-wave rectifier, with a circuit as in Figure 2.5, and with same $V_s = 120\,V$, $f_s = 50\,Hz$, *but assume a purely resistive load* $R = 50\,\Omega$. *Assume that firing angle is 80°. Sketch the load voltage and compare with Figure 2.5. Explain whether such circuit would be feasible in practice. Calculate the value of required snubber* L_s.

Solution
Figure 2.6 shows the circuit and the waveforms. It can be seen that the load current is in phase with voltage and the load voltage has no negative segments.

The problem with this circuit is that the current derivative at turn-on is very high and in particular at high firing angles. This high di/dt might destroy the thyristor, so a dI/dt protective snubber would be needed.

In order to calculate Ls the current equation is studied:

$$i_L(t) = \frac{V_s}{R}\left(1 - e^{R/L_s t}\right)$$

Assuming that the supply voltage is constant for the duration of switching transient, and considering worst case peak voltage, and therefore $V_s = 169\,V$, *while the current derivative is:*

$$di_L/dt = \frac{V_s}{R}\frac{R}{L_s}e^{R/L_s t},$$

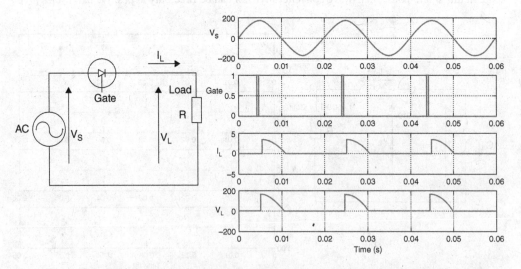

Figure 2.6 Thyristor in a single-phase half-wave converter with resistive load in Example 2.1.

and the initial current derivative for t = 0 is:

$$\mathrm{d}i_L/\mathrm{d}t(t=0) = \frac{V_s}{L_s}$$

Therefore to limit the current derivative to di/dt = 100 A/μs, an inductor of at least $L_s > 1.69\,\mu H$ is needed.

2.3   Losses in HVDC Thyristors

Losses in a semiconductor component occur as a product of the current through the device and the voltage across the device. The losses are dissipated as heat and, in large HVDC converters, the total requirement for heat removal can be significant, approaching several megawatts. High-voltage direct-current converters typically use special liquid cooling systems, which have an effect on losses, costs and system reliability.

The main losses in HVDC converter thyristors include:

- conduction losses;
- turn-off losses;
- snubber losses;
- reverse-leakage current loss;
- forward-leakage current loss;
- gate driver loss.

The principal loss components in HVDC converters include conduction and turn-off losses. Figure 2.7 shows the shape of ON-state curves for a thyristor. The voltage across thyristor can be expressed as:

$$v_T = V_{T0} + R_{on}i_T \qquad (2.1)$$

where V_{T0} is the threshold voltage (at zero current), which is typically $1 < V_{T0} < 1.5$ V depending on thyristor voltage rating; R_{on} is the on-state resistance, which is typically $0.5 < R_{on} < 2$ mΩ, depending on current rating; and i_T is the anode instantaneous current.

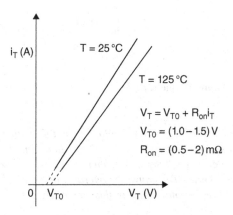

Figure 2.7   Typical on-state characteristic for a high-power thyristor.

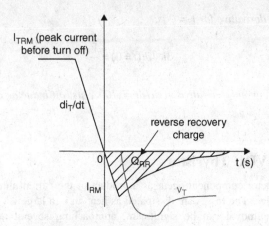

Figure 2.8 Thyristor turning OFF.

The on-state loss power can be determined by integrating the product of the current and the voltage:

$$P_{con} = \frac{1}{T}\int_0^T v_T i_T = \frac{1}{T}\int_0^T \left(V_{T0}i_T dt + R_{on}i_T^2 dt\right) = V_{T0}I_{TM} + R_{on}I_T^2 \tag{2.2}$$

where T is the period, I_{TM} is the average thyristor current, which can be determined by integrating instantaneous current or using the duty ratio δ (i.e. conducting period as a percentage of the full cycle) and I_T is the RMS value of thyristor current. In a six-pulse bridge each thyristor conducts for 120° and therefore $\delta = 0.33$.

A typical turning-off curve of a thyristor is shown in Figure 2.8. The current 'overshoots' to a small negative value to recover charge in the P-N junctions, and the element switches off after a short period of time. The peak reverse current I_{RM} depends on the current falling derivative (di_T/dt) and the peak conducting current I_{TRM}. There are detailed methods to calculate the reverse-recovery loss but simplest method is to use the reverse-recovery charge Q_{RR} or turn-off energy loss E_{off}, which is supplied in the thyristor manufacturer data sheets. The reverse recovery loss power is:

$$P_{off} = \int_0^{t_{off}} i_T v_t dt$$

$$P_{off} = E_{off} f_s \tag{2.3}$$

where f_s is the thyristor switching frequency.

Example 2.2
A six-pulse, 2000 A, 500 kV HVDC converter employs thyristors with the characteristics shown in Figure 2.9. These thyristors have a 6500 V, and 2800 A rating and 170 devices are used in each valve. The blocking voltage across each thyristor is therefore 2941 V. Assume that the series inductor is designed to limit di/dt to 10 A/μs. The converter operates in a typical six-pulse pattern with 120° conducting intervals. Calculate the total losses in this converter.

Solution
From Figure 2.9, $V_{T0} = 1V$, $R_{on} = (1.7 - 1)/2000 = 0.00035\,\Omega$, $E_{on} = 2.2\,Ws/pulse$, $E_{off} = 43\,Ws/pulse$,

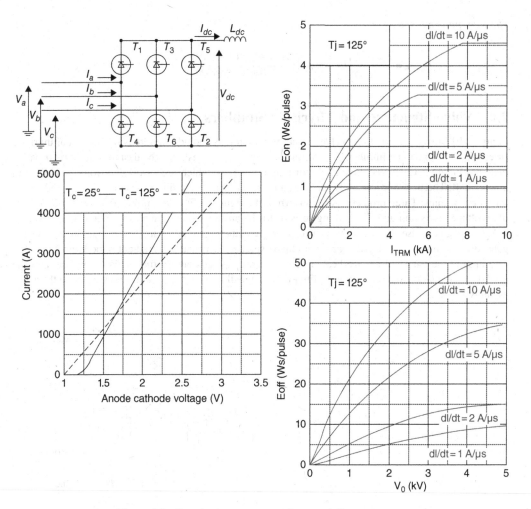

Figure 2.9 Test thyristor on-state and turn on/off energy curves.

The total ON-state loss is:

$$P_{con} = 2n3\left(V_{T0}I_{TM} + R_{on}I_T^2\right) =$$

$$P_{con} = 2 \times 170 \times 3\left(1 \times \frac{2000}{3} + 0.00035 \times \left(\frac{2000}{\sqrt{3}}\right)^2\right) = 1.144\,\text{MW}$$

The turn-on loss is:

$$P_{on} = E_{on}f_s = 6 \times 170 \times 2.2 \times 50 = 0.112\,\text{MW}$$

The turn-off loss is:

$$P_{off} = E_{off}f_s = 6 \times 170 \times 43 \times 50 = 2.19\,\text{MW}$$

The total percentage loss is:

$$P_{loss\%} = \frac{P_{con} + P_{on} + P_{off}}{I_{dc} \times V_{dc}} = \frac{(1.144 + 0.112 + 2.19) \times 10^6}{2000 \times 500000} \times 100 = 0.34\%$$

2.4 Valve Structure and Thyristor Snubbers

Figure 2.10 shows the converter valve structure, which may include hundreds of individual thyristor assemblies, grouped in a number of valve racks. Figure 2.11 illustrates the design of a thyristor valve rack and Figure 2.12 shows valve racks forming six valves, which are suspended from the ceiling in a valve hall. The thyristor assembly includes a thyristor, driver, passive-protection circuits and monitoring electronics. Three protections are shown in the figure: L_s for *di/dt* protection, RC for *dv/dt* and overvoltage protection and R_{dc} grading resistor for balancing voltages across switches in a valve.

When a gate pulse is sent, the thyristor starts conducting initially in the PN region around the gate connection and the conducting area gradually spreads. If the current gradient is too large, the initial area around the gate will be overheated before current spreads across full thyristor surface, and the thyristor may be destroyed. In order to limit *di/dt,* a small inductor is used in series with the thyristor, as shown by L_s.

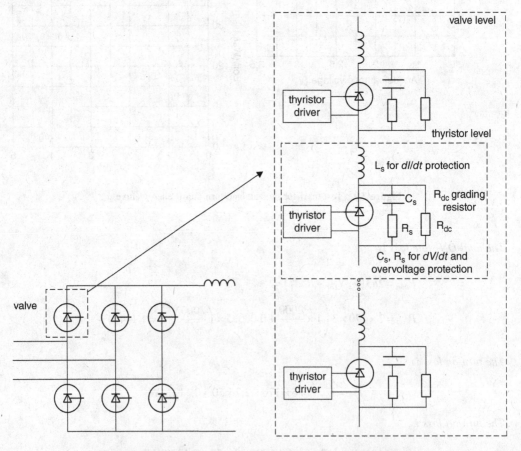

Figure 2.10 Thyristor valve structure and protection for *dv/dt* and *di/dt*.

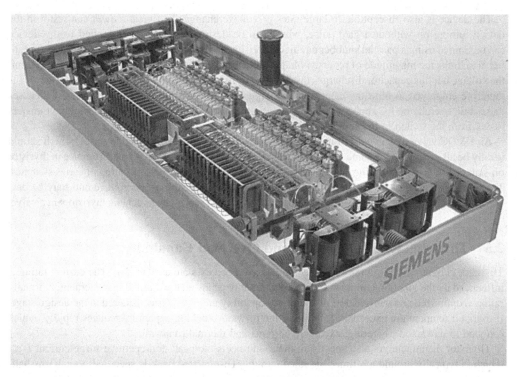

Figure 2.11 Thyristor valve rack assembly. Reproduced with permission of Siemens.

Figure 2.12 Thyristor valve hall at Yunnan-Guangdong HVDC station. Reproduced with permission of Siemens.

The device is also susceptible to large rates of voltage change. Too great a *dv/dt* can result in the device turning on without a gate pulse, which can be destructive. The *dv/dt* applied to the device can be limited using a parallel snubber circuit comprising an RC combination in series. The same snubber also limits the magnitude of reverse voltage at the turn-off instant. A capacitor is sufficient to limit the voltage but this capacitor discharges through the thyristor at the next turn on. A series resistor, R_s, is therefore employed to limit the capacitor discharge current at the next turn on. This resistor causes losses but, overall, the RC snubber reduces losses in the switch, and therefore the RC snubber transfers losses from the switch to the snubber resistor.

An HVDC valve may consist of hundreds of individual thyristors connected in series, which should ideally be all stressed to the same voltage during all operating conditions. A small difference in thyristor on-state resistance or switching speed will cause voltage unbalance across the string of series switches. The thyristor with the slowest switching-on speed will be subjected to overvoltage and may be destroyed. The grading resistors are used to help equalize voltage sharing between the thyristors in a valve.

2.5   Thyristor Rating Selection and Overload Capability

Thyristor current rating is commonly specified as average ON-state current, I_{TM}. The current rating is influenced by the junction temperature and therefore it depends on the thermal management. Normally rating is optimized and there will be no overload capability unless this is considered in the design stage. If thyristor temperature exceeds rated values, the forward-blocking capability reduces rapidly, which can lead to unwanted triggering, excessive currents and thermal runaway.

Thyristor manufacturers also specify in their datasheets the peak nonrepetitive surge current I_{TSM}, which is typically around ten times the rated current. This current peak is specified on a 10 ms half-sine pulse at rated temperature, however it is nonrepetitive. The temperature of the junction will rise significantly during such a high pulse and the thyristor will not be able to withstand further blocking voltage. On detection of overcurrent, the protection system will block the thyristor driver for a period of time until the temperature is sufficiently reduced to resume normal operation. The manufacturers also specify limiting load integral $I^2 t$, which can be used to calculate nonrepetitive peak pulse for a different duration.

The thyristor voltage rating is specified as maximum repetitive peak forward voltage V_{DRM} and maximum repetitive peak reverse voltage V_{RRM}. Large thyristors are typically manufactured as symmetrical components and these two values are therefore identical. Note that the thyristor will be destroyed if V_{DRM} or V_{RRM} are exceeded even for a very short time. Because of voltage variations during normal operation and the presence of harmonics, typical operating voltage stress for a thyristor will be selected at around 50% of the maximum repetitive forward/reverse voltage.

3

Six-Pulse Diode and Thyristor Converter

3.1 Three-Phase Uncontrolled Bridge

A three-phase full bridge (Graetz bridge) diode converter is shown in Figure 3.1. This is the simplest three-phase topology, which will illustrate three-phase AC/DC conversion. This converter operates in the same way as a thyristor converter with zero delay angle.

The AC system is assumed to be symmetrical and balanced and the voltages are defined as:

$$v_a = V\cos(\omega t)$$
$$v_b = V\cos\left(\omega t - \frac{2}{3}\pi\right)$$
$$v_c = V\cos\left(\omega t + \frac{2}{3}\pi\right)$$

(3.1)

where V is the line-neutral peak magnitude voltage. Note that the three switches are connected to the positive DC pole and the remaining three switches to the negative pole where the label numbers correspond to the sequence of conduction. The diodes will start conducting when anode voltage is higher than cathode. Therefore diodes conduct when the respective phase voltages are at highest value, as shown in Figure 3.2, for a test system consisting of diode converter between AC system $V_{LL} = 410\,\text{kV}$ and DC source of 500 kV. Each diode conducts for 1/3 of a cycle (120°). At any time one diode conducts on the positive rail and one on the negative.

The DC-side inductor ensures that DC current stays approximately constant for one pulse, and therefore DC current commutates from one switch to another every 60°. The commutation occurs every 120° on the positive rail and on the negative rail, however commutation instants on the negative rail are lagging by 60°.

The diode bridge average DC voltage can be calculated by averaging the surface below the V_{dc} curve:

$$V_{dc0} = 2\frac{3}{2\pi}\int_{\pi/6}^{\pi/6+2\pi/3} V\cos\omega t\,d(\omega t)$$

(3.2)

High-Voltage Direct-Current Transmission: Converters, Systems and DC Grids, First Edition.
Dragan Jovcic and Khaled Ahmed.
© 2015 John Wiley & Sons, Ltd. Published 2015 by John Wiley & Sons, Ltd.

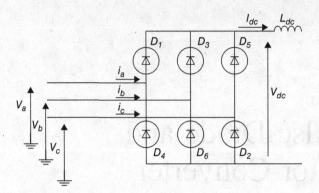

Figure 3.1   Diode six-pulse AC/DC converter.

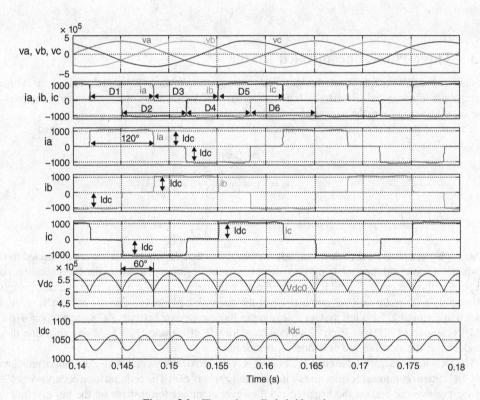

Figure 3.2   Three-phase diode bridge plots.

$$V_{dc0} = \frac{3\sqrt{3}}{\pi}V = \frac{3\sqrt{6}}{\pi}\mathrm{V} = \frac{3\sqrt{2}}{\pi}\mathrm{V_{LL}} \tag{3.3}$$

where V is the line-neutral RMS voltage and $\mathrm{V_{LL}}$ is the line-line RMS voltage. The above DC voltage V_{dc0} is called the ideal no-load voltage. It corresponds to the voltage of a thyristor rectifier with zero firing angle. This is also the maximum DC voltage that a six-pulse thyristor converter can achieve.

The AC current in each phase consists of 120° long squares per each half cycle. The peak magnitude of fundamental component of AC current is obtained using the Fourier series:

$$I = 2\frac{1}{\pi}\int_{\pi/6}^{\pi/6+2\pi/3} I_{dc}\cos(\omega t)d(\omega t)$$

$$I = 2\frac{\sqrt{3}}{\pi}I_{dc}$$

(3.4)

The RMS value of fundamental component of AC current is from Eq. (3.4):

$$I = \frac{\sqrt{6}}{\pi}I_{dc}$$

(3.5)

3.2 Three-Phase Thyristor Rectifier

This section considers three-phase bridge topology as in the previous section, but thyristor delay angle is considered and a commutation overlap (resulting from transformer inductance) is included. Figure 3.3 shows the converter topology.

Figure 3.4 shows the voltage and current waveforms, assuming similar parameters as in Figure 3.2 but thyristors are employed and converter is interfaced using $L_t = 0.1$ H. The operation is similar to that in Figure 3.2, however a firing-delay angle α is introduced. The delay angle is measured from the instant of positive thyristor forward voltage (intersection of two phase voltages), which corresponds to 30° on phase a voltage.

Figure 3.4 also shows the commutation overlap μ, resulting in a DC voltage dip each time the commutation occurs. Neglecting initially the commutation overlap, the average DC voltage can be obtained considering surface below the DC voltage curve:

$$V_{dc} = 2\frac{3}{2\pi}\int_{\pi/6+\alpha}^{\pi/6+\alpha+2\pi/3} V\cos\omega t\, d(\omega t)$$

(3.6)

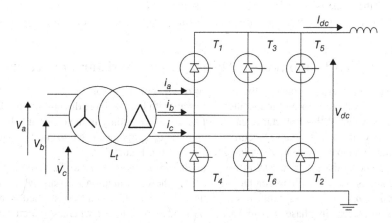

Figure 3.3 Thyristor six-pulse AC/DC converter with a transformer.

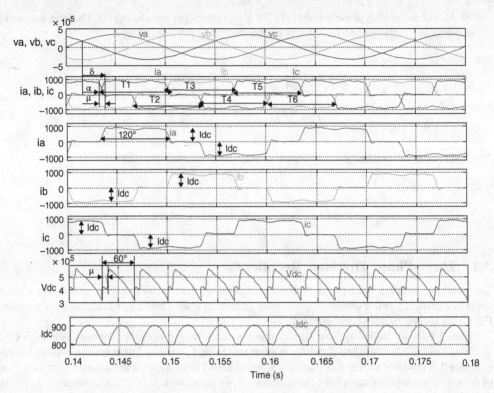

Figure 3.4 Thyristor six-pulse AC/DC converter with transformer ($L_t = 0.1$ H) and ignition delay ($\alpha = 30$).

$$V_{dc} = \frac{3\sqrt{3}}{\pi} V \cos\alpha = \frac{3\sqrt{6}}{\pi} V \cos\alpha = \frac{3\sqrt{2}}{\pi} V_{LL} \cos\alpha \qquad (3.7)$$

The above DC voltage is also expressed as:

$$V_{dc} = V_{dc0} \cos\alpha \qquad (3.8)$$

where V_{dc0} is the diode bridge ideal DC voltage defined in Eq. (3.3). This formula illustrates that the converter DC voltage is controllable through the firing delay angle, but the control gain is nonlinear.

3.3 Analysis of Commutation Overlap in a Thyristor Converter

The commutation overlap occurs in the presence of an inductance on the AC side of the thyristor converters. Typically transformer inductance is present with high-voltage direct-current (HVDC) converters and the value of this inductance can be quite large, commonly in the order of 0.1–0.2 pu. This inductance prevents instantaneous DC current commutation from one switch to another. The result is a commutating overlap, causing a DC voltage dip, as seen in Figure 3.4.

Figure 3.5 shows the electrical circuit for commutation from valve T_1 to T_3, assuming that T_3 has received gate signal while T_1 is conducting. During the commutation overlap, which lasts in the period $\alpha < \omega t < \delta$, three valves conduct simultaneously. The DC current is assumed to be constant. The outgoing current in phase a (and valve T_1) gradually reduces, whereas the current in phase B (and valve T_3) gradually increases, as shown in the time-domain converter variables during the commutation process in Figure 3.6. Table 3.1 shows how the main variables evolve during the commutation

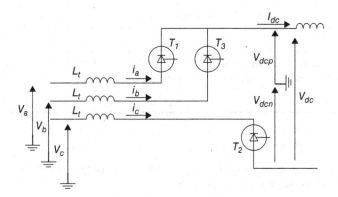

Figure 3.5   Converter equivalent circuit during commutation.

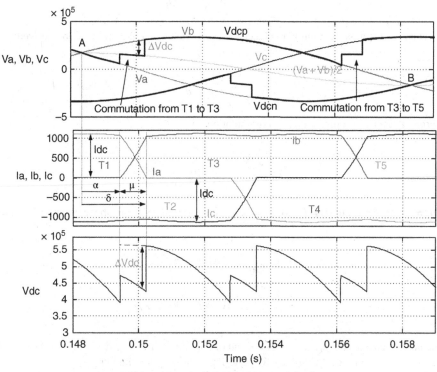

α – Firing delay angle (start of commutation)
δ – Extinction angle (end of commutation)
μ – Commutation overlap angle
ΔVdc – DC voltage drop because of commutation overlap

Figure 3.6   Commutation from valve T_1 to T_3 (phase A to phase B) in rectification mode.

Table 3.1   Variables during the commutation period.

Variable	i_a	i_b	V_{dcp}
$\omega t = \alpha$ (beginning of commutation)	I_{dc}	0	v_a
$\alpha < \omega t < \delta$ (during commutation)	$\frac{\sqrt{6}V_{g\_LL}}{2L_t} = \frac{di_a}{dt}$	$\frac{\sqrt{6}V_{g\_LL}}{2L_t} = \frac{di_b}{dt}$	$(v_a + v_b)/2$
$\omega t = \delta$ (end of commutation)	0	I_{dc}	v_b

process. Note that the converter voltage (V_{g_LL}) is measured on the grid side of the transformer and it is appropriately scaled for the transformer ratio.

The commutation can only happen between points A and B, while $V_b > V_a$. In rectification mode, commutation happens close to point A, whereas in inversion the converter is operated with large firing angles, close to point B. This section considers rectification only.

With reference to the circuit in Figure 3.5, the Kirchhoff's voltage equation along the commutation loop ($\alpha < \omega t < \delta$) is:

$$v_a - v_b = L_t \frac{di_a}{dt} - L_t \frac{di_b}{dt} \tag{3.9}$$

From the above formula, assuming a balanced system, the current equation can be derived:

$$\frac{\sqrt{2}V_{g\_LL}}{2L_t} = \frac{di_a}{dt} \tag{3.10}$$

Integrating, the phase a current during commutation is:

$$i_a = \frac{\sqrt{2}V_{g\_LL}}{2\omega L_t} (\cos\alpha - \cos\omega t) \tag{3.11}$$

Since $i_a = 0$ for $\omega t = \alpha$ and $i_a = I_{dc}$ for $\omega t = \delta$, as seen in Table 3.1, the above equation becomes:

$$I_{dc} = \frac{\sqrt{2}V_{g\_LL}}{2\omega L_t} (\cos\alpha - \cos\delta)$$

$$I_{dc} = \frac{\sqrt{2}V_{g\_LL}}{2\omega L_t} (\cos\alpha - \cos(\alpha + \mu)) \tag{3.12}$$

Equation (3.12) enables the prediction of the end of commutation (angle $\delta = \alpha + \mu$), if the start of commutation (control angle α), AC voltage V_{LL} and DC current I_{dc} are known:

$$\cos(\alpha + \mu) = \cos\alpha - \frac{I_{dc} 2\omega L_t}{\sqrt{2}V_{g\_LL}} \tag{3.13}$$

This analytical result is important and it is used with inverter controllers to predict the extinction angle, as discussed in Chapter 5.

The commutation process has the effect of reducing DC voltage by the value V_{dc_com}, as seen in Figure 3.6. This voltage drop can be calculated considering the surface area below the curve as:

$$V_{dc\_com} = \frac{1}{2\pi/3} \int_{\alpha}^{\delta} \left[v_b - \frac{v_a + v_b}{2} \right] d(\omega t)$$

$$V_{dc\_com} = \frac{V_{dc0}}{2} (\cos\alpha - \cos\delta) \tag{3.14}$$

$$V_{dc\_com} = \frac{3}{\pi} I_{dc} \omega L_t$$

From the above equation, the conclusion is derived that the commutation voltage drop depends on the direct current and has an effect equivalent to a fictitious resistance $R_c = 3\omega L_t/\pi$. Using Eqs (3.8) and (3.14):

$$V_{dc} = V_{dc0} \cos\alpha - \Delta V_{dc}$$

$$V_{dc} = V_{dc0} \frac{\cos\alpha + \cos\delta}{2} \tag{3.15}$$

$$V_{dc} = \frac{3\sqrt{2}}{\pi} V_{g\_LL} \cos\alpha - \frac{3}{\pi} \omega L_t I_{dc}$$

Example 3.1
A six-pulse HVDC converter is connected to 220 kV AC grid using a 220 kV/330 kV, 800 MVA, $X_t = 12\%$ transformer.

- *Determine the commutation overlap if the rectifier is operating at 20° firing angle and an 1800 A DC current.*
- *Determine the commutation overlap if the converter is operating at 15° and with same DC current (tap-changer adjustment).*
- *Discuss how an increase in transformer reactance will affect the commutation overlap.*

Solution
Transformer inductance is:

$$L_t = X_t \frac{V_{cll}^2}{S_t} \frac{1}{2\pi 50} = 0.12 \frac{330\,000^2}{800\,000\,000} \frac{1}{2\pi 50} = 0.052\,H$$

The turn-off angle and commutation overlap are calculated as:

$$\alpha + \mu = a\cos\left[\cos(\alpha) - \frac{I_{dc} 2\omega L_t}{\sqrt{2} V_{cll}} \right]$$

$$\alpha + \mu = a\cos\left[\cos(20) - \frac{1800 \times 2 \times 314.15 \times 0.052}{\sqrt{2} \times 330,000} \right]$$

$$\alpha + \mu = 35.5°$$

$$\mu = 15.5°$$

With firing angle of 15°:

$$\alpha + \mu = 32.9°$$

$$\mu = 17.9°$$

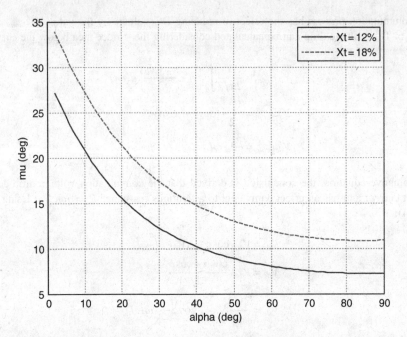

Figure 3.7   Commutation angle as the function of operating angle in Example 3.1.

Using Eq. (3.12), Figure 3.7 shows the commutation angle as the function of the operating angle for two values of transformer inductance. Larger transformer leakage reactance will generate a larger commutation angle.

3.4   Active and Reactive Power in a Three-Phase Thyristor Converter

In order to get accurate expression for converter AC current and the power factor consider the power-balance equation:

$$P_{ac} = P_{dc0}$$
$$3\text{VIcos}(\varphi) = V_{dc}I_{dc} \tag{3.16}$$

where φ is the power factor angle. Using Eqs (3.15) and (3.16):

$$\text{Icos}(\varphi) = \frac{\sqrt{6}}{\pi}I_{dc}\frac{\cos\alpha + \cos\delta}{2} \tag{3.17}$$

The expression for AC current in Eq. (3.5) is obtained for $\mu = 0$, but it is accurate to within 5% even for a very large μ, and therefore it is justifiable to approximate:

$$\text{I} \approx \frac{\sqrt{6}}{\pi}I_{dc} \tag{3.18}$$

Replacing Eq. (3.18) in Eq. (3.17), the following results:

$$\cos(\varphi) \approx \frac{\cos\alpha + \cos\delta}{2} \tag{3.19}$$

or using Eq. (3.15):

$$\cos(\varphi) \approx \frac{V_{dc}}{V_{dc0}} \tag{3.20}$$

The power-factor angle is therefore directly dependent on the firing angle and the overlap angle. It is concluded that the power factor is better if converter operates at low firing angles. For this reason the line-commutated converter (LCC) HVDC controllers optimize operating conditions to minimize a firing angle of around 15–20°, which leaves sufficient room for control action in each direction. If the power factor angle is known, the reactive power can be calculated as:

$$Q = P\tan(\varphi) \tag{3.21}$$

3.5 Inverter Operation

Figure 3.8 shows the converter DC voltage as a function of the firing angle, using Eq. (3.8), and the same test system as in previous sections. As the firing angle increases over 90°, the DC voltage becomes negative and the converter moves into inversion mode. Current cannot change direction in thyristor converters, so negative DC voltage implies power reversal.

When the commutation overlap is considered, the condition for the inversion mode is from Eq. (3.15):

$$\cos\alpha + \cos\delta = 0$$
$$\alpha = \frac{\pi - \mu}{2} \tag{3.22}$$

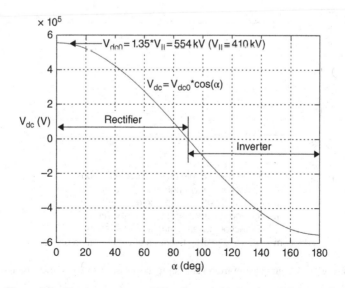

Figure 3.8 Thyristor converter DC voltage as the function of firing angle.

Figure 3.9 shows the plot of converter variables in inversion mode. The firing angle is $90 < \alpha < 180$, and the following angles are commonly used for inversion mode:

- Ignition advance angle $\beta = 180 - \alpha$, and therefore $\cos(\beta) = -\cos(\alpha)$.
- Extinction advance angle $\gamma = 180 - \delta$, $\cos(\gamma) = -\cos(\delta)$. Note that the overlap angle is $\mu = \delta - \alpha = \beta - \gamma$.

The extinction angle is very important for inverter operation because it defines the safe period for thyristor reverse recovery. Referring to the plots in Figure 3.9, thyristor T_1 (connected to phase a) will turn off in the interval between points A and B. After turning off, it must recover the forward blocking state before point B. At point B it will become forward biased. Thyristor T_3 (connected to phase b) firing must also happen before point B, while it is forward biased. Therefore the thyristor T_3 must be fired sufficiently early, because commutation overlap and the reverse recovery time must be completed before point B. The control challenge at the inverter is how to determine firing angle α in order to

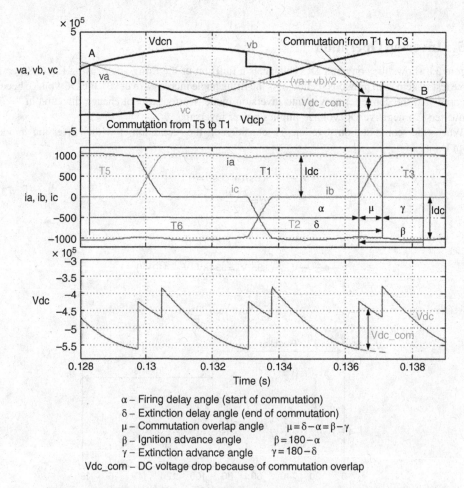

α – Firing delay angle (start of commutation)
δ – Extinction delay angle (end of commutation)
μ – Commutation overlap angle $\mu = \delta - \alpha = \beta - \gamma$
β – Ignition advance angle $\beta = 180 - \alpha$
γ – Extinction advance angle $\gamma = 180 - \delta$
Vdc_com – DC voltage drop because of commutation overlap

Figure 3.9 Commutation from valve T_1 to T_3 (phase A to B) in inversion mode.

achieve a balance between a sufficiently large γ and to avoid large reactive power caused by an excessively large γ. For a desired γ it is possible to calculate required firing angle α in steady state but this is very difficult in transient conditions.

Writing Eq. (3.15) with the above inverter angles:

$$V_{dc} = -V_{dc0}\cos\beta - V_{dc\_com}$$

$$V_{dc} = -V_{dc0}\frac{\cos\beta + \cos\gamma}{2} \tag{3.23}$$

Replacing further Eq. (3.3) in Eq. (3.23), and considering that the negative sign is usually omitted with inverter equations:

$$V_{dc} = \frac{3\sqrt{2}}{\pi}V_{g\_LL}\cos\beta + \frac{3}{\pi}\omega L_t I_{dc} \tag{3.24}$$

Rearranging further Eq. (3.23), DC voltage is obtained as the function of the extinction angle:

$$V_{dc} = \frac{3\sqrt{2}}{\pi}V_{g\_LL}\cos\gamma - \frac{3}{\pi}\omega L_t I_{dc} \tag{3.25}$$

The above two equations are important because they describe the effect of different control strategies at the inverter. It is seen in Eqs. (3.24)–(3.25) that the commutating resistance has a different effect (sign) on V_{dc} depending on whether control angle β or γ is kept constant:

- If firing angle $\beta = const$ in Eq. (3.24), then a DC current increase implies a DC voltage increase. This will have stabilizing effect on the HVDC system, as discussed in Chapter 6.
- If $\gamma = const$ in Eq. (3.25) (β is manipulated in order to maintain $\gamma = const$ in an appropriate feedback loop) then a DC current increase reduces the inverter DC voltage. This will have destabilizing effect on the HVDC system, as discussed in Chapter 6.

The Eq. (3.12) for commutation overlap for inverter operation becomes:

$$I_{dc} = \frac{\sqrt{2}V_{g\_LL}}{2\omega L_t}(\cos\gamma - \cos\beta)$$

$$I_{dc} = \frac{\sqrt{2}V_{g\_LL}}{2\omega L_t}(\cos\gamma - \cos(\gamma + \mu)) \tag{3.26}$$

Example 3.2
An HVDC inverter uses thyristors that have 700 μs turn-off time. Determine the safe extinction angle for this converter. Analyse how the extinction angle depends on the turn-off time.

Solution
The minimum γ equals the turn-off time in 20 ms period (50 Hz system);

$$\gamma_{min} = t_{off} \times 50 \times 360 = 12.6°$$

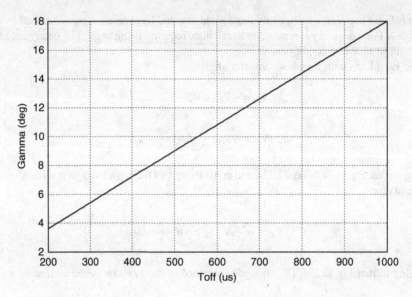

Figure 3.10   Gamma angle for a range of thyristor turn off times in Example 3.2.

The actual extinction angle in the HVDC converter would be maintained at around 15–18° to allow for some operating margin. Figure 3.10 shows the γ_{min} for a range of typical turn-off times.

4

HVDC Rectifier Station Modelling, Control and Synchronization with AC Systems

4.1 HVDC Rectifier Controller

The principal thyristor converter control equation is given in Eq. (3.15), which demonstrates that DC voltage can be manipulated directly by varying firing angle α. For a multibridge high-voltage direct-current (HVDC) rectifier, the DC voltage becomes:

$$V_{dcr} = B\frac{3\sqrt{2}}{\pi}V_{g\_LLr}\cos\alpha_r - B\frac{3}{\pi}\omega L_{tr}I_{dc} \tag{4.1}$$

where subscript 'r' stands for rectifier and B represents the number of series-connected six-pulse bridges. Most often $B = 2$ or 4. It is a common practice to connect multiple bridges in series (appropriately phase shifted) in order to reduce harmonics. The DC voltage waveform in Figure 3.2 has six pulses per cycle and therefore the dominant harmonic on the DC side is the sixth harmonic. A 12-pulse system can be created if two six-pulse converters with 30° phase shifted transformer secondaries are connected in series. Such a 12-pulse system has a dominant 12th harmonic on the DC side and 11th and 13th on the AC side. Note that each bridge might have an independent controller to improve flexibility and reliability.

The HVDC system DC current is expressed using rectifier and inverter voltages as:

$$I_{dc} = \frac{V_{dcr} - V_{dci}}{R_{dc}} \tag{4.2}$$

where R_{dc} is the total DC-side resistance and V_{dci} is the inverter-side DC voltage. The DC current, therefore, can be controlled either by using rectifier or inverter DC voltage but most commonly it is controlled on the rectifier side, as analysed in Chapter 6.

Figure 4.1 shows a simplified controller schematic for a rectifier converter. It includes a feedback PI controller for DC current in the inner loop. As DC voltage responds according to the *cosine* function of

High-Voltage Direct-Current Transmission: Converters, Systems and DC Grids, First Edition.
Dragan Jovcic and Khaled Ahmed.
© 2015 John Wiley & Sons, Ltd. Published 2015 by John Wiley & Sons, Ltd.

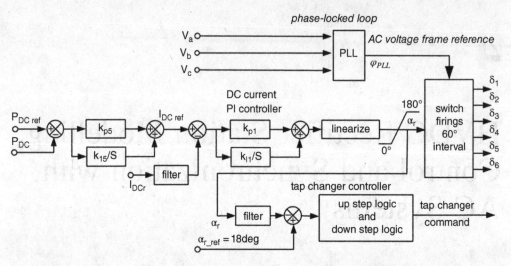

Figure 4.1   Rectifier controller.

the firing angle as shown in Eq. (4.1), the system is nonlinear. The system gain depends on the operating point, which can cause feedback control problems. In order to eliminate nonlinear gain a linearization element is introduced in the controller (shown as 'linearize'). This element performs an inverse *cosine* function which is further studied in Chapter 7. The phase-locked loop (PLL) provides information on the AC bus phase angle to the controller, as discussed below. The switch firings provide six pulses, which are 60° spaced (one to each valve), and delayed from AC voltage zero crossings by angle α_r.

The nominal rectifier operating angle is typically $\alpha_r = 15$–$20°$. It is desirable to have a low operating angle to reduce reactive power and harmonics but a sufficient margin must be allowed to compensate for disturbances to the AC voltage. In some operating conditions, high values for the firing angle will result – for example when DC current is low or when AC voltage is high. A slow-acting transformer tap changer is provided in order to optimize the valve-side AC voltage. The tap-changer controller will measure the firing angle and make adjustments to the tap-changer position in order to increase or decrease the transformer ratio. The tap changer is a motorized slider on the transformer primary, which is capable of adjusting transformer ratio typically in steps of 1.25%, and the total adjustment range is around 40%. This is a very slow-operating controller (it has a time constant of the order of minutes), which ensures that converter variables are optimized but it does not react to fast disturbances. A deadband of around 2% is normally used to prevent unwanted up/down hunting and wearing of mechanical components.

4.2   Phase-Locked Loop (PLL)

The main role of a PLL in HVDC systems is to give a reference signal for a converter controller, which is synchronized with the AC commutation voltage. As the operating conditions in the AC system change, the zero crossings of AC voltages will move. A thyristor can only be fired when it is forward biased, and in inversion mode it should be fired sufficiently early to allow reverse recovery for a given AC voltage magnitude. Precise information about the AC voltage position is therefore required for converter firing control.

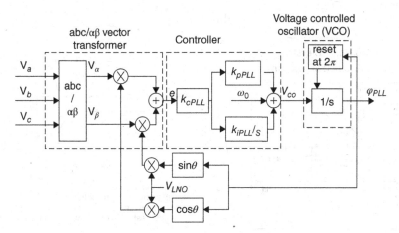

Figure 4.2 Transvector PLL.

The earliest HVDC systems used zero crossing detection on AC voltages for firing synchronization. This synchronization was vulnerable to harmonic instabilities. Figure 4.2 shows the transvector PLL, which is commonly used with the latest HVDC. This PLL tracks positive sequence of three-phase AC voltage in order to derive the phase angle reference signal. It includes three segments:

- 3/2 phase transformer ($ABC \rightarrow \alpha\beta$);
- controller, which ensures tracking;
- voltage-controlled oscillator (VCO), which converts frequency into a position angle (a resettable integrator generating sawtooth waveform).

The input to the PLL is the measured voltage signal from the three-phase AC voltage (v_a, v_b, v_c), as shown in Figure 4.2. The PLL output is a reference sawtooth waveform with frequency and phase angle closely following the AC voltage. If the inner controller in the PLL is tuned for fast tracking of the AC system dynamics then the reference signal will be closely following the changes in the AC system voltage angle. In this case, the actual firing angle will not deviate from the ordered firing angle. In the opposite case, with low PLL controller gains, the reference signal will have only slow dynamic changes with very loose tracking of the AC system dynamics.

Figure 4.3 shows how the actual firing angle depends on the ordered firing angle and the AC system dynamics. The firing angle ordered from the controller is α_c. The solid line shows the AC voltages in the nominal state. When the AC system is perturbed, as shown by the dotted line, the actual position of the voltage crossings will change. In this perturbed state the actual firing angle seen by the converter (α) becomes different from that ordered from the controller (α_c) because of the change in the reference point. This happens because PLL internal dynamics are not able to track the AC system position instantaneously. A PLL works like a Kalman filter by indirectly adjusting the output to track the input signal position. This indirect tracking provides excellent immunity from noise on the AC voltage signal. Large PLL gains can improve tracking but the system becomes more sensitive to harmonics and may lose synchronism during large disturbances. The value ω_o is the expected frequency of the AC system ($2\pi f$).

Figure 4.3b uses small-signal block-diagram representation to depict the role of the PLL in deriving the actual converter firing angle. The PLL output (φ_{PLL}) cancels (with some delay) any changes in AC system position (φ_V).

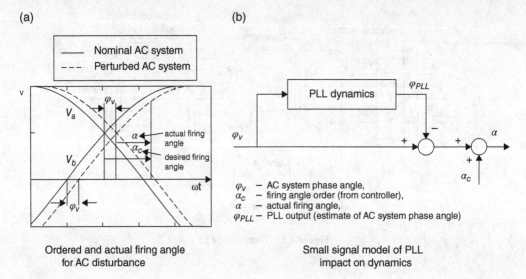

Figure 4.3 Influence of AC system dynamics on the converter firing angle. (a) Time domain and (b) small signal model.

The α and β components of AC voltages for transvector type PLL in Figure 4.2, are defined as:

$$v_\alpha = \frac{2}{3}v_a - \frac{1}{3}v_b - \frac{1}{3}v_c \tag{4.3}$$

$$v_\beta = \frac{1}{\sqrt{3}}(v_b - v_c) \tag{4.4}$$

The control error is defined as:

$$e = v_\alpha v_{\sin} + v_\beta v_{\cos} \tag{4.5}$$

After substituting Eqs (4.3)–(4.4) in Eq. (4.5):

$$e = V^2 \sin(\varphi_{PLL} - \varphi_V) \tag{4.6}$$

It can be seen that the PLL controller will respond to the difference between the actual phase position of the AC system and the PLL output-phase angle. However the error signal also depends on the AC voltage magnitude V^2. This is not desirable because, for low AC voltages, the PLL controller gain will be low and it will respond slowly. Some new PLL systems use adaptive gain to compensate for the dependency on AC voltage magnitudes.

4.2.1 Master Level HVDC Control

Figure 4.4 shows the topology of master-level controller at the rectifier. This control is at a higher (and slower) level, which gives DC current reference to the primary DC current control shown

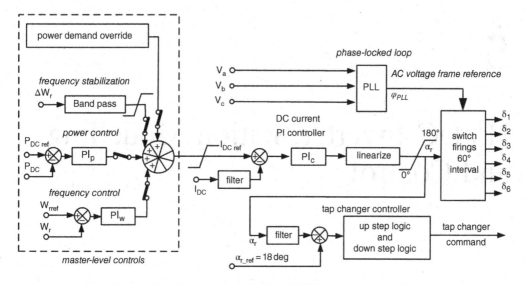

Figure 4.4   Master-level and low-level HVDC controls at rectifier.

in Figure 4.1. There are several possible control modes but not all modes are used at each HVDC system:

- Power control is the primary control mode, which is employed in most HVDC systems.
- Frequency stabilization may be used to improve stability in a particular bandwidth where known weakly damped oscillatory modes may exist in the AC system. The frequency range may include unwanted machine-machine regional oscillations (0.1–5 Hz), or interarea modes (0.01–0.1 Hz). This signal is typically limited to ±3–5% of the current order and added to the power controller output. If stabilization is required for an HVDC remote station (inverter) then the controller should compensate for delays incurred by signal transfer. This function is similar to power oscillation damping on generators.
- Frequency control may be possible with HVDC, for example if it operates in isolated mode, in which case power control is disabled. If a line-commutated converter (LCC) HVDC is supporting isolated AC system operation, it is required to provide synchronous condenser or a static VAR compensator (SVC) for reactive power support at the terminal bus. This is rarely used as the normal mode of operation with LCC HVDC but it may be included as an emergency control configuration.
- The power-demand override includes a range of limiters and rate limiters for DC power controller. These limiters are active during startups, postfault recoveries and other events.

5

HVDC Inverter Station Modelling and Control

5.1 Inverter Controller

5.1.1 Control Structure

The basic equation for the inverter terminal is similar to that for the rectifier – see Eq. (4.1):

$$V_{dci} = B\frac{3\sqrt{2}}{\pi}V_{g\_LLi}\cos\beta + B\frac{3}{\pi}\omega L_{ti}I_{dci} \tag{5.1}$$

where subscript 'i' stands for inverter. Note that sign with the commutation overlap drop is positive (opposite to rectifier), and normally inverter angles β and γ are used. The angle $\alpha = 180 - \beta$ is the only control variable whereas several different control goals exist. The primary control aim is to prevent commutation failure as this would cause system collapse. In addition, the goal is to maintain inverter DC voltage close to optimal values and also to ensure system operation in case that the rectifier terminal loses control capability. Figure 5.1 shows the basic inverter control topology. There are three control systems – γ control, DC/AC voltage control and DC current control – which compete through the '*minimum*' element and the one that gives minimal angle defines the operating mode.

5.1.2 Extinction Angle Control

In the inversion mode, as demonstrated in Figure 3.9, it is important to keep a safe extinction angle, γ, in order to prevent commutation failure. It is also desirable to maintain γ as low as possible in order to minimize reactive power consumption, harmonics and losses. The minimum γ required for safe commutation with high-power thyristors is around 10°–15°. Typically, γ reference is set to around 15–20°, which enables safe commutation under some expected disturbances.

The control challenge with inversion mode is that controller can only influence the beginning of thyristor conduction, which is the firing angle β. After thyristor is triggered there is a commutation overlap period (typically 10–30°) while current reduces in the outgoing thyristor and current increases in the

High-Voltage Direct-Current Transmission: Converters, Systems and DC Grids, First Edition.
Dragan Jovcic and Khaled Ahmed.
© 2015 John Wiley & Sons, Ltd. Published 2015 by John Wiley & Sons, Ltd.

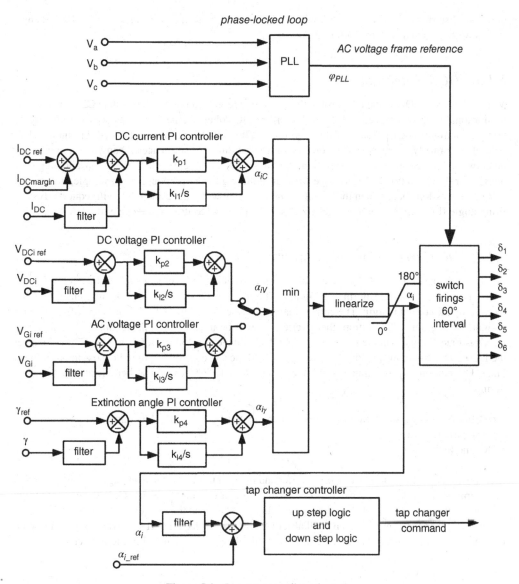

Figure 5.1 Inverter controller schematic.

triggered thyristor. Control is not possible during commutation overlap. However, the length of commutation overlap depends on the external conditions like AC voltage and DC current and cannot be predicted at the firing instant. It is therefore necessary to allow a certain safety margin in the γ reference value.

Gamma minimum control is also called constant extinction angle (CEA) control. This is a default control mode at the inverter. Typically γ is controlled in a feedback manner as shown in Figure 5.1 and by using thyristor voltage sensors, which rapidly measure the extinction angle for each thyristor. The minimum angle for a six-pulse bridge is passed to the controller and compared with the reference angle to make adjustments in the firing angle. The CEA control, however, is the least stable control mode for inverter and for this reason CEA control is not used on HVDC systems with weak AC systems or during disturbances. However CEA will always run in the background as the largest allowed angle, α_i, in order to prevent commutation failure.

5.1.3   DC Voltage Control

Many HVDCs use DC voltage control at the inverter side, which is more stable than CEA. Typically a proportional integral (PI) feedback control regulates DC voltage at the desired value. Alternatively AC voltage control can be used as shown in Figure 5.1, although this is rarely employed. In Figure 5.1 the controller normally operates in DC voltage control but during disturbances it can move to one of the other two modes. During normal operation $\alpha_{i\gamma}$ will be only a few degrees larger than α_{iV}, allowing some margin for the regulation of DC voltage without switching to CEA mode. If, for example, inverter AC voltage depressions occur, then the actual extinction angle will reduce and CEA will request advanced firing angle (lower $\alpha_{i\gamma}$) in order to keep γ at safe levels, and control will change to CEA mode.

5.1.4   DC Current Control at Inverter

The DC current controller (CC) mode is also employed at the inverter side as a backup control mode, which is not active during normal operation. The inverter CC receives current reference, which is reduced by the current margin. The current margin is typically around 10–15% of the rectifier reference current. During normal operation, the inverter CC will be in saturation, demanding firing angle α_{iC}, which is much larger than the α_{iv} demanded by DC voltage controller. However, during disturbances, the DC current can reduce below I_{DCref}-$I_{DCmargin}$ and the CC controller will reduce α_{iC} in order to stabilize DC current through transient recovery. The inverter CC control is beneficial and necessary during:

- rectifier AC system disturbances (faults), when the rectifier controller hits minimum angle limit and loses current control;
- DC line faults.

In a bidirectional HVDC system, the inverter and rectifier terminals can exchange roles very quickly. Both terminals will typically have the same control structure as in Figure 5.1. Only the current margin setting and the sign of DC voltage reference are different between the two terminals. If the AC systems are different then the controller gains will also differ. The system can reverse power in a fast and seamless manner just by manipulating the current margin, as will be discussed in the next chapter.

5.2   Commutation Failure

In a normal inverter operation it is necessary that each valve receives reverse blocking voltage for a safe turnoff time (extinction angle) and after the commutation overlap is completed, in order for the thyristor to regain forward blocking capability. This may not happen if, for example, commutation overlap is longer than normal or does not complete before the next forward voltage. In such case the thyristor will not gain a blocking state and it will continue to conduct for the full next cycle. This is an unwanted operating condition because the thyristor on the opposite pole on same converter leg will be fired in the next 120° and this will create a short circuit across the DC voltage.

Figure 5.2 illustrates a commutation failure. Firstly, AC voltage magnitude drops to 0.5 pu at 0.155 s. This causes an increase in the DC current and the commutation overlap is significantly prolonged at the next firing instant. T_3 does not take DC current and T_1 continues to conduct. This is a commutation failure. At the next firing instant T_4 (a thyristor on the same phase connecting to the opposite pole)

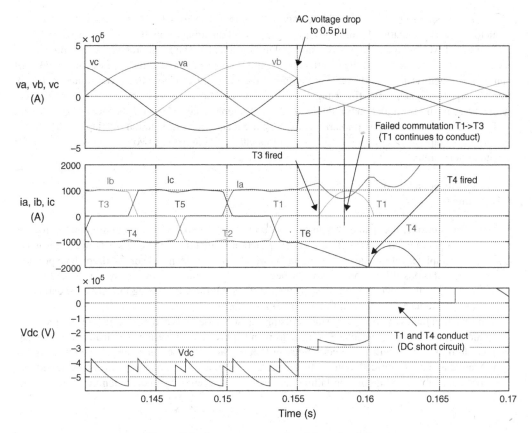

Figure 5.2   Inverter commutation failure simulation.

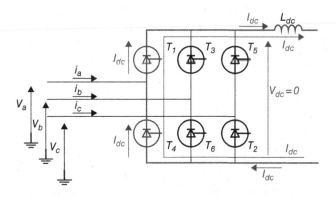

Figure 5.3   DC current path after commutation failure.

is fired and DC voltage is therefore shorted through T_1 and T_4. The shorted inverter DC voltage further increases DC current and this exacerbates the situation. Figure 5.3 shows the current path after commutation failure.

Commutation failure and DC short circuit are not destructive for HVDC; the system will recover in few cycles assuming that the AC voltage has recovered. The commutation failure is a localized DC phenomenon where DC voltage collapses and DC current increases but it is limited by smoothing inductors. The commutation failure does not cause a direct disturbance on the AC side: there is no high AC current or AC voltage drop (the AC side will not see a fault). The main problem with commutation failure is that the power flow to the inverter AC system is interrupted and, if the AC system is weak, this can cause a significant disturbance. Furthermore, the recovery process after a commutation failure implies large transient inverter firing angles, which draw significant reactive power from the AC system while the active power injection is reduced. These conditions make AC system recovery more difficult.

The commutation failure is primarily caused by the increased commutation overlap (μ). The commutation overlap is defined by Eq. (3.26), and this enables establishing two main conditions for the commutation failure:

- inverter AC voltage drop;
- DC current increase.

The CEA controller is capable of adjusting operating angle α_i to avoid commutation failure for any steady-state operating conditions (like gradual reduction in inverter AC voltage). However, it is not possible to completely eliminate commutation failure in case of sudden AC voltage drops. The problem arises because it is possible to control only the beginning of commutation overlap but once a thyristor is fired the conduction is solely determined by the circuit conditions. Increasing the reference extinction angle can only reduce the probability of commutation failure.

Commutation failure is commonly studied using statistical methods and it is expressed as a probability, which depends on the percentage voltage drop on the inverter AC system. Typically, a sudden AC voltage drop of 5–15% will cause commutation failure. Single-phase faults are as onerous as three-phase faults as far as commutation failure onset is concerned. AC systems with larger impedance are more vulnerable to commutation failure because AC voltage swings more when the load changes.

6

HVDC System V-I Diagrams and Operating Modes

6.1 HVDC-Equivalent Circuit

The equivalent circuit of the DC side of a HVDC system is shown in Figure 6.1. There are two equivalent controllable DC sources, which represent the two converters. If the inverter is operated in a constant extinction angle, the inverter equivalent resistance becomes negative as shown in Figure 6.1b.

The DC current is proportional to the difference between rectifier and inverter DC voltages, as shown in Eq. (4.2). Analysing this equation, DC current can be controlled either using a rectifier or an inverter. In most HVDC systems, a rectifier controls DC current while an inverter controls DC voltage as this method results in the optimal HVDC design. Some HVDC systems with extremely weak inverters operate with DC current control at the inverter side.

6.2 HVDC V-I Operating Diagram

Figure 6.2 shows the HVDC system static operating point as the intersection of the rectifier and the inverter V-I operating curves. The rectifier is typically in the constant current mode and the curve is a vertical line in the HVDC V-I diagram. Three different curves are shown for the inverter, representing three common modes.

The inverter curve $\beta_i = const$ provides a good, natural stabilizing response. Assuming that DC current is perturbed so that it increases, the HVDC operating point would move along $\beta_i = const$ curve and therefore the inverter DC voltage increases. This increase in the inverter DC voltage implies a DC current reduction from Eq. (4.2), thus counteracting (stabilizing) the original disturbance.

Figure 6.3 shows a more detailed V-I operating diagram for an HVDC system. The rectifier is in constant control of the current, but the rectifier voltage can only be increased until the firing angle reaches the minimum limit. Typically a small margin of $\alpha_{min} = 2°$ is provided to prevent the thyristor triggering while it is reverse biased. The inverter in this diagram includes three modes. In normal

High-Voltage Direct-Current Transmission: Converters, Systems and DC Grids, First Edition.
Dragan Jovcic and Khaled Ahmed.
© 2015 John Wiley & Sons, Ltd. Published 2015 by John Wiley & Sons, Ltd.

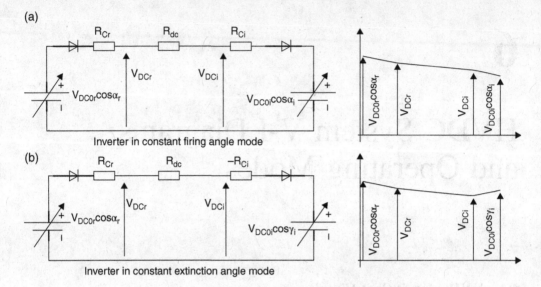

Figure 6.1 HVDC system equivalent electrical circuit.

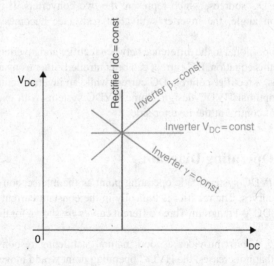

Figure 6.2 HVDC operating point as intersection between rectifier and inverter curves.

operation, DC voltage control is employed, which is a horizontal curve and establishes a stable operating point with rectifier constant current control.

Figure 6.4 shows the operating point when the rectifier AC voltage reduces. If the rectifier AC voltage reduces sufficiently, the rectifier will operate along an $\alpha = const = 2°$ curve and the intersection with the inverter curve may not be defined. For this reason the inverter also has a constant current mode but the reference current is reduced by the current margin.

If the inverter AC voltage reduces, the inverter controller moves to constant extinction angle mode in order to prevent commutation failure, as shown in Figure 6.5. This operating mode ensures that the minimum extinction angle is preserved under all external circuit conditions.

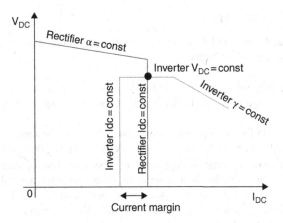

Figure 6.3 HVDC full V-I operating diagram (inverter in V_{dc} control mode).

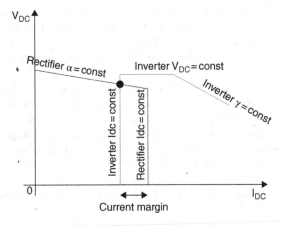

Figure 6.4 HVDC operating point for reduced rectifier AC voltage (inverter in CC mode).

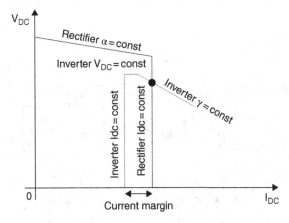

Figure 6.5 HVDC operating point for reduced inverter AC voltage (inverter in constant extinction angle (CEA) mode).

6.3   HVDC Power Reversal

The control topology at the two HVDC terminals is identical because the rectifier and inverter roles are interchangeable. There are two differences:

- the terminal that receives the current margin operates as an inverter;
- the sign of the DC voltage reference is different (but stays unchanged) at the two terminals.

Figure 6.6 shows the HVDC test system, which is used in all further simulations. This is a 500 MW monopolar bidirectional HVDC with a typical medium length (100–300 km) DC cable parameters.

Figure 6.7 illustrates power reversal on the test HVDC system. The top three graphs show rectifier variables and the lower three graphs show inverter (the roles are interchanged at 1.9 s). At 1.1 s, the power reversal process is initiated by sending the current margin to the rectifier and removing the current margin from the inverter controller. At the rectifier side, the DC current reference drops by 0.1 pu while at the inverter controller the DC current reference increases by 0.1 pu. The rectifier controller now experiences too high current and keeps increasing the firing angle in an attempt to bring the current to the reference level. Simultaneously, the inverter controller experiences too low current and keeps reducing the firing angle, trying to increase the current to the reference level. During this process the rectifier firing angle increases to over 90° and it starts inverting while the inverter's firing angle reduces to below 90° and it moves into rectification. The second graph shows that the DC voltage polarity reverses at both sides. The slope of the curves is identical in this case as the two controllers have the same proportional integral (PI) gains.

The third graph shows the three controller firing angles, which are competing in the *minimum* element. The rectifier current controller cannot lower the current to a new setting and it will saturate as the DC voltage controller takes over control at 1.9 s. As seen in the third graph, for inverter variables, the inverter current controller takes over from the DC voltage controller as soon as the current margin is removed. It is noted that the DC current stays approximately constant during the reversal process. In modern HVDC systems, the power-reversal process can be shorter than in Figure 6.7, and possibly within 0.2 s. In some systems the speed of power reversal will be limited, considering limitations on other equipment like cables and filters and limitations on the AC system ramp rates.

Figure 6.8 shows the full V-I diagram including the positive and negative power direction. The diagram is symmetrical around the abscissa axis and illustrates the interchangeable roles of rectifier and inverter.

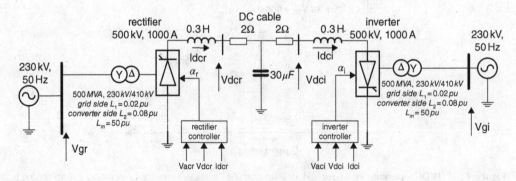

Figure 6.6   HVDC test system.

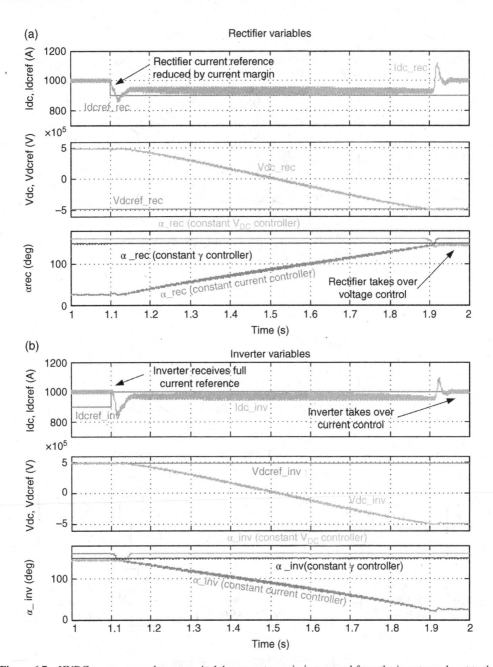

Figure 6.7 HVDC power reversal process. At 1.1 s current margin is removed from the inverter and sent to the rectifier. Power reversal is completed at 1.95 s.

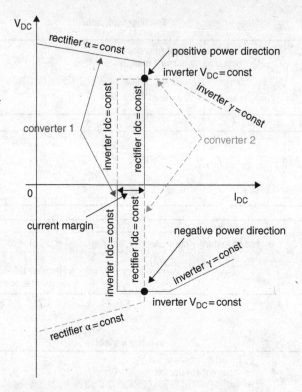

Figure 6.8   HVDC V-I diagram with positive and negative power direction.

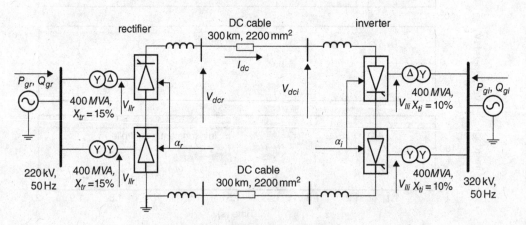

Figure 6.9   HVDC system in Example 6.1.

Example 6.1

An HVDC is being considered as an interconnection between two AC systems as shown in Figure 6.9. The HVDC ratings are: 600 MW, 400 kV (specified at rectifier side), 1500 A. It is required that the rectifier operates at an 18° nominal angle and that the inverter extinction angle is controlled at 16°. The copper DC cable is selected as 2200 mm², with a total distance of 300 km. Assume that the copper specific resistance is $R_{ocu} = 1.73 \times 10^{-8}$ $\Omega m^2/m$. Determine:

- *the transformer ratio for the rectifier and inverter transformers;*
- *the inverter operating angle β;*
- *DC cable losses;*
- *the reactive power drawn at the rectifier side;*
- *the reactive power drawn at the inverter side.*

Solution
Rectifier side:
Transformer inductance:
The transformer V_{llr} can be obtained iteratively as Vllr = 167.8 kV.

$$L_{tr} = X_{tr}\frac{V_{llr}^2}{S_t}\frac{1}{2\pi 50} = 0.15\frac{167\ 800^2}{400\ 000\ 000}\frac{1}{2\pi 50} = 0.0333\ H$$

Valve-side AC voltage:

$$V_{dcr} = B\frac{3\sqrt{2}}{\pi}V_{g\_LLr}\cos\alpha - B\frac{3}{\pi}\omega L_{tr}I_{dc}$$

$$V_{g\_LLr} = \left[V_{dcr} + B\frac{3}{\pi}\omega L_{tr}I_{dc}\right]\frac{\pi}{B3\sqrt{2}\cos\alpha}$$

$$V_{g\_LLr} = \left[400\ 000 + 2\frac{3}{\pi}314.15\times 0.0333\times 1500\right]\frac{\pi}{2\times 3\sqrt{2}\cos(18)} = 167.4\ kV$$

Transformer ratio is: $n_r = 220/167$.
Ideal DC voltage

$$V_{dc0r} = B\frac{3\sqrt{2}}{\pi}V_{g\_LLr} = B\frac{3\sqrt{2}}{\pi}167.4 = 452.1\ kV$$

Power factor angle:

$$\phi_r = a\cos\left[\frac{V_{dcr}}{V_{dc0r}}\right] = a\cos\frac{400\ 000}{452\ 100} = 27.77°$$

Reactive power:

$$Q_{gr} = P_{gr}\tan(\varphi_r) = 600\ 000\ 000\tan(27.75) = 316\ MVar$$

Inverter side:
DC cable resistance and loss

$$R_{dc} = 2R_{ocu}\frac{l_c}{S_c} = 2\times 1.73\times 10^{-8}\frac{300\ 000}{2200\times 10^{-6}} = 4.718\ \Omega$$

$$P_{dcloss} = I_{dc}^2 R_{dc} = 1500^2\times 4.718 = 10.616\ MW = 1.77\%$$

Inverter DC voltage:

$$V_{dci} = V_{dcr} - R_{dc}I_{dc} = 400\ 000 - 1500\times 4.718 = 392.92\ kV$$

Transformer inductance:
The transformer Vllr can be obtained iteratively as Vllr = 158 kV.

$$L_{ti} = X_{ti}\frac{V_{lli}^2}{S_{ti}}\frac{1}{2\pi 50} = 0.1\frac{158\ 000^2}{400\ 000\ 000}\frac{1}{2\pi 50} = 0.0199\ H$$

Valve side AC voltage:

$$V_{dci} = B\frac{3\sqrt{2}}{\pi}V_{LLi}\cos\gamma - B\frac{3}{\pi}\omega L_{ti}I_{dc}$$

$$V_{LLi} = \left[V_{dci} + B\frac{3}{\pi}\omega L_{ti}I_{dc}\right]\frac{\pi}{B3\sqrt{2}\cos\gamma}$$

$$V_{LLi} = \left[392\ 920 + 2\frac{3}{\pi}314.15\times 0.0199\times 1500\right]\frac{\pi}{2\times 3\sqrt{2}\cos(16)} = 158.2\ kV$$

The transformer ratio is: $n_i = 320/158$.
The inverter firing angle:

$$\gamma + \mu = a\cos\left[\cos(\gamma) - \frac{I_{dc}2\omega L_{ti}}{\sqrt{2}V_{LLi}}\right]$$

$$\gamma + \mu = a\cos\left[\cos(16) - \frac{1500\times 2\times 314.15\times 0.0815}{\sqrt{2}\times 158.2}\right]$$

$$\beta = \gamma + \mu = 28.64°,\ \ \mu = 12.64°$$

$$\alpha_i = 180 - \beta = 151.36°$$

Ideal inverter DC voltage

$$V_{dc0i} = B\frac{3\sqrt{2}}{\pi}V_{LLi} = B\frac{3\sqrt{2}}{\pi}158.2\ kV = 427\ kV$$

Power factor angle:

$$\phi_i = a\cos\left[\frac{V_{dci}}{V_{dc0i}}\right] = a\cos\frac{392920}{466000} = 23.16°$$

Reactive power:

$$Q_{gi} = V_{dci}I_{dc}\tan(\varphi_i) = 392\ 920\times 1500\times \tan(23.16) = 252.1\ MVar$$

7

HVDC Analytical Modelling and Stability

7.1 Introduction to Converter and HVDC Modelling

The analysis, design and testing of power electronics systems without computer simulation is extremely time consuming and therefore expensive. In particular, with high-power converters and HVDC testing is very expensive, even when scaled prototypes are employed. In the first decades of HVDC development, low-power hardware models were used. However with the development of computer simulation in the late 1990s, good accuracy and fast models became available. Nowadays, most HVDC design and development work is done on simulators. When proven converter topologies are employed for new projects, the design is primarily based on extensive simulation studies and this is often adequate to proceed directly to commissioning tests.

High-voltage direct-current systems embedded inside national grids involve dynamic phenomena with several widely different values of time constants, and therefore multiple modelling methods exist:

Detailed switching transients modelling.
  In the fastest timeframe are power electronics dynamic phenomena at switch level. Here the transients between ON and OFF states are studied in great detail with nonlinear functions perhaps also temperature dependent, enabling accurate representation of switching losses and driver design. The simulation step is in the order of *nanoseconds,* and commonly a single switch or a single converter is studied. Typical commercial software would include PSPICE and SABER. Such modelling is not addressed in this book.
Modelling with switchings.
  The second group of models accurately presents all switching instances and calculates all variables between switching events. This type of analysis presumes ideal switching transients (instant change between OFF and ON). There will typically be at least 100–500 calculations between each switching instant, which implies a simulation step of around 50 μs for line-commutated converter (LCC) HVDC or 1–10 μs for voltage source converter (VSC) HVDC. The switch voltage drop and its

High-Voltage Direct-Current Transmission: Converters, Systems and DC Grids, First Edition.
Dragan Jovcic and Khaled Ahmed.
© 2015 John Wiley & Sons, Ltd. Published 2015 by John Wiley & Sons, Ltd.

snubbers are embedded in the analysis. This is essential for conduction losses assessment, and for accurate harmonic analysis. The control design can also be quite accurately performed. Typical commercial simulation software in this category includes SIMULINK/SimPowerSystems, PSCAD/EMTDC, EMTP and others which are widely used by manufactures and developers. These models can be used for two-terminal HVDC, for other converters in transmission system applications and in general for grids with a small number of converters. In cases with a large number of converters, like a DC grid, which may include more than 20 VSCs, the model will be significantly complex and the simulation speed may become unacceptably slow. These models are always represented in an ABC coordinate frame with all three-phase oscillating variables. As this simulation supports only trial and error methods, any multivariable optimization is time consuming. This type of simulation is used for benchmarking the accuracy of analytical and phasor models in this manuscript.

Analytical dynamic modelling of converters.

At the next step, in order to simplify HVDC models and improve simulation speed, the actual switching waveforms are approximated with fundamental frequency or harmonic sinusoidal signals. Such a method is called analytical modelling in this book (it is also called average-value modelling), where the essence of the approach is to consider the average behaviour of the switched circuit over a period. The nonlinear phenomena may be retained or linearized but all dynamic elements are preserved as much as possible. Usually there is at least one order of magnitude improvement in simulation speed compared with the simulation switching models because typical simulation steps are in the range of 20–100 μs. These models are suitable for studying systems with numerous converters, like DC grids. The analytical models are typically represented in an ABC frame with oscillating variables, but they can be converted to a DQ rotating frame, as is explained in Appendix B. If linear models are required, then modelling should be performed in a rotating DQ frame where all variables are DC signals. Linearized models (also known as small-signal models) facilitate eigenvalue studies and control design. Note that DQ frame modelling eliminates oscillating variables and gives further improvement in simulation speed.

Phasor modelling.

With very large national transmission systems, the dominant dynamics are in the low frequency range (below 5 Hz) and depend primarily on the rotor inertia modes of large generators. The power flow and voltage magnitudes are of interest and all electrical circuits are represented using reactances at fundamental frequency and phasors. This modelling approach is called phasor modelling, as HVDC systems are represented with phasor (vector) equations that neglect most of the main-circuit dynamics. Such models are very convenient for the initial converter dimensioning and for power-flow interaction studies with AC systems. Phasor models can be derived from DQ dynamic models.

7.2 HVDC Analytical Model

Figure 7.1 shows the schematic for the complete HVDC system model with all key interaction variables between subunits. The subunits have been analysed in the previous chapters and a complete model can be assembled in an ABC frame. The rectifier and inverter coupling blocks provide links between AC currents and DC variables, which are given in Eqs (3.18) and (3.19). The DC system and AC system models are given in boxes as they can be represented in different levels of detail and different coordinate frames depending on the nature of the model:

- With fundamental frequency modelling, like those used for power-flow studies, the AC system will be given as a fundamental frequency phasor model, which has limited accuracy (bandwidth below 5 Hz).
- If more detailed transient studies are required, with frequencies of interest above 5 Hz, then full dynamic modelling of AC and DC systems is needed.

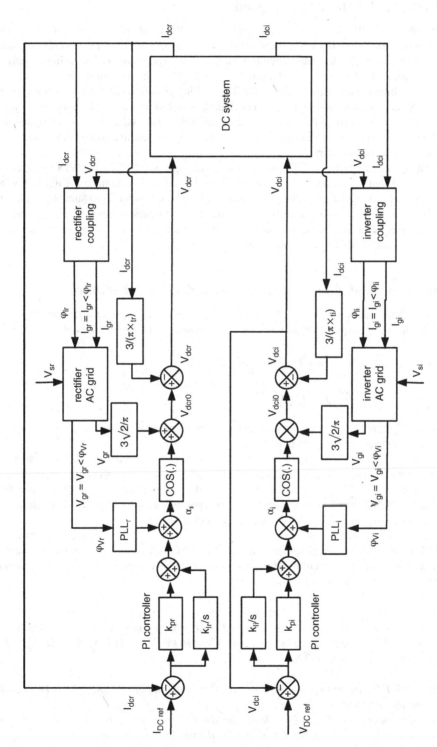

Figure 7.1 HVDC dynamic model schematic.

7.3 CIGRE HVDC Benchmark Model

In the early 1990s, CIGRE developed a representative HVDC benchmark model that has become widely used as the test system for new control strategies. The model is sufficiently simple for studies at research centres and all the main circuit parameters are specified in detail. Importantly, it captures some crucial HVDC operating characteristics and difficulties. On the downside, the model is not meant for bidirectional operation and the CIGRE working group did not discuss control systems.

Figure 7.2 shows the 1 GW, 500 kV CIGRE benchmark model. It includes a 'T' model for a typical 100–300 km DC cable with smoothing reactors on the DC side. Each AC system is represented with equivalent source behind a first-order impedance. At each AC bus the reactive power supply is included but also tuned and high-pass filters.

The modelling in this chapter will follow the basic topology of the CIGRE HVDC benchmark. A comprehensive small-signal linearized model will be derived for the complete CIGRE HVDC-HVAC system. The state-space modelling approach will be adopted as the most suitable method for presenting the dynamics of complex systems. These modelling principles are suitable to accurately represent HVDC responses in the frequencies of control bandwidth, say below 50–100 Hz. The known HVDC stability issues in this frequency range will be presented.

7.4 Converter Modelling, Linearization and Gain Scheduling

Neglecting the commutation overlap in the first instance, the basic converter equation is given in Eq. (3.5). Linearizing this equation around the operating point, the DC voltage deviation ΔV_{dc} is:

$$\Delta V_{dc} = \frac{3\sqrt{2}}{\pi} V_{g\_LL0} \sin\alpha_0 \Delta\alpha + \frac{3\sqrt{2}}{\pi}\cos\alpha_0 \Delta V_{g\_LL}$$
$$\Delta V_{dc} = C_1 \Delta\alpha + C_2 \Delta V_{g\_LL} \tag{7.1}$$

where subscript '0' represents the variable value at the operating point under consideration. The symbol 'Δ' denotes deviation around the operating point.

The coefficients C_1 and C_2 are constant at a particular operating point but their value will change as the operating point changes. Note that the operating point can change significantly, as the power-transfer level changes and, in particular, during large disturbances (outages). It may be seen from Eq. (7.1) that the coefficient C_1 is very important as it represents gain for the controller acting on α and therefore closed loop stability will be affected as C_1 changes.

Control designers would like coefficient C_1 to be constant across the full operating range; however, from Eq. (7.1) C_1 depends on actual operating angle α_0 and AC voltage V_{LL0}. For this reason, a nonlinear compensator is commonly introduced in the controller, which compensates for the inherent nonlinear nature of C_1 (gain scheduling), as seen in Figure 4.1.

Example 7.1

Consider a three-phase thyristor converter connected to AC system and feeding a DC load as shown in Figure 7.3. The following parameters are given: $V_s = 410\,kV$, $V_{dci} = 514\,kV$, $R = 1\,\Omega$, $L_{dc} = 1\,H$, $R_s = 0.01\,\Omega$, $L_t = 0.1\,H$, $I_{dcref} = 1000\,A$.

a. *Simulate a 10% DC current step response at two different operating DC voltages: 514 and 20 kV. Comment on the responses.*
b. *Develop a nonlinear controller to provide similar dynamic response across converter operating range. Draw a transfer function block diagram for the whole system.*
c. *Assume that the AC system impedance reduces to $L_t = 0.01\,H$. Explain the impact on converter dynamics.*

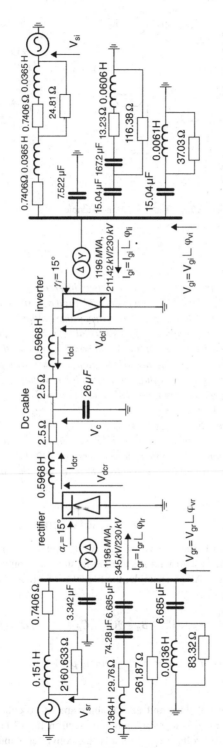

Figure 7.2 CIGRE HVDC benchmark model, 1 GW, 500 kV (rectifier short-circuit ratio (SCR) = 2.5, 84°, inverter SCR = 2.5, 75°).

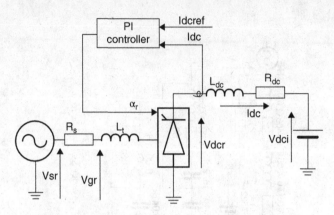

Figure 7.3 Test system in Example 7.1.

Solution

a. *Figure 7.4a shows a 10% DC current step response at two different operating DC voltages. It may be seen that the response at $V_{dc0} = 20\ kV$ is much faster with high overshoot, since the firing angle is larger in this case and therefore C_1 is larger (larger control gain). In the extreme cases the gain variation can cause instabilities and in practical systems special gain linearization circuits are used.*

b. *It is necessary to develop a linearizing control block that compensates for the variations in the control loop gain. The control system with the linearizing function is shown in Figure 7.5. The step response simulation in Figure 7.4b shows that the controller now has a similar response at any DC voltage. In the actual system a lookup table will be used instead of the feedback compensating loop to avoid stability problems.*

c. *In the block diagram in Figure 7.5, it is seen that L_t will affect the closed loop dynamics through the coefficient C_2. However it is not possible to directly compensate for the C_2 variation using the control path. Therefore variation in AC system parameters will affect dynamic responses and the control design strategy is to use a worst-case scenario to determine the controller gains.*

7.5 AC System Modelling for HVDC Stability Studies

A complex AC system like that in Figure 7.2 is commonly represented as an n-th order state-space model. The phase a model for the system with n-states, m-inputs and p-outputs is firstly derived in the static ABC frame for one phase. As an example, the model for rectifier AC system phase a is:

$$sx_a = A_a x_a + B_a u_a$$
$$y_a = C_a x_a + D_a u_a \quad A_a \in \mathfrak{R}^{n \times n}, B_a \in \mathfrak{R}^{n \times m}, C_a \in \mathfrak{R}^{p \times n}, D_a \in \mathfrak{R}^{p \times m} \tag{7.2}$$

$$x_a = \begin{bmatrix} x_1 \\ \cdot \\ x_n \end{bmatrix}_a, \quad u_a = \begin{bmatrix} i_{gr} \end{bmatrix}_a, \quad y_a = \begin{bmatrix} v_{gr} \end{bmatrix}_a, \quad m = 1, \ p = 1$$

where s is a Laplace operator ($s = d(.)/dt$), and all variables are AC oscillating signals. The LCC converter model has AC voltage magnitude as the input, as seen in Eq. (7.1), and therefore the AC model output in Eq. (7.2) will be AC voltage v_{gr}. Similarly, the converter output is AC current as seen in Eq. (3.18), and therefore the AC model input in Eq. (7.2) is the converter current i_{gr}. With VSC converters, as will be shown in Part II of this book, the inputs and outputs to the AC system are reversed.

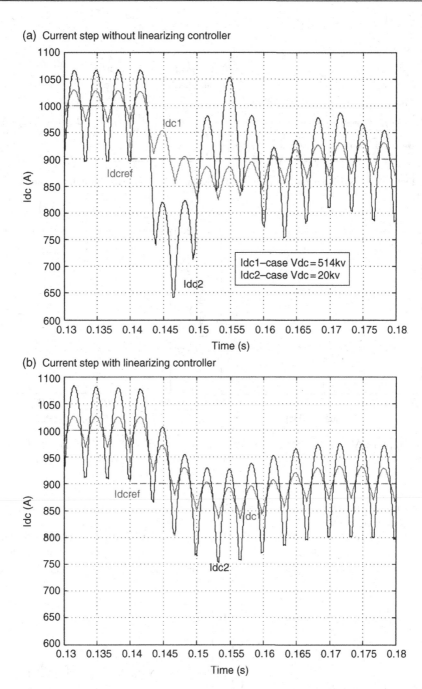

Figure 7.4 DC current controller step response for two different operating points (Vdc = 514 and 20 kV) in Example 7.1.

Figure 7.6 shows the frequency response of the CIGRE rectifier AC system (in an ABC frame), which clearly indicates the resonant peak around 90–100 Hz. The AC system parameters are purposely selected to locate the first resonance at low frequencies in order to excite interaction problems between the AC and DC sides in the CIGRE model. Most AC systems will have impedance characteristics with

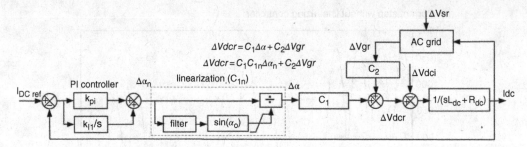

Figure 7.5 Block diagram for the system in Example 7.1 with gain scheduling controller.

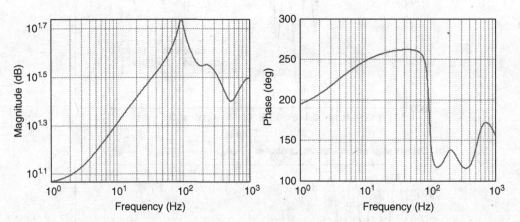

Figure 7.6 Rectifier AC system ABC frame model frequency response (SCR = 2.5, 84°) for the HVDC CIGRE benchmark.

the first resonance above the third harmonic but the system impedance can change with operating conditions (line trippings).

At the next stage a complete three-phase ABC frame model is assembled by appending the remaining two phases. Using the methods presented in Appendix B, and employing Park's ABC-DQ transformation under the assumption of a balanced and symmetrical system, the AC system model in DQ frame becomes:

$$
\begin{aligned}
sx_{dq0} &= A_{dqr}x_{rdq0} + B_{rdc}u_{rdq0} \\
y_{dcr} &= C_{dcr}x_{rdq0} + D_r u_{rdq0}
\end{aligned}
\qquad
x_{dq0} =
\begin{bmatrix}
\begin{bmatrix} x_1 \\ \cdot \\ x_n \end{bmatrix}_d \\
\begin{bmatrix} x_1 \\ \cdot \\ x_n \end{bmatrix}_q \\
\begin{bmatrix} x_1 \\ \cdot \\ x_n \end{bmatrix}_0
\end{bmatrix}
,\;
u_{rdc} =
\begin{bmatrix}
[i_{gr}]_d \\
[i_{gr}]_q \\
[i_{gr}]_0
\end{bmatrix}
,\;
y_{dcr} =
\begin{bmatrix}
[v_{gr}]_d \\
[v_{gr}]_q \\
[v_{gr}]_0
\end{bmatrix}
\qquad (7.3)
$$

$$A_{dqr} = \begin{bmatrix} A_a & \omega I_{n \times n} & 0_{n \times n} \\ -\omega I_{n \times n} & A_a & 0_{n \times n} \\ 0_{n \times n} & 0_{n \times n} & A_a \end{bmatrix}, \quad B_{rdc} = \begin{bmatrix} B_a & 0_{n \times m} & 0_{n \times m} \\ 0_{n \times m} & B_a & 0_{n \times m} \\ 0_{n \times m} & 0_{n \times m} & B_a \end{bmatrix}, \quad m = 1$$

$$C_{dcr} = \begin{bmatrix} C_a & 0_{p \times n} & 0_{p \times n} \\ 0_{p \times n} & C_a & 0_{p \times n} \\ 0_{p \times n} & 0_{p \times n} & C_a \end{bmatrix}, \quad D_r = \begin{bmatrix} D_a & 0_{p \times m} & 0_{p \times m} \\ 0_{p \times m} & D_a & 0_{p \times m} \\ 0_{p \times m} & 0_{p \times m} & D_a \end{bmatrix}, \quad p = 1$$

In the above DQ frame, rotating at frequency ω, all the variables are DC signals and they can be directly coupled with the DC-side model. The DQ frame model also enables linearization, which is important for small signal stability studies. Note that the frequency of all variables in DQ frame is shifted by $\pm \omega$.

Table 7.1 shows the set of dominant eigenvalues for the rectifier AC system of the CIGRE benchmark model in the DQ frame. The first two eigenvalues are derived from the original oscillatory mode at around 95 Hz in the static ABC frame.

Figure 7.7 shows the frequency response for the inverter AC system, which confirms the existence of the first resonance around 100 Hz. Table 7.2 shows the CIGRE inverter AC system dominant eigenvalues, where it is apparent that the first resonant peak is also at around 100 Hz but with much better damping. The eigenvalues 5 and 6 are located at frequency of exactly 50 Hz or 314 rad/s, which indicates that they originate from real eigenvalues in the ABC frame.

Table 7.1 Rectifier AC system eigenvalues (in rotating DQ frame).

	Eigenvalue	Damping ratio	Frequency
1	$-42.53 + 282.12j$	0.085	282 rad/s (44.9 Hz)
2	$-42.53 - 282.12j$	0.085	282 rad/s (44.9 Hz)
3	$-42.53 + 910.4j$	0.046	910 rad/s (144.9 Hz)
4	$-42.53 - 910.4j$	0.046	910 rad/s (144.9 Hz)
5	$-43.79 + 314.15j$	0.139	314 rad/s (50 Hz)
6	$-43.79 - 314.15j$	0.139	314 rad/s (50 Hz)
7	$-505.4 + 1079j$	0.468	1079 rad/s (172 Hz)
8	$-505.4 - 1079j$	0.468	1079 rad/s (172 Hz)

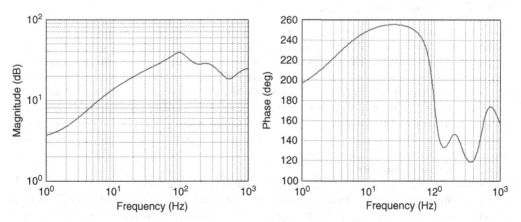

Figure 7.7 Inverter AC system ABC frame model frequency response (SCR = 2.5, 75°) for the HVDC CIGRE benchmark.

Table 7.2 Inverter AC system eigenvalues (in rotating DQ frame).

	Eigenvalue	Damping ratio	Frequency
1	$-127.3 + 293.6j$	0.432	294 rad/s (46.8 Hz)
2	$-127.3 - 293.6j$	0.432	294 rad/s (46.8 Hz)
3	$-127.3 + 921.9j$	0.138	922 rad/s (146.8 Hz)
4	$-127.3 - 921.9j$	0.138	922 rad/s (146.8 Hz)
5	$-43.8 + 314.15j$	0.139	314 rad/s (50 Hz)
6	$-43.8 - 314.15j$	0.139	314 rad/s (50 Hz)
7	$-530.8 + 1141.4j$	0.465	1141 rad/s (182 Hz)
8	$-530.8 - 1141.4j$	0.465	1141 rad/s (182 Hz)

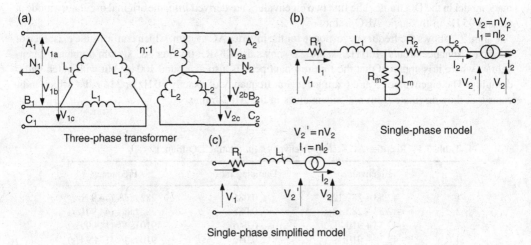

Figure 7.8 Transformer model. (a) Three-phase transformer, (b) single-phase model and (c) single-phase simplified model.

7.6 LCC Converter Transformer Model

The line-commutated HVDC converter is connected to the AC grid through an isolating transformer. Although, theoretically, an HVDC converter can be connected to the grid using just an inductor, a transformer is used in all systems. The single-phase equivalent circuit of the isolating transformer at the low-frequency range is presented in Figure 7.8. The primary winding resistance and leakage inductance are R_1 and L_1, respectively. Similarly, the secondary winding referred to primary, resistance and leakage inductance are R_2 and L_2, respectively. L_m represents magnetizing inductance of the windings and R_m represents magnetic (hysteresis) losses and core (eddy currents) losses. The value of the shunt magnetizing impedance is much larger than the leakage impedance. Thus, the magnetizing impedance can be neglected and the model is approximated as in Figure 7.8c. The zero-sequence variables are not transferred through the transformer.

With LCC converters there is no need to include transformer impedance in the AC system model. The transformer impedance produces equivalent DC-side voltage drop as discussed in section 3.3 and therefore the converter AC voltage is the voltage at the grid-side terminals of the transformer. The dynamics of the transformer are included by increasing the value of the DC side smoothing reactors by the equivalent DC side inductance.

Considering the link between converter AC and DC current, the transformer reactance transferred to the DC side is:

$$L_{t\_dc} = L_t(4 - 3\mu/\pi) \tag{7.4}$$

Therefore the equivalent smoothing reactance on the DC side is:

$$L_r = L_{dc} + L_{t\_dc} \tag{7.5}$$

where L_{dc} represents the DC smoothing reactance. Figure 7.9 shows the DC-side model for HVDC.

7.7 DC System Model

7.7.1 DC Cable/Line Modelling

In the dynamic studies concerned primarily with the low frequencies, DC cables are represented as a lumped T or π model as shown in Figure 7.9. The per kilometre data for overhead lines (OHL) and for typical DC cables are shown in Table 7.3. If more detailed studies are required, then a distributed parameter model is used.

The DC side of the CIGRE system (DC cable and smoothing reactors) can be modelled with the following dynamic equations:

$$sI_{dcr} = -R_r I_{dcr}/L_r - V_c/L_r + V_{dcr}/L_r$$
$$sI_{dci} = -R_i I_{dci}/L_i - V_c/L_i + V_{dci}/L_r \tag{7.6}$$
$$sV_c = 1/C_{dc}I_{dcr} + 1/C_{dc}I_{dci}$$

Figure 7.10 illustrates the CIGRE system DC-side frequency response where it is assumed that the input is V_{dcr} and the output is I_{dci}. It shows that there is a sharp resonant peak around 60 Hz, which implies

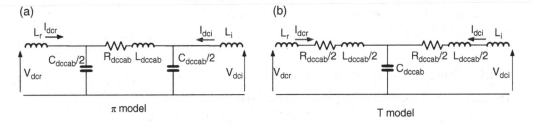

Figure 7.9 DC side model, including smoothing reactors and DC cable. (a) π model and (b) T model.

Table 7.3 Typical DC line and cable data.

Line data R	R (Ω/km)	L (mH/km)	C (μF/km)	G (μS/km)	Max current (A)
DC OHL $\pm$ 400 kV	0.0114	0.9356	0.0123	—	3500
DC OHL $\pm$ 200 kV	0.0133	0.8273	0.0139	—	3000
DC cable $\pm$ 400 kV	0.0095	2.1120	0.1906	0.048	2265
DC cable $\pm$ 200 kV	0.0095	2.1110	0.2104	0.062	1962

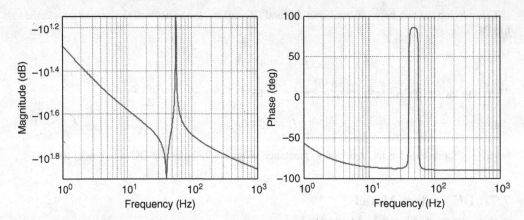

Figure 7.10   DC system frequency response for HVDC CIGRE benchmark.

lingering oscillations around this frequency in all step responses. Many HVDC cables combined with smoothing reactors will have a first resonance in the frequency range 50–100 Hz, which is more-or-less damped depending on the resistance. This resonant peak cannot be damped by DC current HVDC control loops since they typically have bandwidth below 10–20 Hz.

7.7.2   Controller Model

The controller model at both the rectifier side and the inverter side is commonly included in the DC system model. The HDVC controller model comprises the following dynamic equations:

- two equations for phase-locked loop (PLL) dynamics at the rectifier side;
- two equations for PLL dynamics at the inverter side;
- a proportional integral (PI) controller for DC current at the rectifier;
- a PI controller for DC voltage/γ at the inverter (if used);
- dynamics of feedback filters.

7.7.3   Complete DC System Model

The state-space representation for the complete DC side model becomes:

$$
\begin{aligned}
sx_{dc} &= A_{dc}x_{dc} + B_{dcr}u_{dcr} + B_{dci}u_{dci} + B_{dcinp}u_{dcinp} \\
y_{rdc} &= C_{rdc}x_{dc} + D_{rdc}u_{dcr} \\
y_{idc} &= C_{idc}x_{dc} + D_{idc}u_{dci} \\
y_{outdc} &= C_{outdc}x_{dc}
\end{aligned}
\tag{7.7}
$$

where the states are as defined in Eq. (7.6) and with additional states from the above list of control equations. The inputs originate from the rectifier-side AC system, from the inverter-side AC system and the external reference inputs:

$$
u_{dcr} = \begin{bmatrix} V_{grd} & V_{grq} \end{bmatrix}, \quad u_{dci} = \begin{bmatrix} V_{gid} & V_{giq} \end{bmatrix}, \quad u_{dcinp} = \begin{bmatrix} I_{dcrref} \end{bmatrix}
\tag{7.8}
$$

and the outputs are connecting to the rectifier AC system, the inverter AC system and also include the external outputs:

$$y_{rdc} = [I_{grd} \ I_{grq}], \quad y_{idc} = [I_{gid} \ I_{giq}], \quad y_{outdc} = [I_{dcr}] \tag{7.9}$$

7.8 HVDC-HVAC System Model

If the detailed AC system model from Eq. (7.3) and the DC system model from Eq. (7.7) are included in the overall systems model in Figure 7.1, a linear state-space HVDC model can be assembled. This model is suitable for eigenvalue studies of system dynamics and control system design.

The complete model for the HVDC and the two AC systems in the CIGRE system is derived in state space as:

$$sx_s = A_s x_s + B_{inp} u_{inp}$$
$$y_{out} = C_{out} x_s + D_{inp} u_{inp} \tag{7.10}$$

where the system matrix contains three submatrices representing the DC system (A_{dc}), the rectifier AC system (A_{dqr}) and the inverter AC system (A_{dqi}):

$$A_S = \begin{bmatrix} A_{dc} & B_{dcr}C_{dcr} & B_{dci}C_{dci} \\ B_{rdc}C_{rdc} & A_{dqr} & 0 \\ B_{idc}C_{idc} & 0 & A_{dqi} \end{bmatrix}, \quad x_s = \begin{bmatrix} x_{dc} \\ x_{rdq} \\ x_{idq} \end{bmatrix} \tag{7.11}$$

and where B_{dcr} is the input matrix for the DC subsystem that takes outputs from the rectifier AC system, C_{dcr} is the output matrix of rectifier AC subsystem that provides inputs to the DC system, and similarly other coupling matrices can be defined. For the CIGRE benchmark model the matrix dimensions are:

- the DC system A_{DC} is 13th order (depending on the control filters and artificial states);
- the rectifier AC system A_{dqr} is 16th order;
- the inverter AC system A_{dqi} is 16th order.

7.9 Analytical Dynamic Model Verification

Table 7.4 shows the controller parameters that are used in this chapter to study the dynamics of the CIGRE model. There are several possible options for inverter control modes, and the constant firing

Table 7.4 Controller parameters for CIGRE benchmark HVDC system.

	Rectifier	Inverter
Control mode	DC current control	Constant beta
Nominal firing angle	22°	40°
Proportional gain (rad)	1.099	—
Integral gain (rad/s)	91.58	—
Feedback filter constant (s)	0.002	—
PLL K_{cPLL}	1	1
PLL K_{pPLL}	10	10
PLL K_{iPLL} (s^{-1})	50	50

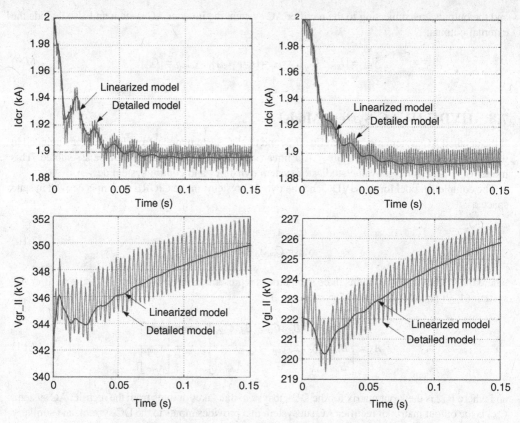

Figure 7.11   Verification of the small-signal LCC HVDC model.

angle ($\beta = const$) is selected for modelling in this chapter. Constant firing angle should give some 'middle-ground' conclusions comparing with constant gamma or constant DC voltage.

Figure 7.11 shows verification of the linearized small-signal dynamic model. The detailed model for benchmarking accuracy is the switching model (on simulation platform from EMTP family) with a 50 μs simulation step and with all nonlinear elements. It is seen that the linearized state-space model is accurate for rectifier and inverter variables and for AC and DC systems. All internal model states have similar accuracy. All parameters in the model have physical meaning and can be used in various dynamic (parametric) studies.

7.10   Basic HVDC Dynamic Analysis

7.10.1   Eigenvalue Analysis

Table 7.5 shows the set of 18 dominant eigenvalues for the CIGRE HVDC model. The damping ratio and frequency for each eigenvalue are also shown. The system is stable but there are two eigenvalues (1, 2) with very poor damping. If model reduction techniques are employed it can be shown that the minimal model order for accurate dynamic representation of this system is 8–10.

7.10.2   Eigenvalue Sensitivity Study

The last column in Table 7.5 shows the results of eigenvalue sensitivity analysis. An eigenvalue sensitivity study examines the dependence of eigenvalue migration on the model parameters (a derivative

Table 7.5 Dominant eigenvalues for the CIGRE HVDC benchmark model.

	Eigenvalue	Damping ratio	Frequency	Eigenvalue sensitivity
1	$-37.343 + 437.23j$	0.085	437 rad/s (69.6 Hz)	Rec, Rec
2	$-37.343 - 437.23j$	0.085	437 rad/s (69.6 Hz)	Rec, Rec
3	$-90.311 + 398.73j$	0.226	398 rad/s (63.4 Hz)	Inv, Inv
4	$-90.311 - 398.73j$	0.226	398 rad/s (63.4 Hz)	Inv, Inv
5	$-118.84 + 195.9j$	0.606	196 rad/s (31.2 Hz)	Rec, Rec
6	$-118.84 + 195.9j$	0.606	196 rad/s (31.2 Hz)	Rec, Rec
7	$-5.62 + 5.38j$	1.04	5.38 rad/s (0.86 Hz)	Inv, Inv
8	$-5.62 - 5.38j$	1.04	5.38 rad/s (0.86 Hz)	Inv, Inv
9	$-5.49 + 5.049j$	1.09	5.0 rad/s (0.796 Hz)	Inv, Inv
10	$-5.49 - 5.049j$	1.09	5.0 rad/s (0.796 Hz)	Inv, Inv
11	-43.46	—	0	Rec, Rec
12	-151.01	—	0	Rec, Rec
13	$-107.0 + 981.7j$	0.109	982 rad/s (156.4 Hz)	Rec, Rec
14	$-107.0 - 981.7j$	0.109	982 rad/s (156.4 Hz)	Rec, Rec
15	$-154.4 + 1002.0j$	0.154	1002 rad/s (159.6 Hz)	Inv, Inv
16	$-154.4 - 1002.0j$	0.154	1002 rad/s (159.6 Hz)	Inv, Inv
17	$-1014.0 + 236.8j$	4.282	237 rad/s (37.7 Hz)	Inv, Inv
18	$-1014.0 - 236.8j$	4.282	237 rad/s (37.7 Hz)	Inv, Inv

of the eigenvalue with respect to the model parameter), which is obtained using the product of the left and right eigenvectors. This analysis indicates which parameter mostly affects a particular eigenvalue. Only AC system parameters are studied because AC topology can change (load connection, AC line switching, filter connection, etc.) whereas DC-system topology generally does not change. The two parameters that mostly impact particular eigenvalues are noted and if they belong to the rectifier AC system, then *Rec* is recorded in the last column, while *Inv* is recorded for the inverter.

The dominant eigenvalues, 1 and 2, are the result of coupling the dominant rectifier AC system mode shown in Table 7.1 with the DC system dynamics. The inverter AC system is represented in the overall dynamics predominantly with eigenvalues of 3 and 4.

7.10.3 *Influence of PLL Gains*

The PLL gains are significant for the overall dynamics and there is some freedom in adjusting these gains. The gains in Table 7.4 are chosen to be relatively low in order to prevent stability problems with very weak AC systems. However, if PLL gains are too low then the controller may not be able to track the AC system position after large disturbances like AC faults. Figure 7.12 shows the eigenvalue locus as the rectifier PLL gains (k_{cPLL} in Figure 4.2) are increased two times. The original eigenvalues (from Table 7.5) are shown with diamonds. It shows that the dynamics will improve with increased PLL gains but further increases (five times and over) will cause overall stability to deteriorate. It is also interesting that PLL gains predominantly affect eigenvalues 7–11 (low frequency) and have no impact at higher frequencies.

Figure 7.13 shows the eigenvalues as the inverter PLL gains are increased four times. Some improvement is possible for $0 < k_{cPLL} < 1.5$ but stability deteriorates for gains $k_{cPLL} > 2$. For large PLL gains a new oscillatory mode is created and the system response will have poorly damped oscillations around 3–8 Hz. The eigenvalue 11 also rapidly moves towards an imaginary axis, which will substantially slow the controller response. The above conclusions apply only for the operation around nominal operating point. Further studies are required for other operating points, like contingencies with low AC voltage.

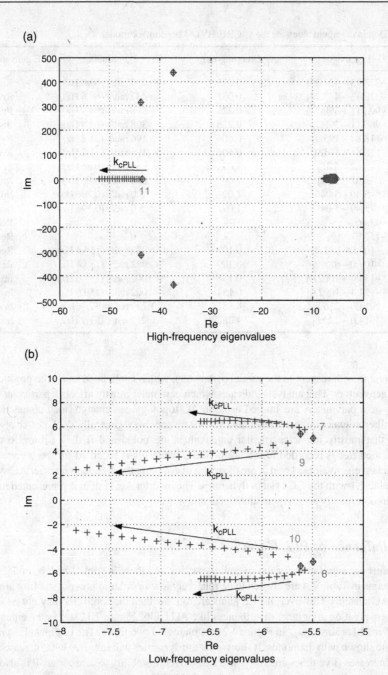

Figure 7.12 Eigenvalue locus as rectifier PLL gains are increased two times ($\Diamond$ shows original gains). (a) High-frequency eigenvalues and (b) low-frequency eigenvalues.

(a)

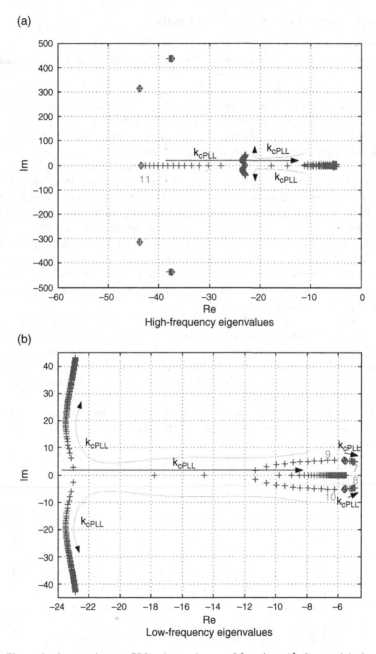

Figure 7.13 Eigenvalue locus as inverter PLL gains are increased four times (◊ shows original gains). (a) High-frequency eigenvalues and (b) low-frequency eigenvalues.

7.11 HVDC Second Harmonic Instability

The second harmonic instability is an HVDC control instability phenomenon that has been experienced on several practical systems. It is the result of unfavourable positive feedback and natural impedance resonance associated with AC and DC systems.

When an AC system is disturbed there will be prolonged voltage oscillation at the lowest resonant frequency, like, for example 95 Hz in Figure 7.6. This oscillation will have two resulting oscillations on the DC side at frequencies: $95 - 50 = 45$ Hz and $95 + 50 = 145$ Hz, because of converter frequency transformation (discussed in Appendix B). The magnitude of DC current oscillation at 45 Hz will depend on the DC impedance and also on the controller response.

The HVDC DC cables frequently have high impedance at around 50 Hz as shown in Figure 7.10. The DC current controller has a bandwidth well below 50 Hz (perhaps up to 20 Hz) but it introduces some phase lag and this may contribute to reducing the Bode stability margin. The DC current at 45 Hz, in turn, generates 95 Hz AC current according to converter interaction equations. This AC current then contributes to amplify the original AC voltage oscillation at 95 Hz and the feedback loop is complete.

The second harmonic resonance is a control issue that results in dynamic instability with oscillations of progressively increasing magnitude. As oscillations increase, the HVDC protection will trip the system.

A particularly difficulty scenario occurs if the AC side resonance is exactly at the second harmonic (100 Hz). This scenario happened on several occasions with the early HVDC systems when AC line tripping reduced the first resonance frequency.

Alternatively, second-harmonic instability may be initiated even without resonance conditions on the AC system, as second harmonics are quite common on the AC side. A small second harmonic on AC systems can result from system asymmetry, various nonlinear loads and switching events. The zero sequence component on the AC side also resonates with the 50 Hz component on the DC side. On several HVDC systems, the transformer energization produced DC inrush currents, which resulted in the initiation of second harmonic instability.

Figure 7.14 shows the eigenvalue location for the CIGRE benchmark model as the rectifier controller proportional gain is increased ($\Diamond$ represents eigenvalues with original gains). The branches '1' and '2'

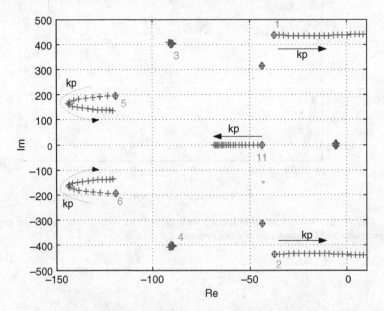

Figure 7.14 Eigenvalue movement as the rectifier controller proportional gain is increased for the CIGRE benchmark model ($\Diamond$ shows original gains).

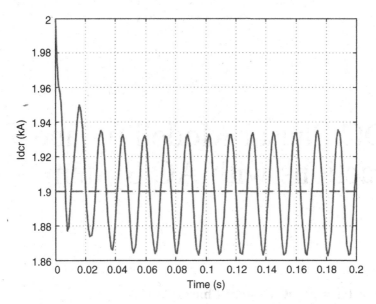

Figure 7.15 Second harmonic instability simulation, obtained on CIGRE benchmark system with rectifier proportional gain increased 2.2 times.

cross an imaginary axis for k_p moderately increased (increased only twice), which is a poor control robustness. The dominant eigenvalues cross the stability boundary at a frequency of around 430 rad/s (68 Hz), indicating oscillatory instability, which is a confirmation of the second harmonic resonance. Figure 7.15 illustrates the second harmonic instability in the time domain.

7.12 Oscillations of 100 Hz on the DC Side

It is important not to confuse the second harmonic instability, which gives 50 Hz on the DC side, with 100 Hz oscillations on the DC side. Some second harmonic on the DC side is quite common and it is typically stable and not increasing. The DC side 100 Hz is caused by a negative sequence on AC side, which is commonly the result of unbalanced/unsymmetrical AC system. Typically long untransposed AC lines feeding HVDC cause some DC-side second harmonic, which is essentially a harmonic phenomenon.

Oscillations of 100 Hz on the DC side cannot be mitigated through a normal feedback control loop but special control injections of oscillating signals can be used to alleviate this issue.

8

HVDC Phasor Modelling and Interactions with AC System

8.1 Converter and DC System Phasor Model

This chapter is concerned with the inherent interaction mechanisms between DC and AC systems at fundamental frequency. Dynamics are neglected. This analysis will aid understanding of the power-transfer limits on AC grids, the reactive power flow with HVDC converters and, importantly, the dependence of the AC voltage profile on the DC power flow and various operating modes.

It is not necessary to involve the complete HVDC system in this study. Each terminal will be considered separately, focusing on the interaction with the local AC system, while the remote terminal will be simplified as a DC voltage or DC current source.

The inverter side will be considered first because interaction problems are more important with inverter AC systems. The analysis is very similar for the rectifier converter and it is given in brief in the last section of this chapter.

There are only four phasor equations that connect the DC and AC sides of a LCC and they will be repeated here for completeness. The basic equation that links the inverter DC voltage and AC voltage is given in Eq. (3.25):

$$V_{dc} = \frac{3\sqrt{6}}{\pi} V_g \cos\gamma - \frac{3}{\pi}\omega L_t I_{dc} \tag{8.1}$$

All the AC variables are given in RMS (phase-neutral for voltages) in this chapter. From the above equation, the AC system voltage magnitude V_g is the only AC variable that influences the DC system. If the converter is in constant extinction angle (CEA) control mode, the angle gamma is known and the actual firing angle β ($\beta = 180 - \alpha$) can be obtained from Eq. (3.26):

$$I_{dc} = \frac{\sqrt{6}V_g}{2\omega L_t} (\cos\gamma - \cos\beta) \tag{8.2}$$

High-Voltage Direct-Current Transmission: Converters, Systems and DC Grids, First Edition.
Dragan Jovcic and Khaled Ahmed.
© 2015 John Wiley & Sons, Ltd. Published 2015 by John Wiley & Sons, Ltd.

Alternatively, the same equation is used to determine γ if the converter angle β is known. The further two equations show the influence of DC side variables on the AC current. The magnitude of the AC current is given by Eq. (3.5):

$$I_g = \frac{\sqrt{6}}{\pi} I_{dc}$$ (8.3)

The power factor angle φ is:

$$\cos(\varphi_I) \approx \frac{\cos\gamma + \cos\beta}{2}$$ (8.4)

and this angle provides the position of the AC current because the coordinate frame is linked with the voltage V_g.

8.2 Phasor AC System Model and Interaction with the DC System

Figure 8.1 shows the simplified circuit of an HVDC converter connected to an AC grid. A converter is considered as an AC current source with a current phasor:

$$\overline{I}_g = I_g \angle \varphi_I$$ (8.5)

where the phasor components are given in the section above.

The use of reactances assumes that only the fundamental frequency is of interest and all dynamics are neglected. The AC system is defined by the remote voltage magnitude V_s (assumed constant), and an equivalent impedance $z_s = R_s + jX_s$. The AC system will commonly have many lines, generators and loads, while the series impedance z_s represents equivalent system behaviour at the point of common coupling (PCC). It is accepted that properly sized reactive power correction capacitors are supplied

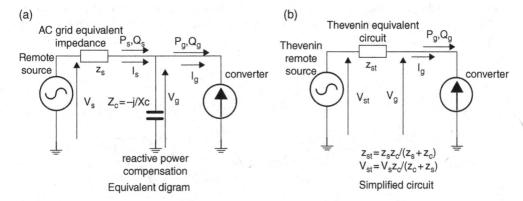

Figure 8.1 LCC HVDC converter connection with AC system.

$z_c = -j/X_c$. The reactive power equipment will typically supply full converter reactive power demand, which is around 50–60% of the rated power. The transformer impedance $z_t = jX_t$ is also included:

$$X_t = X_{tpu}\frac{V_s^2}{S_t}$$
(8.6)

where X_{tpu} is the per unit transformer impedance, which is normally 0.1–0.2 pu, and S_t is the transformer rating. With LCC converters, the converter AC voltage is assumed at the grid side of the transformer and the transformer impedance effect is therefore included in the DC model.

The AC system can be further simplified using Thevenin equivalent representation (Voltage V_{st} and impedance z_e), as shown in Figure 8.1b. The vector equation for AC system is:

$$\overline{I_g} = \frac{\overline{V_{st}} - \overline{V_g}}{z_{st}}$$
(8.7)

and the phasor diagram is shown in Figure 8.2. Assuming that the coordinate frame is linked with the converter voltage V_g, the AC current becomes:

$$I_g \angle \varphi_I = \frac{V_{st} \angle \varphi_{Vs} - V_g}{z_{st} \angle \varphi_Z}$$
(8.8)

The phasor equation above gives two scalar equations. When separated into real and imaginary components and summing the squares of components:

$$\left(R_{st}I_{gd} - X_{st}I_{gq} + V_g\right)^2 + \left(X_{st}I_{gd} + R_{st}I_{gq}\right)^2 = V_s^2$$
(8.9)

Rearranging Eq. (8.9) the equation for the grid voltage is found, assuming that the converter current is known:

$$V_g = -R_{st}I_{gd} + X_{st}I_{gq} + \sqrt{V_s^2 - \left(X_{st}I_{gd} + R_{st}I_{gq}\right)^2}$$
(8.10)

However, as the AC current also depends on the AC voltage magnitude as in Eqs (8.2)–(8.3), it is not possible to solve the power flow directly. It is therefore recommended that an iterative process or one of the nonlinear solvers be used to determine power flow from the above set of equations.

In Figure 8.2 the phasor diagram is shown for two different DC power injection levels. As DC current increases, the AC current magnitude and the current angle increase (converter AC current is always

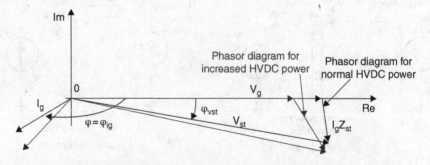

Figure 8.2 Phasor diagram of AC system connected to inverter.

lagging terminal voltage). As AC impedance is predominantly inductive, the increased reactive power consumption has the effect of reducing the converter AC voltage V_g. The active power will therefore be increasing at a slower rate than DC current and may be reducing after a certain operating point. At the point where AC power reduces as the DC current continues increasing, the system becomes unstable. This phenomenon is similar to voltage instability with AC systems, as given by the 'nose curve'.

8.3 Inverter AC Voltage and Power Profile as DC Current is Increasing

The assumptions for the inverter mode study are:

1. The magnitude of the remote source voltage is constant $V_s = const$. The coordinate frame is linked with PCC voltage V_g and therefore angle of the remote voltage will be changing.
2. No tap-changer action is considered (constant tap position) as a tap changer is very slow.
3. No capacitor bank switching is considered (constant capacitive support) as capacitor-switching events are slower than the events in this study.
4. There are no controls in the AC system (or they have no time to respond) as the events considered here are faster.
5. The inverter operates in constant (minimum) γ mode (CEA).
6. The DC current is an independent variable, which is assumed to be well controlled by the rectifier. The power flow is solved to reach a stable operating point for each value of DC current.

The test system for interaction studies is similar to those in the previous chapters but a unity transformer ratio is assumed as shown in Table 8.1. The short-circuit ratio (SCR) is studied in depth in the next chapter.

Figure 8.3 shows the inverter variables as DC current increases. The DC and AC voltages reduce as the DC power is increased. The AC voltage is high at low power, which is the result of excess reactive power. This AC system can transmit a maximum of around 610 MW of active power, which is called maximum available power) (MAP). Any higher DC current injection would cause system instability as AC power would reduce and AC voltage would collapse. This form of instability is important for HVDC systems and it has been experienced in some installations with weak AC grids. The system in Figure 8.3 can operate safely within the rated power range of 500 MW; however, if the strength of the AC system changes (tripping of lines or generators) then the MAP point can possibly move below the rated power level.

It is also seen that DC voltage and AC voltage have high values at low loading and therefore CEA control is not the preferred mode at low power levels.

Table 8.1 HVDC test system for fundamental frequency studies.

DC system power rating	500 MW
AC system voltage	$V_{sll} = 420$ kV
DC voltage (at inverter)	$V_{dci} = 496$ kV
Transformer rating,	$S_t = 500$ MVA
Transformer impedance	$X_{tpu} = 0.1$ pu
Transformer ratio	$n_t = 1$
Nominal extinction angle	$\gamma = 18°$
Minimal extinction angle	$\gamma = 15°$
AC impedance	$z_s = 7.96 + j79.6 \, \Omega$ (SCR = 4, X/R = 10)
Reactive power supply	50% of Pdc (Q = 250 MVAr). $X_c = -0.0014 \, \Omega$

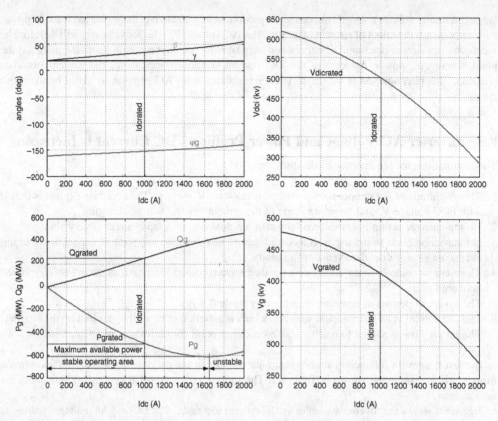

Figure 8.3 Interaction between inverter DC and AC system for DC current increase, assuming CEA mode (γ = constant).

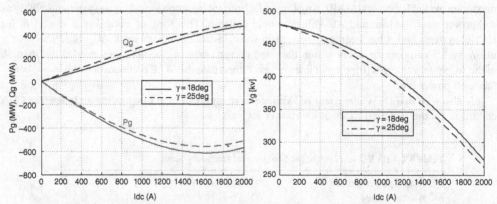

Figure 8.4 Interaction between DC and AC systems for two different γ angles. CEA mode.

8.4 Influence of Converter Extinction Angle

In this section, the impact of the extinction angle value on the converter curves is investigated. A larger extinction angle may be desired in order to improve resilience from commutation failure.

Figure 8.4 shows the active and reactive power as the function of DC current for two different extinction angles. It is evident that the larger extinction angle has less favourable properties because the peak power is lower and the stability margin is lower. Since the MAP power is around 540 MW with $\gamma = 25°$, which is only 8% higher than the rated power, such systems may not be satisfactory in practice.

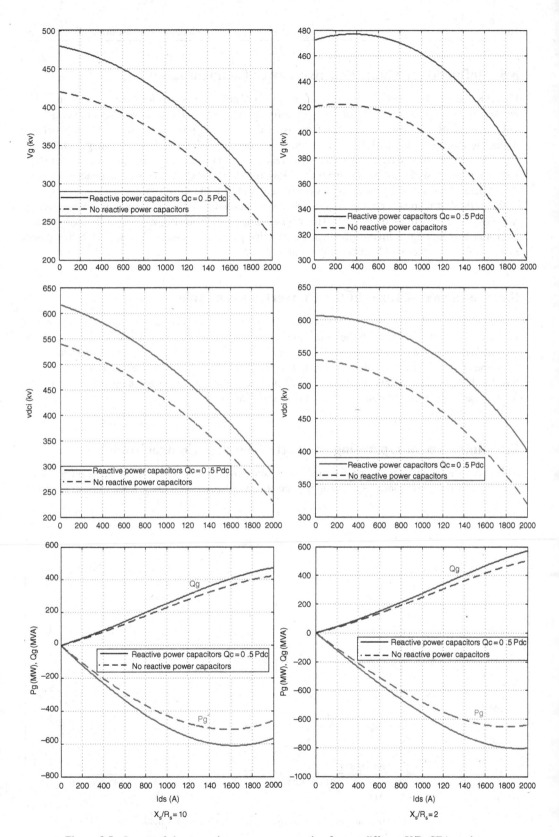

Figure 8.5 Impact of shunt reactive power compensation for two different X/R. CEA mode.

8.5   Influence of Shunt Reactive Power Compensation

Figure 8.5 shows the interaction between the inverter DC and AC systems for the cases with and without reactive power capacitors, X_c, and for two different ratios, X/R. The reactive power compensation will increase the grid voltage and therefore also DC voltage significantly and this has a beneficial impact in terms of increased power transfer. Note, however, that the no-load voltage with shunt capacitors is very high. The voltage curve with reactive power compensation also has a steeper slope indicating the possibility of more dynamic stability issues. Shunt capacitors will, in general, increase AC system series impedance, causing more voltage swing for the same power change.

When the X/R ratio is high, the system behaves as an inductive line and therefore voltage largely depends on the reactive power flow. The voltage drop at high load is high regardless of whether converter is rectifying or inverting. When the resistive part is increased (X/R low), the active power injection form inverter will help boost the grid voltage.

8.6   Influence of Load at the Converter Terminals

There will frequently be some local load at the converter PCC terminals, although more commonly at the inverter terminals. Figure 8.6 shows the interaction curves for two cases: where the load is equal to the rated DC power and where it is equal to twice the rated power. A load at the inverter terminals will consume converter power locally and this is beneficial because it avoids problems caused by power transfer through large system impedance. At rectifier terminals the effect is the opposite because the load demands more power transfer through the series impedance.

8.7   Influence of Operating Mode (DC Voltage Control Mode)

The inverter normally operates in CEA mode or constant DC voltage control mode. Some other control modes are also used on some systems but they are not considered in this study. Regardless of the actual

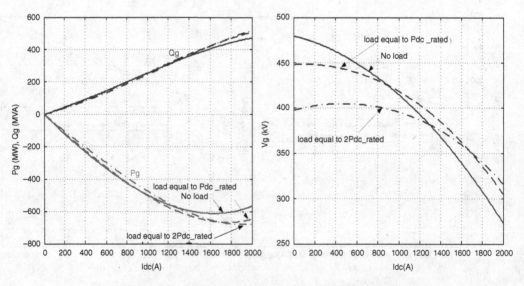

Figure 8.6   Impact of load at inverter AC bus. CEA mode.

inverter operating mode, the CEA control is always active in the background and it is activated when required to prevent commutation failure.

Figure 8.7 shows the interaction variables when constant DC voltage control is used. It is observed that the system moves to the CEA control around 1100 A in order to prevent gamma falling below 15°.

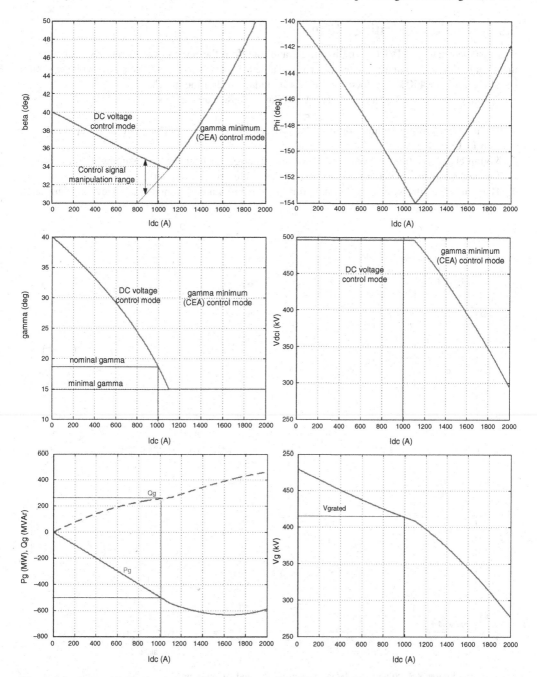

Figure 8.7 Interaction between inverter DC and AC system as DC current increases, assuming DC voltage control mode (V_{dci} = constant), also showing mode transition to CEA mode.

It is evident that DC voltage control mode produces better AC power, lower AC voltage drop and in general better stability, compared with the CEA mode. Any disturbance on the AC system will be corrected faster if the inverter is in DC voltage control mode.

8.8   Rectifier Operating Mode

It is assumed that the HVDC is bidirectional and therefore the test system for the rectifier mode is identical to that for the inverter mode shown in Table 8.1; however, the remote source voltage should be adjusted ($V_{sll} = 450\,kV$), which can be accommodated using the tap changer.

Assumptions for rectifier mode study are similar to those for inverters:

1. The magnitude of the remote source voltage is constant $V_s = const$. The coordinate frame is linked with PCC voltage V_g.
2. No tap changer action is considered (constant tap position) because tap changing is too slow.
3. No capacitor bank switching is considered (constant capacitive support).
4. There are no controls in the AC system (or they have no time to respond) as the events considered here are faster.

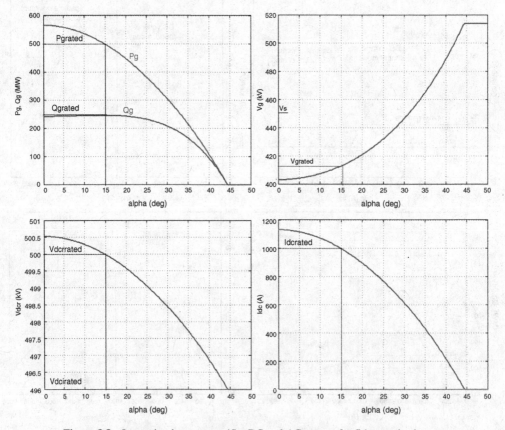

Figure 8.8   Interaction between rectifier DC and AC system for firing angle change.

5. The inverter operates in constant DC voltage mode ($V_{dci} = const$). A DC line of 4 Ω total resistance is assumed. Therefore DC current can be calculated.
6. Rectifier firing angle (α) is an independent variable. The power flow is solved to reach a stable operating point for each value of α.
7. The inverter DC current control would normally be activated at low DC current, but this is not considered. It is assumed that the current order is communicated between converters sufficiently quickly.

Figure 8.8 shows the locus of steady-state operating points for a rectifier as the firing angle is increasing. A low firing angle implies high power according to the cosine function, but α cannot be reduced below zero. No instability is observed, unless the AC system becomes very weak as studied in the next chapter.

9

HVDC Operation with Weak AC Systems

9.1 Introduction

The characteristics of AC systems have a significant impact on the operation of HVDC systems. Commonly, the AC systems that cause operating issues are labelled as weak, but this notation has two different aspects:

- high-impedance systems, which are associated with voltage stability and power transfer limitations;
- low-inertia systems, which are linked with frequency deviation concerns.

The high-impedance systems are mostly responsible for HVDC stability concerns and they are frequently associated with the notion of weak systems. The HVDC operational experience has shown that, in particular, the relative AC strength compared with HVDC power is an important indicator of HVDC operating problems.

9.2 Short-Circuit Ratio and Equivalent Short-Circuit Ratio

9.2.1 Definition of SCR and ESCR

Expanding in more depth on the same AC system representation as in Figure 8.1, the impedance vector is:

$$\overline{z_s} = R_s + jX_s = z_s \angle \varphi_Z \tag{9.1}$$

which is also characterized by the ratio X_{ratio}:

$$X_{ratio} = \frac{X_s}{R_s} = \tan \varphi_Z \tag{9.2}$$

High-Voltage Direct-Current Transmission: Converters, Systems and DC Grids, First Edition.
Dragan Jovcic and Khaled Ahmed.
© 2015 John Wiley & Sons, Ltd. Published 2015 by John Wiley & Sons, Ltd.

For transmission systems, typically $5 < X_{ratio} < 20$, while it can be much lower for distribution grids. Referring to the topology in Figure 8.1, the short-circuit ratio (SCR) is introduced as:

$$SCR = \frac{V^2_{s\_LL}}{z_s P_{dc}} \tag{9.3}$$

where P_{dc} is the converter rated power.

The AC system is normally characterized by the SCR and the ratio X_{ratio}, which enables calculation of AC system parameters:

$$R_s = \frac{V^2_{s\_LL}}{SCR P_{dc}} \frac{1}{\sqrt{1 + X^2_{ratio}}} \qquad X_s = \frac{V^2_{s\_LL}}{SCR\, P_{dc}} \frac{X_{ratio}}{\sqrt{1 + X^2_{ratio}}} \tag{9.4}$$

The SCR can also be expressed using the short-circuit level (SCL). The SCL is defined as the power that the system in Figure 8.1 can deliver when converter terminals are short circuited, and this power will also be considered as the base power in this chapter:

$$SCL = \frac{V^2_{s\_LL}}{z_s} = S_{base} \tag{9.5}$$

Further, the voltage and current bases can be defined:

$$V_{base\_LL} = V_{s\_LL}, \quad I_{base} = \frac{V_{s\_LL}}{\sqrt{3}z_s} = \frac{S_{base}}{\sqrt{3}V_{base\_LL}} \tag{9.6}$$

The system in Figure 8.1 is quite simple and SCL can be used here to determine SCR. In general, however, the terms SCR and SCL are used for completely different purposes. In the SCR definition, z_s represents impedance to the nearest bus with fixed AC voltage under normal conditions that can take full converter power. As an example, a large voltage source converter (VSC) or STATCOM or wind farm can establish a firm AC voltage bus under normal operation. On the other hand, z_s in the SCL definition in (9.5) is the impedance to the rotating machine that can sustain the rated AC voltage under fault conditions. The SCL is associated with fault studies and such a remote source is required as the machine speed stays unchanged and constant excitation will give firm back EMF (electromotive force) for the short fault duration. If the SCL is used in the SCR definition, the conclusions on stability may become overly conservative.

The AC system strength is classified as shown in Table 9.1. It is extremely difficult to operate HVDC with AC systems having SRC < 2. The SCR can be increased by strengthening the AC system, for example by adding transmission lines or generation plants or by increasing system voltage. These measures are expensive and some other methods should be considered as discussed in the sections below.

Table 9.1 SCR classification.

SCR	Classification	Operating difficulties
SCR > 3	Strong system	No operating problems
2 < SCR < 3	Weak AC system	Operating difficulties can be expected. Some special controls are required
SCR < 2	Very weak AC system	Serious operating difficulties can be expected. Very few HVDC systems operate with such low SCR

In order to include the influence of the reactive power capacitors and shunt filters on the converter bus, the equivalent short-circuit ratio (ESCR) is introduced:

$$ESCR = \frac{SCL - Q_f}{P_{dc}} \tag{9.7}$$

where Q_f is the reactive power supplied by the shunt capacitors/filters. The reactive power capacitors improve the power factor and improve the voltage profile but they also reduce the SCR and may reduce the stability limits.

With weak AC systems the impedance z_s is large and this implies a large voltage swing at the converter terminals as the loading is changing. The maximum power transfer is determined by the system impedance and the voltage level, where larger impedance implies a lower maximum power transfer limit.

Example 9.1
An HVDC system of 2 GW is connected to an AC system as shown in Figure 9.1. Determine:

a. SCR,
b. SCR assuming that one line is out of service,
c. ESCR with both lines in service,
d. Comment on the above results, regarding the HVDC operating difficulties.

Solution

a. Both lines in service.

$$\overline{z_1} = 5 + j2\pi f 0.05, \quad z_2 = 5 + j2\pi f 0.05$$

$$\overline{z_3} = \frac{\overline{z_1 z_2}}{\overline{z_1} + \overline{z_2}} = 2.5 + j7.854 = 8.24 \angle 72.34°$$

$$SCR = \frac{V_{sll}^2}{z_3 P_{dc}} = \frac{220\,000^2}{8.24 \times 2e9} = 2.94$$

This is a system at the border between strong and weak.

b. One line out of service.

$$SCR = \frac{V_{sll}^2}{z_1 P_{dc}} = \frac{220\,000^2}{16.48 \times 2e9} = 1.47$$

This is a very weak system and HVDC will not be able to operate.

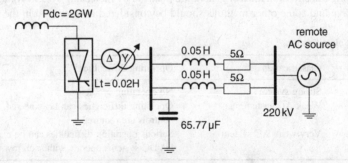

Pdc=2GW

Lt=0.02H

0.05 H   5Ω

0.05 H   5Ω

65.77 µF

220 kV

remote
AC source

Figure 9.1   System in Example 9.1.

c. ESCR calculation.

$$Qc = \frac{V_s^2}{1/(2\pi fC)} = 1 \text{ GVAr};$$

$$ESCR = \frac{SCL - Q_f}{P_{dc}} = \frac{\dfrac{220\,000^2}{8.24 \times 2e9} - 1e9}{2e9} = 2.44$$

d. Comments.

This example illustrates how shunt reactive power reduces the system strength. Shunt capacitors improve power factor of the line current but they may create dynamic HVDC operating difficulties.

9.2.2 Operating Difficulties with Low SCR Systems

A weak AC system causes operating issues when connected to inverters whereas connection with rectifiers is less critical. It is unfortunate that weak AC systems are most often connected to inverters because HVDC systems are typically installed to supply power to areas with deficient generation. A rectifier is most frequently connected to stronger AC systems like remote generation regions.

Some of the problems that occur with high-impedance AC systems include:

- voltage instability;
- small-signal control instability;
- commutation failures;
- harmonic resonances;
- temporary overvoltages.

The first two problems will be discussed further in some depth in the following sections.

Commutation failure is among the foremost concerns with weak AC systems but it is linked with voltage instability. The primary cause of commutation failure is the depression in the AC system voltage. The more AC voltage deviates, the higher is the probability of commutation failure. Weak AC systems have large impedance and this leads to large voltage swings, which increase commutation failure probability. Moreover, recovery from commutation failure becomes particularly difficult with weak AC systems. During the recovery period the converter operating angle is high demanding more reactive power, while active power transfer is low, and to exacerbate problem the depressed AC voltage implies that reactive power supply from capacitors is low.

The AC system impedance has characteristic resonant peaks, which typically fall at higher frequencies and do not cause harmonic magnification issues. However, with very weak AC systems the first resonant peak will be shifted to lower frequencies and may coincide with low harmonics. If AC harmonics are magnified they can cause control problems with HVDC firing circuit resulting in feedback instability as studied in Chapter 8.

9.3 Power Transfer between Two AC Systems

A theoretical study of power transfer between two AC systems is now provided as an example. As a line-commutated converter (LCC) is a current source for the AC grid, the study considers one AC voltage source and an ideal current source as shown in Figure 9.2. The aim is to explore theoretical limits for power exchange between an AC system and a converter through an equivalent line, given by impedance z_s.

The basic equations for power-flow analysis are given in Eqs (8.7)–(8.10). Assuming that the converter current magnitude is fixed at I_{base}, which is theoretically the largest possible current, while the current angle φ_{Ig} is an independent variable, the active power P_g and point of common coupling (PCC) voltage V_g curves can be obtained as shown in Figure 9.3.

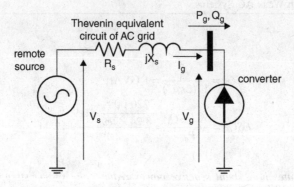

Figure 9.2 AC system and converter as a current source for phasor interaction studies.

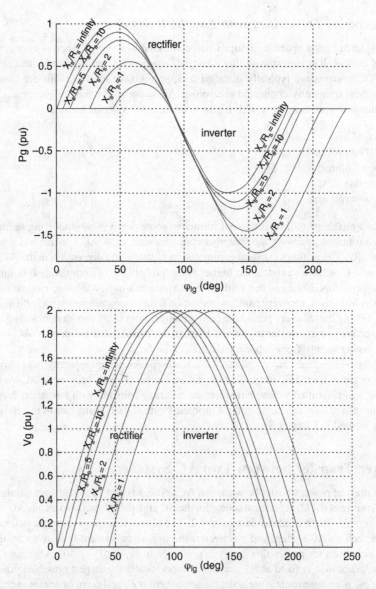

Figure 9.3 Maximum power transfer between an AC system and a LCC converter.

The case X_{ratio} = *infinity*, corresponds to a textbook case of power exchange between two AC systems through an inductive line. In this case, the maximum power transfer is obtained for a 90° phase shift between the voltages giving a 45° phase shift between any voltage and the current. In such case the maximum power transfer is equal to the short-circuit power (P_{dc} = *SCL*). Therefore theoretical minimum is *SCR* = *1*, and this applies both to rectifier and inverter operation. The converter voltage in such case would be *1.41 × V_s*.

If the X_{ratio} reduces, then the maximum power transfer reduces for rectifier and increases for inverter as seen in Figure 9.3. As an example, X_{ratio} = 5 gives minimal theoretical *SCR* = 1/0.75 = 1.33 for

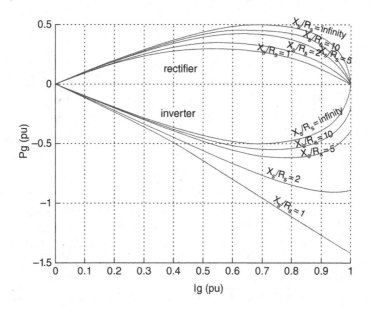

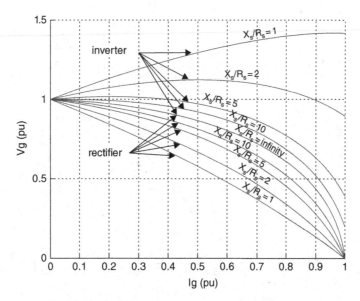

Figure 9.4 Maximum power transfer between an AC system and a LCC converter, assuming that converter current is in phase with the converter voltage.

rectifier and $SCR = 1/1.2 = 0.83$ for inverter. However, to achieve these values, the current angle would need to be exactly 48° for the rectifier or 138° for the inverter, which is difficult in practice. A typical LCC converter operates with current lagging voltage by around 30–40°.

Figure 9.4 shows the maximum power transfer between an AC system and an LCC converter, assuming that the current is in phase with the voltage. Such operating condition might occur if a static synchronous compensator (STATCOM) or a VSC compensates LCC reactive power at the PCC. It is seen that the minimal $SCR = 2$ is achieved for both rectifier and inverter in case $X_{ratio} = infinity$.

Figure 9.5 shows maximum power transfer for a more practical case where the converter current is lagging voltage by 30° (corresponding to $Q_g = 0.58 P_g$). It demonstrates that the theoretically minimal

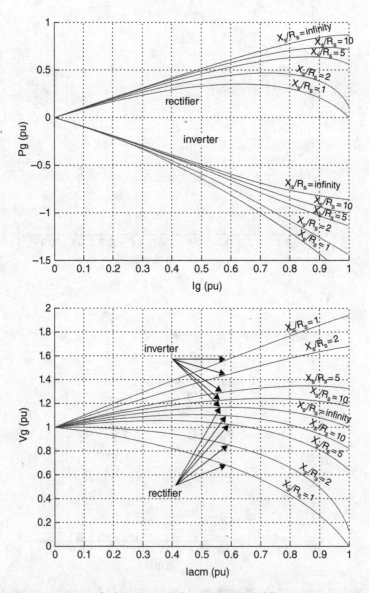

Figure 9.5 Maximum power transfer between an AC system and a LCC converter, assuming that converter current is lagging converter voltage by 30°.

$SCR = 1.25$ for both rectifier and inverter when X_{ratio} = infinity. When $X_{ratio} = 5$, the theoretically minimal $SCR = 1.5$ for rectifier and 0.8 for inverter. However the maximum allowed voltage swing at the PCC bus should also be considered in order to derive general conclusions.

9.4 Phasor Study of Converter Interactions with Weak AC Systems

Figure 9.6 shows the interaction between HVDC and AC system, similarly as in Figure 8.4, but the SCR is varied from strong ($SCR = 6$) to weak (SCR = 2.3). It demonstrates that the theoretically minimal system strength for the 500 MW test system is around $SCR = 2.5$, in order to enable the maximum available power (MAP) higher than the rated power. In practice, however, a margin of at least 10–20% above the theoretical minimum would be needed.

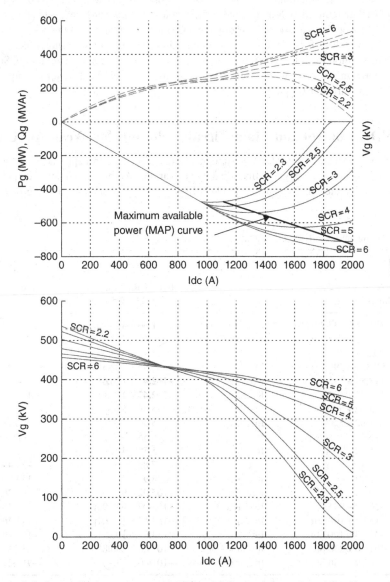

Figure 9.6 Interaction between converter DC and AC system for various AC grid strengths.

The voltage graph shows that weak AC systems cause a very steep voltage slope as DC power is varied. This amplifies stability problems since any external AC voltage drop (because of a remote fault) further depresses the power curve. The voltage swing between no-load and full-load condition can be significant. Note also that reactive power supply from shunt capacitors reduces according to a square function of AC voltage, and this further depresses AC voltage.

9.5   System Dynamics (Small Signal Stability) with Low SCR

Table 9.2 shows the system eigenvalues for a reduced SCR at the rectifier and inverter sides, considering the CIGRE HVDC benchmark model studied in Chapter 7. In both cases stability deteriorates with lower SCR. Lower SCR at the rectifier side predominantly affects eigenvalues 5 and 6 (referring to labels from Table 7.5), which indicates the presence of oscillations at around 30 Hz. Stability deteriorates very fast with reduced SCR at the inverter side, and instability can be expected around the second harmonic (119 Hz on the AC side or 59 Hz on the DC side). The real part of the dominant eigenvalues 3,4 reduces around 30% for such a small SCR reduction. It is important to note that these dynamic issues arise in addition to all the problems discussed in previous sections related to low SCR. In practice, when operating problems with weak AC systems arise, it might be difficult to pinpoint the exact cause that led to converter tripping.

9.6   HVDC Control and Main Circuit Solutions for Weak AC Grids

If an HVDC is connected to a weak AC system then some special control approaches are required. The simplest solutions include a larger extinction angle or various inverter stabilization control loops. The inverter can be operated in DC voltage mode, which gives better stability. On the downside, these

Table 9.2   Dominant CIGRE HVDC eigenvalues with reduced SCR.

	Original eigenvalues Rec SCR = 2.5, Inv SCR = 2.5	Eigenvalue sensitivity	Rectifier reduced SCR Rec SCR = 2.0, Inv SCR = 2.5	Inverter reduced SCR Rec SCR = 2.5, Inv SCR = 2.0
1	−37.343 + 437.23j	Rec, Rec	−41.60 + 414.29j	−33.19 + 443.37j
2	−37.343 − 437.23j	Rec, Rec	−41.60 − 414.29j	−33.19 − 443.37j
3	−90.311 + 398.73j	Inv, Inv	−81.96 + 397.33j	**−63.94 + 374.78j**
4	−90.311 − 398.73j	Inv, Inv	−81.96 − 397.33j	**−63.94 − 374.78j**
5	−118.84 + 195.9j	Rec, Rec	**−102.47 + 176.65j**	−124.87 + 176.81j
6	−118.84 + 195.9j	Rec, Rec	**−102.47 − 176.65j**	−124.87 − 176.81j
7	−5.62 + 5.38j	Inv, Inv	−5.39 + 5.43j	−6.13 + 5.93j
8	−5.62 − 5.38j	Inv, Inv	−5.39 − 5.43j	−6.13 − 5.93j
9	−5.49 + 5.049j	Inv, Inv	−5.55 + 4.85j	−5.49 + 5.01j
10	−5.49 − 5.049j	Inv, Inv	−5.55 − 4.85j	−5.49 + 5.01j
11	−43.46	Rec, Rec	−47.82	**−29.49**
12	−151.01	Rec, Rec	−148.9	−155.73
13	−107.0 + 981.7j	Rec, Rec	−105.02 + 937.95j	−107.26 + 982.51j
14	−107.0 − 981.7j	Rec, Rec	−105.02 − 937.95j	−107.26 − 982.51j
15	−154.4 + 1002.0j	Inv, Inv	−154.63 + 1002.1j	**−127.70 + 955.15j**
16	−154.4 − 1002.0j	Inv, Inv	−154.63 − 1002.1j	**−127.70 − 955.15j**
17	−1014.0 + 236.8j	Inv, Inv	−1014.0 + 236.8j	**−955.29 + 267.11j**
18	−1014.0 − 236.8j	Inv, Inv	−1014.0 − 236.8j	**−955.29 − 267.11j**

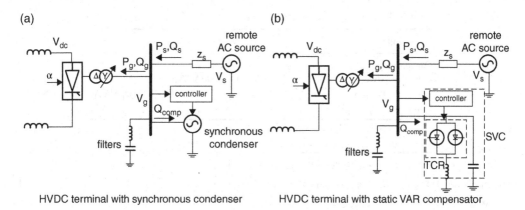

Figure 9.7 Synchronous condensers and an SVC as solutions for HVDC with a weak AC systems.

approaches always imply a larger firing angle, which in turn demands more reactive power, increases losses, produces more valve stresses and generates more harmonics.

One possible control solution is to employ DC current control at the inverter terminal (as used at the Korean Jeju Island HVDC link, which enabled operation with very weak AC system ($SCR < 2$)). As a negative consequence of this approach the nominal firing angles are larger and both rectifier and inverter are rated for somewhat higher voltages and currents.

Synchronous condensers are employed with many HVDC links to avoid low SCR problems as shown schematically in Figure 9.7. Synchronous condensers provide positive shunt impedance in parallel with the AC system and this improves SCR. They also increase the system inertia, which helps with transient responses. They can be configured to control reactive power or the AC voltage in a feedback manner which indirectly improves stability of the AC system. Synchronous condensers will generally significantly reduce voltage overshoots and improve recovery responses including commutation failure recovery but their response is rather slow and settling time is 200–400 ms.

As a very effective method of maintaining converter AC voltage, the static VAR compensators (SVCs) or STATCOM may be employed. Static VAR compensators are capable of providing reactive power, which is rapidly controllable, and they can be controlled to maintain AC voltage at a constant level with time constants of 50–200 ms (20–100 ms for STATCOMs). However an SVC is essentially a shunt capacitor and therefore it will reduce ESCR, which might negatively affect system dynamics. The overvoltage peaks are larger and commutation failure recovery times are generally larger, than those with synchronous condensers. STATCOMs have the advantage of fastest response and reactive power supply which is independent of the system voltage magnitude.

9.7 LCC HVDC with SVC (Static VAR Compensator)

This section will examine the phasor interaction curves when a SVC is employed at the PCC. It is assumed that the SVC capacitive circuit will provide the full reactive power requirement for the inverter (250 MVAr for the test HVDC). The inductive circuit of SVC has a 350 MVAr rating in order to provide 100 MVAr of inductive power at a light HVDC load. Table 9.3 shows the parameters of the SVC used with the test system.

Figure 9.8 shows the inverter operating curves of the test system when fixed capacitors are used and comparatively when SVC is used for reactive power compensation. In this case a weak AC system ($SCR = 2.0$) is considered to better illustrate the benefits. The same inverter test system from previous sections is in DC voltage control mode.

Table 9.3 SVC parameters in the test case.

Rated voltage	$V_{SVC} = 300\,kV$
Capacitive power	$Q_c = -300\,MVAr$
Inductive power	$Q_l = 400\,MVAr$
Total reactive power	$-300\,MVAr < Q_{comp} < 100\,MVAr$
Converter rating	$400\,MVA$

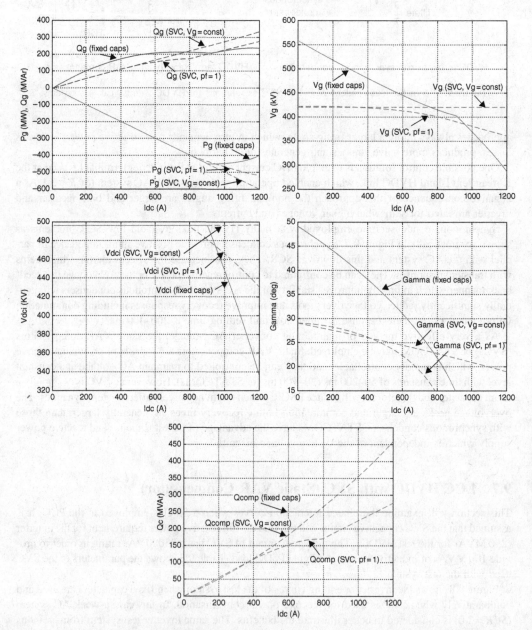

Figure 9.8 Impact of an SVC at the HVDC inverter terminal with a weak AC system (SCR = 2.0).

Two cases of SVC control are considered:

- the SVC regulates the voltage V_g;
- the SVC follows the converter reactive power Q_g, giving the unity power factor as seen from the grid.

The first graph in Figure 9.8 illustrates that an inverter with fixed capacitors would not be able to deliver rated power of 500 MW. It shows that the SVC can significantly increase the MAP, practically eliminating the danger of instability. The SVC with pf = 1 shows a less favourable response as full power is achieved only if DC current increases to 1100 A.

The voltage graph (V_g) is particularly important because it illustrates that the HVDC terminal with the SVC has significantly better voltage profile for the whole power range. The fast SVC controls can adjust reactive power in the same control bandwidth as the HVDC controls. Without SVC, the voltage swing would be over 40%.

The graph with an extinction angle (γ) proves that SVC support ($V_g = const$ control) enables a much higher γ at the HVDC inverter at high powers, which implies that commutation failure probability will be lower.

The last graph shows the required compensator power. The SVC rating with constant voltage control should be around 60% of the HVDC rating.

Figure 9.9 shows the influence of the SVC at the rectifier terminal with a weak AC system ($SCR = 2.0$). In the power graph it is seen that rectifier with fixed capacitors is not able to deliver the rated power. When SVC in voltage control mode is connected, the maximum power achieved is over 550 MW. The grid voltage (V_g) profile shows the remarkable benefit of using SVC. It is also interesting that, if the SVC is operated to follow rectifier reactive power, then the benefit is rather limited.

Figure 9.10 shows the operating curves of the joint system HVDC plus SVC, as seen from the grid. The variables P_s and Q_s from Figures 9.7 and 8.1 are shown, which is the method more commonly used with VSCs. The operating point can be anywhere within the area, which is a very different shape from the familiar circle-resembling PQ shape with VSC HVDC. The active power is controlled by the HVDC while reactive power is regulated independently by the SVC.

9.8   Capacitor-Commutated Converters for HVDC

Capacitor-commutated converters (CCCs) use series connected capacitors either on the valve side or on the grid side of the converter transformer. They are more often placed on the valve side as shown by C_s in Figure 9.11. The CCC concept has been implemented in two practical projects: 2002, 2200 MW, ±70 kV, back-to-back Garabi (Brazil-Argentina) HVDC and 2003, 200 MW, ±13 kV, back-to-back Rapid City HVDC.

It is convenient that the CCC achieves a reactive power supply that is proportional to the line current and therefore they naturally compensate converter reactive power variation. As the result, only a small external reactive power compensation is required. Another advantage of the CCC is that commutation failure probability is reduced because capacitors provide a more stable commutating voltage (which is less dependent on the grid condition).

On the downside, the CCC introduces higher voltage stress on the valves requiring higher converter costs and higher harmonics. In the case of DC faults, the fault current will run through series capacitors and this requires capacitor overrating.

9.9   AC System with Low Inertia

The AC systems with low inertia, when coupled with HVDC, may cause frequency deviation problems. In general, HVDC operation is not affected by frequency deviation because the phase locked loop

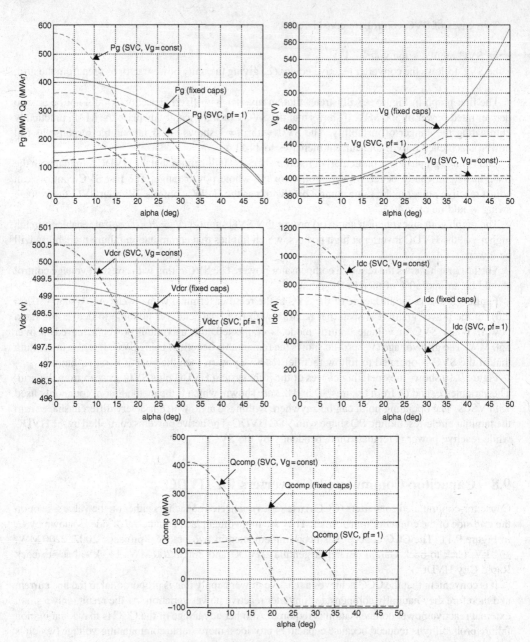

Figure 9.9 Impact of placing an SVC at the HVDC rectifier terminal with a weak AC system (SCR = 2.0).

(PLL) can adequately track AC frequency and phase changes. The frequency deviation, however, is of concern for the AC system, which must comply with frequency standards, and the interaction with HVDC may exacerbate the issue.

A representative test system topology involves an HVDC connected to a single-machine AC system, as shown in Figure 9.12. The machine with inertia may be a generator or a synchronous condenser.

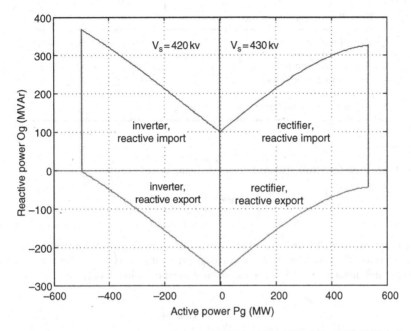

Figure 9.10 PQ diagram of 500 MW HVDC with a −250/+100 MVAr SVC (SCR = 4).

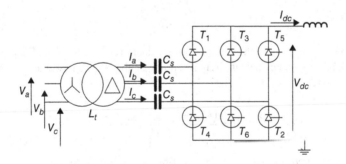

Figure 9.11 HVDC with capacitor commutated converter.

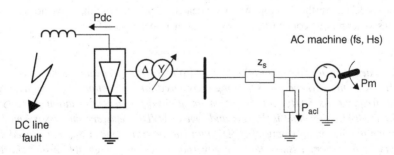

Figure 9.12 Equivalent circuit for systems with low inertia.

Table 9.4 Typical values for inertia constants.

Type of machine	Inertia constant MWs/MVA
Thermal unit, two poles	2.5–6
Thermal unit, four poles	4–10
Hydraulic unit	2–4
Synchronous condenser	1–2

The basic equation for torque balance on the machine is:

$$J\frac{d\omega_m}{dt} = T_m - T_e \tag{9.8}$$

where ω_m is the mechanical speed, T_m is the mechanical torque, T_e is the electrical torque, and J is the moment of inertia of the generator and turbine. Considering small deviations around the operating point and using per unit notation for all variables, an equivalent equation is obtained:

$$2H_s\frac{d\omega_{mpu}}{dt} = T_{mpu} - T_{epu} \tag{9.9}$$

where the per unit inertia constant H_s is defined:

$$H_s = \frac{1}{2}\frac{J\omega_{om}^2}{S_{base}} \tag{9.10}$$

Table 9.4 shows some typical values for the machine inertia constants. Using Eq. (9.9) it is possible to derive expression for frequency deviation Δf depending on the HVDC power and system inertia for the system in Figure 9.12.

$$\Delta f = \frac{(P_{mpu} - P_{aclpu} - P_{dcpu})f_s\Delta t}{2H_s} \tag{9.11}$$

where Δt is the considered time interval (fault-clearing time or HVDC fault-recovery time) and all other variables are defined in Figure 9.12. The fault time is typically known from the protection studies, and this expression therefore facilitates a frequency deviation study for given inertia and power levels. The above formula can also be used to determine the maximum DC power that can be connected to the system under the given frequency deviation limits.

The same model can further be employed with dead networks – those that have no active generation. In such cases synchronous condensers are used and therefore $P_{mpu} = 0$.

Example 9.2
A weak AC system has an equivalent 800 MVA hydro unit with a per unit inertia constant of 3 MWs/ MVA. An HVDC with overhead wires is being considered for connection to this system. Assume that the nominal frequency is 50 Hz and the maximum allowed frequency deviation is 0.5 Hz (0.01 pu). The DC fault self-clearing time is 100 ms and typical HVDC recovery time is assumed to be 200 ms. Determine the maximum rating of HVDC that can be connected, while not exceeding the maximum frequency deviation. Assume that the governor reaction is negligible in the short timeframe considered.

Solution

It is assumed that mechanical power is $P_{mpu} = 1$ pu, and that before the fault electrical power balances the mechanical power P_{mpu}-P_{aclpu}-$P_{dcpu} = 0$. While the DC fault is being cleared for 0.1 + 0.2 s, DC power reduces to zero and the power balance equation becomes P_{mpu}-P_{aclpu}-0 = P_{dcpu}. Therefore:

$$\Delta f = \frac{(P_{dcpu})f_s \Delta t}{2H_s}$$

$$P_{dcpu} = \frac{\Delta f 2H_s}{f_s \Delta t} = \frac{0.5 \times 2 \times 3}{50 \times (0.1 + 0.2)} = 0.2 \text{ pu}$$

$$P_{dc} = P_{dcpu} MVA = 0.2 \times 800 = 160 \text{ MW}$$

Therefore, in the above case, the maximum HVDC rating is 20% of the generator rating. In order to increase HVDC penetration the system inertia could be increased, for example by connecting synchronous condensers.

10

Fault Management and HVDC System Protection

10.1 Introduction

With thyristor HVDC systems, the HVDC controls are generally capable of resolving most transient fault situations in a very short time. On a longer time scale (over 50 ms), if the fault situation persists, various mechanical circuit breakers (CBs) will react. A number of mechanical CBs are associated with an HVDC system but they are used only as a final means of protection – that is, when HVDC controls cannot solve disturbance and permanent isolation is required. The tripping of CBs implies loss of capacity.

Figure 10.1 shows some typical fault locations and the means of protection. The HVDC controls always react to a disturbance first. A tap changer may also react when a modest AC voltage deviation persists for a long time (remote AC fault or tripping).

10.2 DC Line Faults

Line-commutated HVDC is capable of rapidly controlling DC voltage in positive or negative regions and it can therefore react to DC faults according to a fault-control strategy. The fault-control strategy includes limiting the DC current magnitude and changing DC voltage polarity in order to extinguish the DC current.

Figure 10.2 shows system variables (rectifier and inverter) for a DC fault, at 0.45 s using the same 500 MW HVDC test system as in Chapter 6. The fault is cleared by special HVDC controls. Firstly, the normal HVDC control response is studied before 0.57 s in order to analyse shortcomings in dealing with DC faults:

- The rectifier regulates the DC current at reference value (I_{dcref_rec} = 1 pu), which continuously feeds the fault. This requires substantial reduction of rectifier DC voltage and therefore the rectifier firing angle α_r is just below 90°. Reactive power demand is excessive (600 MVAr) because of operating conditions with rated DC current and a very large firing angle.

High-Voltage Direct-Current Transmission: Converters, Systems and DC Grids, First Edition.
Dragan Jovcic and Khaled Ahmed.
© 2015 John Wiley & Sons, Ltd. Published 2015 by John Wiley & Sons, Ltd.

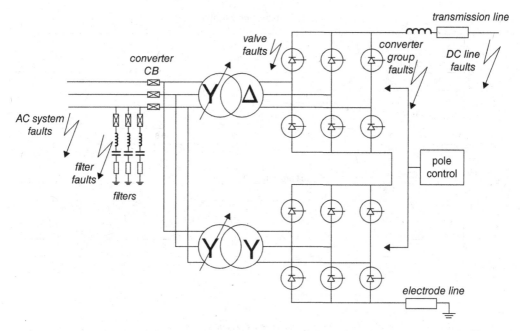

Figure 10.1 Fault locations and protective means with HVDC systems.

- The inverter current reduces below inverter reference value ($I_{dcref\_inv} = I_{dcref\_rec} - I_{margin}$) and the inverter moves to current control mode by reducing inverter firing angle α_i. However, in order to keep the same DC current polarity, the inverter DC voltage polarity changes and the inverter firing angle settles to just below 90°. The inverter station now also operates as a rectifier and feeds the DC fault. Inverter reactive power demand is over 500 MVAr.

The above response is not suitable as a long-term strategy to DC faults since converters still feed DC fault arc as seen in the fault resistor current response. Excessive reactive power demand may also disturb AC systems. It is required to extinguish the fault current (to break the arc) by discharging line energy through converters. A special DC fault-protection control strategy is therefore initiated on detection of DC faults which changes DC voltage profile as shown in Figure 10.3. This special control strategy is as follows:

- the rectifier firing angle is forced into inversion (110–120°), which discharges line energy on the rectifier side;
- the inverter firing angle is limited to inversion mode (110–120°), which discharges line energy on the inverter side.

The above control action is simulated in the model in Figure 10.2, and it becomes active around 0.57 s. Rectifier and inverter angles are both temporarily advanced to 110°, DC voltage is reversed and the fault current drops to zero. This completely extinguishes the fault current and naturally clears the fault. In Figure 10.2, the DC voltage ramp is initiated immediately after the fault is cleared and normal operation resumes after 300–400 ms. In a practical system, restart would be attempted after a delay for deionization. The above sequence is repeated until the DC voltage is recovered to nominal values.

Note that a special discriminatory logic is needed for the detection of DC faults, as the above control strategy is not appropriate in the case of commutation failures (which also bring DC voltage to zero). In DC systems with overhead lines, most DC faults will be transient and they are cleared with fast DC

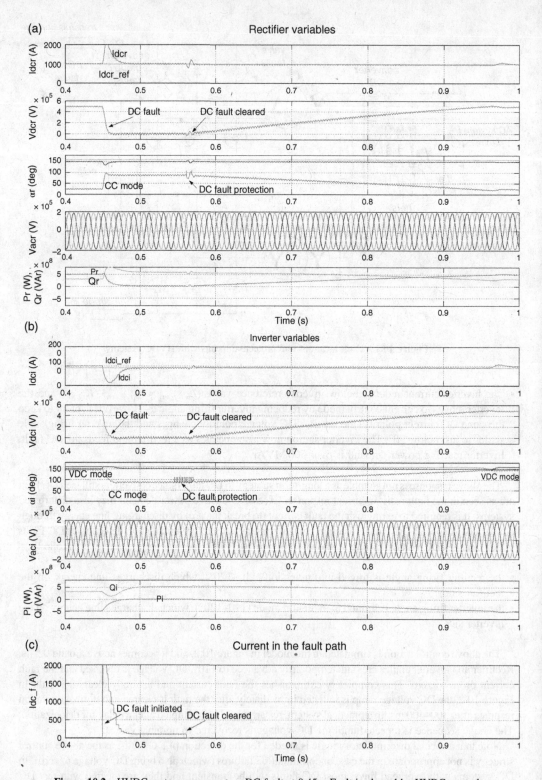

Figure 10.2   HVDC system response to a DC fault at 0.45 s. Fault is cleared by HVDC controls.

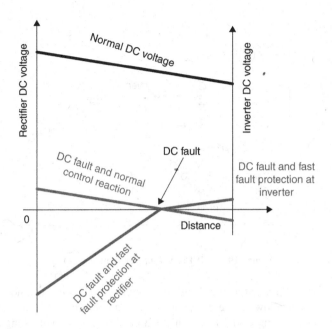

Figure 10.3 Steady-state DC voltage profile for DC faults and effect of DC fault protection control.

control action. However, with cable systems, DC faults are usually permanent and slightly different fault management is used. A special DC fault-protection control strategy is initiated on detection of DC faults, which changes the DC voltage profile as shown in Figure 10.3.

10.3 AC System Faults

10.3.1 Rectifier AC Faults

AC voltage depressions on the rectifier side will result in DC current reduction. The rectifier controller can counteract small AC voltage depressions (a few per cent) by reducing the firing angle until the minimum angle limit is reached ($\alpha_r = 2°$). In case of further rectifier AC voltage depressions, the DC current will be reduced and the inverter will finally move to constant current mode by reducing inverter DC voltage. The HVDC system will now operate at a DC current equal to $I_{dcref\_i} = I_{dcref\_r} - I_{margin}$, which is typically 85–90% of the rated current.

However, if the rectifier AC (and consequently DC) voltage is particularly low it would be inappropriate to operate at 90% of DC current because firing angles are large and reactive power demand becomes excessive. The experience with weak AC systems has shown very difficult recoveries when converters demand large reactive power. For this reason a special voltage-dependent current order limiter (VDCOL) is introduced as shown in Figure 10.4. The VDCOL shape and the whole V-I curves are different for different manufacturers and they are also optimized for particular HVDC installations. The main purpose of VDCOL is to reduce current order for lower DC voltages.

10.3.2 Inverter AC Faults

Inverter AC system faults will always result in commutation failure, except for very small voltage disturbances (a small percentage). A commutation failure is a DC short circuit and results in

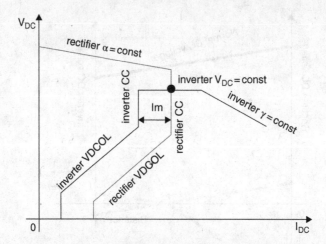

Figure 10.4 HVDC V-I diagram with VDCOL.

temporary loss of HVDC power transfer. However, commutation failure will not be seen as a short circuit on the AC side. Following the detection of commutation failure, the HVDC controls will increase inverter gamma (gamma kick) and initiate gradual power ramp climbing up the V-I curve, using also the VDCOL slope, until normal operating point is reached. The inverter gamma is gradually reduced to nominal values following successful recovery. In some cases repeated commutation failure is encountered and recovery attempts are repeated with progressively lower recovery slopes. Figure 10.5 shows a typical commutation failure recovery process, which can be completed in 200–500 ms. DC voltage drops to zero, which is sure sign of commutation failure. On detecting low extinction angle, inverter moves to CEA mode which establishes normal commutation and power is slowly ramped.

10.4   Internal Faults

The internal faults include thyristor faults, valve faults, bridge faults and a range of other faults inside the valve hall, which in turn can be caused by many factors.

Each valve consists of many individual thyristors connected in series and fired simultaneously. There are several redundant thyristors in each valve, and a valve will operate normally in case of a thyristor failure (a thyristor always fails in a short circuit). The voltage stress on the remaining thyristors will marginally increase but the failed thyristors will be replaced at the next scheduled maintenance.

The most severe internal fault is the valve fault, which can occur as a flashover between valve external connections. Such a fault will bring extreme current and voltage stress across the other valves. The protection response is bypass and converter tripping but the components must be designed to withstand fault conditions for the protection operating time. The bypass involves two steps: firstly two valves on the same converter leg are fired to provide fast redirection of fault current (extinguish fault arc) and to prevent negative DC voltage. In the second stage, if the fault is not cleared, the mechanical bypass switch (BPS) is closed to provide a permanent DC current path.

It should be recalled here that with current source systems, like the line-commutated converter (LCC) HVDC, there is a large inductance on the DC side, which prevents fast current change. Inductors also store energy and, in case of disturbances, a current path must be provided to ensure discharge of inductor energy. For these reasons, the fast converter bypass is normally the first protection step, which is achieved by firing simultaneously two valves on the same converter leg.

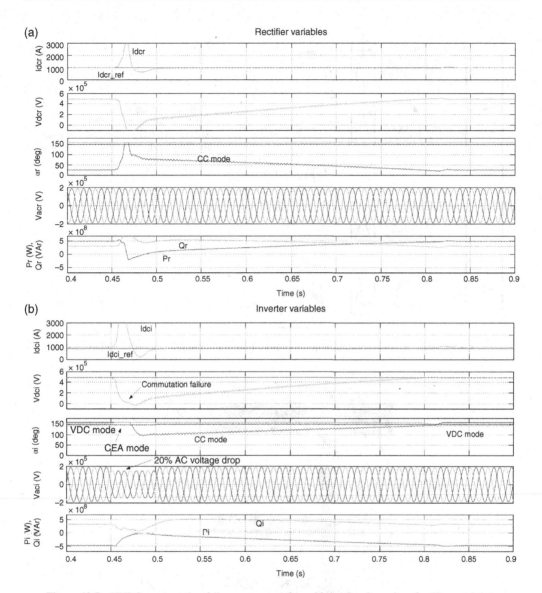

Figure 10.5 HVDC commutation failure recovery (after a 20% AC voltage drop for 50 ms at 0.4 s).

10.5 System Reconfiguration for Permanent Faults

The HVDC controls and the temporary overrating of switches are adequately designed to deal with all transient faults. If the fault condition persists for a longer time, then isolation or reconfiguration is required. As shown in Figures 10.1 and 10.6, each HVDC pole can be isolated using pole CBs and filter CBs.

Figure 10.6 shows the arrangement of DC CBs, but not all of these breakers will be required on all HVDC links. The DC CBs employed for reconfiguration of a bipole HVDC have very low current ratings (not fault levels) and have similar construction to AC CB (common SF$_6$ AC CBs suffice in many cases). Figure 10.7 shows some practical DC circuit breakers. The DC CB technology is further studied in Part III in section 26.4.

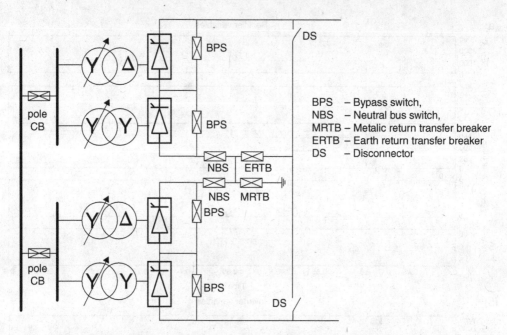

BPS – Bypass switch,
NBS – Neutral bus switch,
MRTB – Metalic return transfer breaker
ERTB – Earth return transfer breaker
DS   – Disconnector

Figure 10.6   Arrangement of DC circuit breakers for system protection and reconfiguration.

Figure 10.7   DC circuit breakers on the Yuannan–Guangdong HVDC. Reproduced with permission of Siemens.

The functions of DC CBs are explained below:

- The bypass switch (BPS) is located across a bridge or pole and it is open in normal operation. In the event of valve or bridge faults it is required to trip the affected bridge and to provide a bypass path for the current. The fast bypass is achieved, firstly, by firing two thyristors on the same converter leg, and this is followed by a permanent bypass using a mechanical BPS, after which the converter is tripped. If the BPS is installed at bridge level, it is possible to operate pole on a single bridge, at half the pole voltage, in order to minimize the loss of power. The BPS enables the converter to be taken in and out of service without power interruption.
- Neutral bus switches (NBSs) are closed and carry rated current in normal operation. In the event of pole faults the current may still flow through a faulted-pole NBS if this current path has lower resistance than the electrode path. In such an event, protection will trip NBS to redirect current to the electrode through a metallic return transfer breaker (MRTB). This enables the isolation of the faulted pole and an uninterrupted power flow through the healthy pole.
- A metallic return transfer breaker (MRTB) is normally closed but carries no current. It will carry a rated current if a pole fault occurs and the NBS is tripped. In most HVDC installations it is not allowed to operate a system with an earth return for a prolonged period of time. If both DC lines are available after a pole fault, then return current is transferred to the DC line on the faulted pole using the following sequence: NBS open, DS close, ERTB close and MRTB open.
- An Earth return transfer breaker (ERTB) is open in normal operation and has the opposite function to an MRTB. It transfers the earth current to the DC line on the faulted pole.

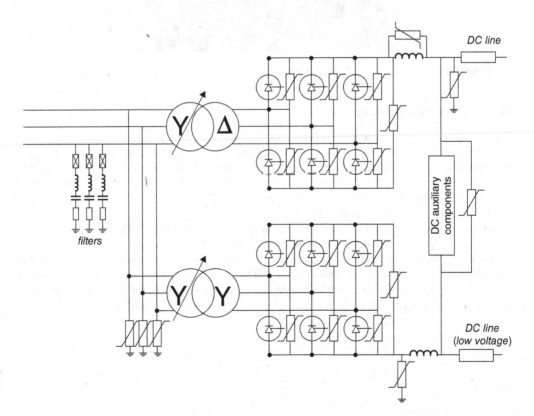

Figure 10.8   Location of surge arresters at an HVDC station.

Figure 10.9   Surge arrester on Yunnan–Guangdong HVDC. Reproduced with permission of Siemens.

10.6   Overvoltage Protection

The HVDC systems include numerous semiconductor-based components, which have high costs and limited voltage withstand abilities and therefore must be protected from overvoltages. The stakes are high with converter systems and it is more expensive to provide large overvoltage margins comparing with other electro-mechanical components. For this reason surge arresters (SAs) are extensively used with HVDC systems, at valve level, bridge level and pole level as transformer protection, as DC cable protection and in other places as shown in Figure 10.8.

Surge arresters will be rated according to insulation coordination strategies. They will always have a protective voltage level below the component rated voltage. An SA protective level that is too low would lead to large leakage current, which can cause SA overheating. The purpose of an SA is to clip the voltage spike but they are not normally designed to damp large energy. If the SA is operated, the current will increase and then converter controls and other protection methods will be activated to manage the fault current. Figure 10.9 shows some practical HVDC surge arresters.

11

LCC HVDC System Harmonics

11.1 Harmonic Performance Criteria

The AC current injected by a line-commutated converter (LCC) has been assumed as an ideal *sine* at fundamental frequency in previous chapters. This current, however, contains harmonics and the quality of converter power is normally evaluated in terms of the following performance parameters: harmonic factor (HF), total harmonic distortion (THD) and distortion factor (DF).

The HF_n, is a measure of individual harmonic magnitude, and it is defined as:

$$HF_n = \frac{V_n}{V_1} \tag{11.1}$$

where V_1 is the RMS value of the fundamental component and V_n is the RMS value of the *nth* harmonic component.

The *THD,* is a measure of closeness between the actual waveform and its fundamental component, and it is defined as:

$$THD = \frac{1}{V_1} \sqrt{\sum_{n=2}^{\infty} V_n^2} \tag{11.2}$$

When calculating the THD the largest harmonic is normally set to 50.

The THD gives the total harmonic content, but it does not indicate the level of each harmonic component. If a common low pass filter is used at the inverter AC terminals, the higher order harmonics would be attenuated more effectively. Knowledge of both the frequency and magnitude of each harmonic is therefore important. The DF indicates the amount of harmonic distortion that remains in a particular waveform after the harmonics of that waveform have been subjected to a second order attenuation. Thus, DF is a measure of effectiveness in reducing unwanted harmonics without having to specify the values of a second order filter and it is defined as:

High-Voltage Direct-Current Transmission: Converters, Systems and DC Grids, First Edition.
Dragan Jovcic and Khaled Ahmed.
© 2015 John Wiley & Sons, Ltd. Published 2015 by John Wiley & Sons, Ltd.

$$DF = \frac{1}{V_1} \sqrt{\sum_{n=2}^{\infty} \left(\frac{V_n}{n^2}\right)^2}$$ (11.3)

The DF of an individual nth harmonic component is defined as:

$$DF_n = \frac{V_n}{V_1 n^2}$$ (11.4)

It is desirable to have as low value as possible for all of the above harmonic performance indicators.

11.2 Harmonic Limits

The harmonics in AC systems cause a range of problems and must be limited. Some common problems resulting from high harmonics include: machine heating and pulsations, insulation stress, overloading of capacitor banks and interference with electronic and telecommunications equipment. With HVDC systems harmonics can also cause control and stability issues. There are also other issues like telephone-line interference and audible noise, which led to cost implications with some HVDC systems but they are not examined in detail.

Table 11.1 shows the voltage harmonic limits according to IEEE and Table 11.2 shows IEC limits.

On the DC side of an HVDC system, there are no loads (customers) and the above limits do not apply. There are, however, two main criteria that influence the design magnitude of DC harmonics:

- Telecommunication interference.
- DC cable insulation limits on harmonics. According to CIGRE brochure 219, for DC extruded cables, the maximum DC voltage distortion should be 3%. Note, however, that many thyristor HVDC system use mass-impregnated DC cables, which have better harmonic tolerance.

Table 11.1 IEEE standard 519 harmonic limits.

Voltage at PCC	Individual harmonic magnitude HF (%)	Total voltage distortion – THD (%)
69 kV and below	3.0	5.0
69–161 kV	1.5	2.5
161 kV and above	1.0	1.5

Table 11.2 IEC standard 61000 3-6 harmonic limits.

Odd harmonics			Harmonics multiples of 3			Even harmonics		
Order	Harmonic voltage (%)		Order	Harmonic voltage (%)		Order	Harmonic voltage (%)	
	MV	HV		MV	HV		MV	HV
5	5	2	3	4	2	2	1.8	1.4
7	4	2	9	1.2	1	4	1	0.8
11	3	1.5	15	0.3	0.3	6	0.5	0.4
13	2.5	1.5	21	0.2	0.2	8	0.5	0.4
			$21 < n < 45$	0.2	0.2			

11.3 Thyristor Converter Harmonics

A thyristor converter behaves as a current source on the AC side. At full power, the AC current for a six-pulse converter bridge, as shown in Figure 3.3 and connected through a Y-Y transformer, can be expressed using the Fourier series (neglecting commutation overlap) as:

$$I_g = 2\frac{\sqrt{3}}{\pi}I_{dc}\left[\sin\omega t - \frac{1}{5}\sin 5\omega t - \frac{1}{7}\sin 7\omega t - \frac{1}{11}\sin 11\omega t + \frac{1}{13}\sin 13\omega t + ...\right] \tag{11.5}$$

which can also be expressed as:

$$I_g = 2\frac{\sqrt{3}}{\pi}I_{dc}\sum_{n=1}^{\infty}\frac{1}{n}\sin(n\omega t), \quad n = 6k \pm 1, \quad k = 1,2,3,... \tag{11.6}$$

It is concluded that all harmonics of the order $6k \pm 1$ are present on the AC side, where k is any integer. The magnitude of harmonics is inversely proportional to the harmonic order.

The AC current generated by a converter with a Y-Δ transformer is:

$$I_g = 2\frac{\sqrt{3}}{\pi}I_{dc}\left[\sin\omega t + \frac{1}{5}\sin 5\omega t + \frac{1}{7}\sin 7\omega t - \frac{1}{11}\sin 11\omega t + \frac{1}{13}\sin 13\omega t + ...\right] \tag{11.7}$$

The Y-Δ transformer therefore gives 180° phase-shifted 5th, 7th harmonics and multiples. The HVDC systems usually use a 12-pulse series connection, which consists of one Y-Y and one Y-Δ converter. The current in a 12-pulse converter system is the sum of Eqs (11.5) and (11.7)

$$I_g = 4\frac{\sqrt{3}}{\pi}I_{dc}\left[\sin\omega t - \frac{1}{11}\sin 11\omega t + \frac{1}{13}\sin 13\omega t - \frac{1}{23}\sin 23\omega t + \frac{1}{25}\sin 25\omega t + ...\right] \tag{11.8}$$

and therefore the 12-pulse system has 11th, 13th and all $12k \pm 1$ harmonics:

$$I_g = 2\frac{\sqrt{3}}{\pi}I_{dc}\sum_{n=1}^{\infty}\frac{1}{n}\sin(n\omega t), \quad n = 12k \pm 1, \quad k = 1,2,3,... \tag{11.9}$$

If a commutation overlap of the angle μ is present, the current 'squares' are rounded and the harmonic magnitude is lower. The harmonic magnitude with overlap I_{nov} is reduced compared to a case with no overlap I_n for n-th harmonic:

$$\frac{I_{nov}}{I_n} = \frac{\sqrt{H^2 + K^2 - 2HK\cos(2\alpha + \mu)}}{\cos\alpha - \cos(\alpha + \mu)}$$

$$H = (\sin(n+1)\mu/2)/(n+1) \tag{11.10}$$

$$K = (\sin(n-1)\mu/2)/(n-1)$$

On the DC side, a thyristor converter appears as a DC voltage source with a DC waveform, shown in Figure 3.4. This voltage source will have $6k$ ($k = 1, 2, ...$) voltage harmonics in the case of a six-pulse bridge. With 12-pulse systems, only $12k$ ($k = 1,2, ...$) harmonics will be present.

The above harmonics are called characteristic harmonics. They occur according to the theoretical study of ideal current/voltage waveforms. The noncharacteristic harmonic (i.e. second, third,...) are not present in theoretical studies but may occur in practical HVDC systems. Some of the causes of noncharacteristic harmonics are:

- an unbalanced AC system caused by small differences in the line impedances in three phases;
- converter transformer asymmetry, and phase difference in transformer ratios;
- converter asymmetry, because of the controller firing angle, or driver or thyristor asymmetry.

AC grid unbalance is quite common cause of non-characteristic harmonics. An unbalance on the AC grid will cause the thyristor converter to produce the following AC harmonics:

$$n = kp \pm 3, \quad k = 0, 1, 2, \ldots \tag{11.11}$$

where p is the pulse number of the thyristor converter. The same unbalance will produce on DC side:

$$n = kp \pm 2, \quad k = 0, 1, 2, \ldots \tag{11.12}$$

In general, a nth harmonic on DC side will produce an $n \pm 1$ harmonic on the AC side.

11.4 Harmonic Filters

11.4.1 Introduction

At each HVDC station there will be harmonic filters on the AC side and perhaps also on the DC side, which are designed to reduce harmonic distortion to within the specified limits. The filters have capacitive reactance at fundamental frequency and they also help to supply the reactive power requirement from converters. Since HVDC reactive power is variable (it is proportional to active power) the capacitors and/or filters will be arranged in switchable banks, which are switched in steps as the HVDC power is varying. The filters are physically large and, together with the switchgear, they can occupy over 50% of the HVDC station footprint.

Most systems use shunt filters but some systems use a combination of shunt and series AC filters. In case of capacitor-commutated converter (CCC) HVDC systems, the HVDC receives full reactive power compensation from series capacitors and therefore shunt filters are small.

Typically, AC-side tuned filters will be used for the two to four lowest order AC harmonics, like 11th and 13th on 12-pulse systems, or 23rd and 25th on 24-pulse terminals. Damped (high-pass) filters will also be employed for higher order harmonics. Figure 11.1 shows the possible filter arrangement for a 12-pulse HVDC terminal while Figure 11.2 shows photographs of some practical AC filters at the HVDC terminal.

An HVDC with overhead lines will commonly have DC-side 12th or 24th filters. The DC cables have high capacitance and they naturally reduce harmonics.

11.4.2 Tuned Filters

The simplest single tuned filter is shown in Figure 11.3, where the converter is shown as a harmonic current source. The other commonly used filters are double tuned and triple tuned. The filter in this figure consists of a series LC circuit and a resistance which is typically just the parasitic resistance of the inductor.

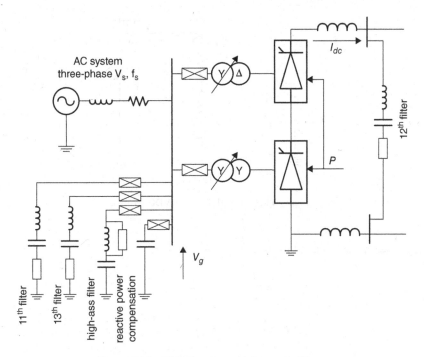

Figure 11.1 HVDC station with harmonic filters.

Figure 11.2 AC filters at Moyle HVDC station (2001). Reproduced with permission of Siemens.

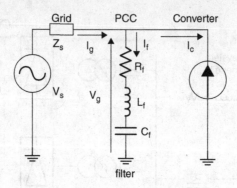

Figure 11.3 Tuned filter at HVDC terminal.

The equivalent impedance of this filter at a frequency ω is:

$$\overline{Z}_f = \frac{V_f}{I_f} = R_f + j\omega L_f + \frac{1}{j\omega C_f}$$

$$\overline{Z}_f = R_f + j\left(\omega L_f - \frac{1}{\omega C_f}\right) = R_f - j\frac{1-\omega^2 L_f C_f}{\omega C_f}$$

(11.13)

The characteristic frequency ω_f (where capacitive impedance equals inductive impedance) of this filter is:

$$\omega_f = \frac{1}{\sqrt{L_f C_f}}$$

(11.14)

and the quality factor q_f is defined as:

$$q_f = \frac{\omega_f L_f}{R_f} = \frac{\sqrt{\frac{L_f}{C_f}}}{R_f}$$

(11.15)

Typical values for the quality factor are in the range $50 < q_f < 100$. The band pass (BP) for the tuned filter is defined as:

$$BP = \pm\frac{1}{2q_f}\omega_f$$

(11.16)

Figure 11.4 shows the Bode plot for two different q_f. Larger q_f (lower R_f) implies better tuned frequency characteristics, which provide lower impedance for current harmonics. The filter impedance at characteristic frequency is:

$$\overline{Z}_{ff} = R_f$$

(11.17)

And therefore the magnitude of the harmonic voltage at frequency ω_f is directly proportional to R_f:

$$V_{gf} = I_{ff} R_f$$

(11.18)

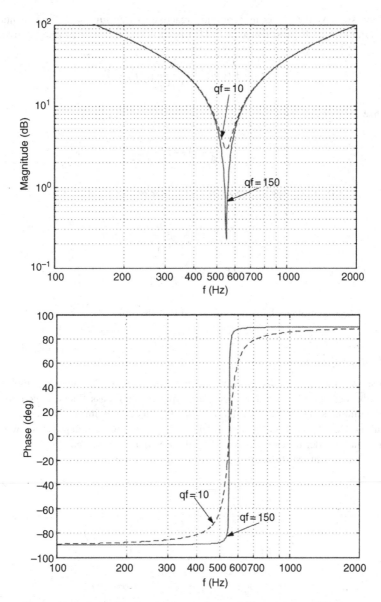

Figure 11.4 Bode plot for 11th harmonic tuned filter with two different q_f.

where I_{ff} is the injected current at ω_f. The filter loss at ω_f frequency is:

$$P_{ff} = I_{ff}^2 R_f \qquad (11.19)$$

It is desired that the impedance at the characteristic frequency is as low as possible in order to reduce the bus-voltage harmonic distortion at this frequency. However, there are physical constraints related to parasitic resistances like inductor size and losses. The filter parameters may also change with the ageing of the components, in which case filters with higher q_f become more detuned.

With HVDC systems the filters also provide reactive power supply at fundamental frequency. Considering Eq. (11.13), the impedance at 50 Hz ($\omega_1 = 2\pi 50$) for the tuned filter is given by:

$$\overline{Z_{f1}} = R_f + j\left(\omega_1 L_f - \frac{1}{\omega_1 C_f}\right) = R_f + jY_{f1} \tag{11.20}$$

and the inverse is determined as:

$$\frac{1}{\overline{Z_{f1}}} = \frac{R_f}{R_f^2 + Y_f^2} - j\frac{Y_{f1}}{R_f^2 + Y_f^2} \tag{11.21}$$

Therefore the active P_f (filter losses) and reactive Q_f filter power at 50 Hz are

$$P_{f1} = 3V_{g1}^2 \frac{R_f}{R_f^2 + Y_f^2} \tag{11.22}$$

$$Q_{f1} = 3V_{g1}^2 \frac{Y_{f1}}{R_f^2 + Y_f^2} \tag{11.23}$$

where V_{g1} is the line-neutral fundamental voltage at the filter bus.

11.4.3 Damped Filters

Figure 11.5 shows the topology for a second-order damped filter. The other commonly used tuned filters are third-order and C-type filters. They are typically high-pass filters, which are designed to eliminate all harmonics above the ones that are directly cancelled with tuned filters. As an example if with 12-pulse HVDC tuned 11th and 13th filters are installed, then a high pass filter will be designed for 24th harmonic in order to damp 23rd, 25th and all higher order harmonics.

The filter impedance at any frequency ω is:

$$\overline{Z_f} = \frac{j\omega R_f L_f}{R_f + j\omega L_f} + \frac{1}{j\omega C_f}$$

$$\overline{Z_f} = \frac{R_f \omega^2 L_f^2}{R_f^2 - \omega^2 L_f^2} + j\frac{\omega C_f\left(\omega R_f^2 L_f - R_f^2 + \omega^2 L_f^2\right)}{R_f^2 - \omega^2 L_f^2} \tag{11.24}$$

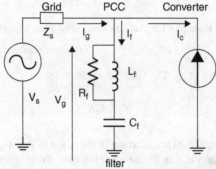

Figure 11.5 Second-order damped filter at HVDC terminals.

The characteristic frequency ω_f of this filter is:

$$\omega_f = \sqrt{\frac{q_f^2 + 1}{q_f^2}} \frac{1}{\sqrt{L_f C_f}} \tag{11.25}$$

and the quality factor is defined as:

$$q_f = \frac{R_f}{\omega_f L_f} \tag{11.26}$$

The quality factor is typically low $1 < q_f < 10$, in order to provide damping at wide range of frequencies. Figure 11.6 shows the Bode plot for a high pass filter designed for 24th harmonic.

Example 11.1
A 500 MW, 250 kV, 2000 A, 12 pulse HVDC terminal employs 11th, 13th and high-pass AC filters. The AC voltage level at the point of common coupling (PCC) is 220 kV. The tuned 11th filter has the following parameters: $C_f = 3.91\,\mu F$, $L_f = 0.0214\,H$, $Rf = 0.7\,\Omega$. Figure 11.7 shows the circuit diagram. Neglecting the AC system impedance (ideally small at 50 Hz and ideally large at 550 Hz), determines:

a. *q_f factor for this filter,*
b. *Reactive power supply by this filter,*
c. *Power loss in the filter considering only 50 and 550 Hz components,*
d. *Voltage distortion at the 11th harmonic,*
e. *Voltage distortion at the 11th harmonic assuming that filter capacitance becomes reduced by 10% because of ageing.*

Solution

a. *Quality factor*

$$q_f = \frac{\sqrt{\dfrac{L_f}{C_f}}}{R_f} = \frac{\sqrt{\dfrac{0.0214}{3.91 \times 10^{-6}}}}{0.7} = 105.6$$

b. *At 50 Hz the filter impedance is:*

$$\overline{Z_{f1}} = 0.7 + j\left(2\pi 50 \times 0.0214 - \frac{1}{2\pi 50 \times 3.91 \times 10^{-6}}\right) = 0.7 - j806.7\,\Omega$$

Therefore

$$Q_{f1} = 3V_{g1}^2 \frac{Y_f}{R_f^2 + Y_{f1}^2} = 3\frac{220000^2}{3} \frac{806.7}{0.7^2 + 806.7^2} = 60\,MVar$$

c. *Losses at 50 Hz*

$$P_{f1} = 3V_{g1}^2 \frac{R_f}{R_f^2 + Y_f^2} = 3\frac{220000^2}{3} \frac{0.7}{0.7^2 + 806.7^2} = 52.066\,kW$$

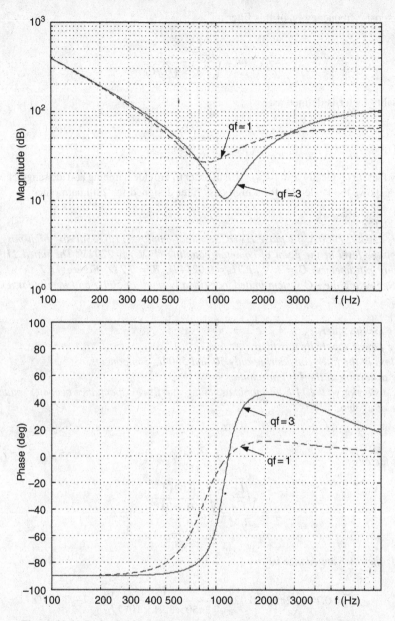

Figure 11.6 Bode plot for 24th harmonic high-pass filter with two different q_f.

At the 11th harmonic, firstly the converter-side AC voltage is needed:

$$V_{g\_LL} = V_{dc} \frac{\pi}{B3\sqrt{2}\cos\alpha} = 250000 \frac{\pi}{2 \times 3\sqrt{2}\cos(15)} = 95.82\,kV$$

The stepping ratio is therefore:

$$n_r = \frac{220.0}{95.8} = 2.296$$

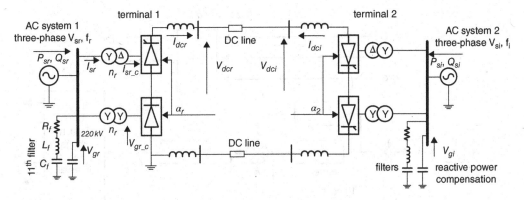

Figure 11.7 HVDC system in Example 11.1.

The RMS fundamental AC current on the grid side, using Eq. (3.5), is:

$$I_{s1} = \frac{B}{n_r}\frac{\sqrt{6}}{\pi}I_{dc} = \frac{2}{2.296}\frac{\sqrt{6}}{\pi}2000 = 1358.4\,A$$

The AC current at 11th harmonic is:

$$I_{s11} = \frac{1}{11}I_{s1} = 123.49\,A$$

The power loss at 11th harmonic is:

$$P_{f11} = I_{s11}^2 R_{f11} = 123.49^2 \times 0.7 = 32.027\,kW$$

The total percentage loss is:

$$P_{loss\%} = \frac{P_{f1}+P_{f11}}{P_{dc}}100 = \frac{52.066+32.027}{500000}100 = 0.0168\%$$

d. *Voltage at 11th harmonic*

$$V_{g11\_ll} = \sqrt{3}I_{g11}R_f = \sqrt{3}123.49 \times 0.7 = 149.73\,V$$

$$V_{g11\_ll\%} = \frac{V_{g11\_ll}}{V_{g1\_ll}} = \frac{149.73}{220000}100 = 0.0168\%$$

e. *The new filter impedance at 11th harmonic*

$$\overline{Z_{f11}} = 0.4 + j\left(2\pi50 \times 11 \times 0.0214 - \frac{1}{2\pi50 \times 11 \times 0.9 \times 3.91 \times 10^{-6}}\right) = 0.7 - j8.21\,\Omega$$

$$Z_{f11} = |\overline{Z_{f11}}| = 8.246\,\Omega$$

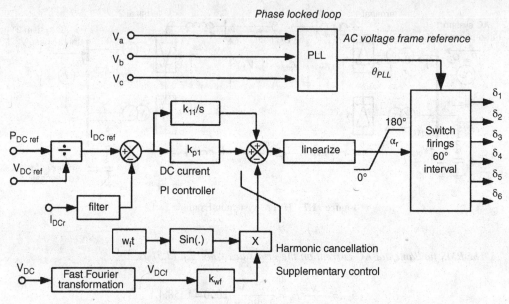

Figure 11.8   HVDC supplementary control for DC harmonic elimination.

$$V_{g11\_ll} = \sqrt{3}I_{s11}Z_{f11} = \sqrt{3}123.49 \times 8.246 = 1.764\,kV$$

$$V_{g11\_ll\%} = \frac{V_{g11\_ll}}{V_{g1\_ll}} = \frac{1764}{220000}100 = 0.8017\%$$

Therefore there is significant increase in 11th harmonic for a 10% change in capacitance.

11.5   Noncharacteristic Harmonic Reduction Using HVDC Controls

Some HVDC systems use special supplementary controls to reduce particular low order noncharacteristic harmonics. This can potentially be an effective and low-cost method but careful studies are required to avoid control instabilities. The scheme can be used with low-order DC-side harmonics that cause problems, like second harmonic resonance.

Figure 11.8 shows the HVDC (rectifier) control schematic with the supplementary control employed for eliminating harmonics at frequency ω_f. As normal HVDC controls have low bandwidth, they are not capable of regulating harmonic voltages. A special oscillator at frequency ω_f is used and the magnitude of the injected oscillation is controlled in a feedback manner using fast Fourier transformation (FFT) processing of the measured DC voltage. The method is used only at a single frequency ω_f because any other frequency would require different and carefully tuned gain k_{wf}.

Bibliography Part I

Line Commutated Converter HVDC

Alstom Grid. HVDC Connecting to The Future, Alstom, 2010.

Burton, R.S., Fuchshuber, C.F., Woodford, D.A, and Gole, A.M. Gole prediction of core saturation instability at an HVDC converter. *IEEE Transactions on Power Delivery*, **11** (4), 1996, 1961–1969.

CIGRE WG 14.07. Guide for Planning DC Links Terminating at AC Systems Locations Having Low Short Circuit Capacities. CIGRE Brochure 68, 1992.

CIGRE WG 14.05. On Voltage and Power Stability in AC/DC Systems. CIGRE Brochure 222, 2003.

CIGRE Working Group 14.03. A Summary of the Report on Survey of Controls and Control Performance in HVDC Schemes. CIGRE Brochure 00, 1994a.

CIGRE Working Group 14.03. DC Side Harmonics and Filtering in HVDC Transmission Systems. CIGRE Brochure 92, 1994b.

CIGRE Working Group 14.03. Guide to the Specification and Design Evaluation of AC Filters for HVDC Systems. CIGRE Brochure 139, 1999.

CIGRE Working Group B2.41. Guide to the Conversion of Existing AC Lines to DC Operation. CIGRE Brochure 583, 2014.

Gole, A.M., Keri, A., Kwankpa, C., Gunther, E.W., Dommel, H.W., Hassan, I., Marti, J.R., Martinez, J.A., Fehrle, K.G., Tang, L., McGranaghan, M.F., Nayak, O.B., Ribeiro, P.F., Iravani, R., and Lasseter, R. Guidelines for modeling power electronics in electric power engineering applications. *Power Delivery, IEEE Transactions on*, **12** (1), Jan 1997, 505, 514.

Hammad, A.E. Analysis of second harmonic instability for the Chateauguay hvdc/svc scheme. *IEEE Transactions on Power Delivery*, **7** (1), 1992, 410–415.

Huang, H., and Ramaswami, V. Design of UHVDC converter station. *Transmission and Distribution Conference and Exhibition: Asia and Pacific*, 2005 IEEE/PES, pp. 1, 6, 2005.

Jovcic, D. Control of HVDC systems operating with long DC cables. AC-DC Power Transmission, 2001. *Seventh International Conference on* (Conf. Publ. No. 485), pp. 113, 118, 28–30 Nov. 2001.

Jovcic, D. Phase locked loop system for FACTS. *IEEE Transactions on Power Systems*, **18** (3), 2003, 1116–1124.

Jovcic, D., Pahalawaththa, N., and Zavahir, M. Analytical modeling of HVDC-HVAC systems. *IEEE Transactions on Power Delivery*, **14** (2), 1999a, 506–511.

Jovcic, D., Pahalawaththa, N., and Zavahir, M. Stability analysis of HVDC control loops. *IEE Proceedings Generation, Transmission and Distribution*, **146** (2), 1999b, 143–148.

Jovcic, D., Pahalawaththa, N., and Zavahir, M. Investigation of the use of inverter control strategy instead of synchronous condensers at inverter side of HVDC system. *IEEE Transactions on Power Delivery*, **15** (2), 2000, 704–709.

Kundur, P. *Power System Stability and Control*, McGraw-Hill, 1994.

High-Voltage Direct-Current Transmission: Converters, Systems and DC Grids, First Edition.
Dragan Jovcic and Khaled Ahmed.
© 2015 John Wiley & Sons, Ltd. Published 2015 by John Wiley & Sons, Ltd.

Naidoo, P., Muftic, D., and Ijumba, N. Investigations into the upgrading of existing HVAC power transmission circuits for higher power transfers using HVDC technology. *Power Engineering Society Inaugural Conference and Exposition in Africa*, 2005 IEEE, pp. 139, 142, 11–15 July 2005.

Nayak, O.B., Gole, A.M., Chapman, D.G., and Davies, J.B. Dynamic performance of static and synchronous compensators at an HVDC inverter bus in a very weak AC system. *Power Systems, IEEE Transactions on*, **9** (3), Aug 1994, 1350, 1358.

Osauskas, C., and Woo, A. Small- signal dynamic modeling of HVDC systems. *IEEE Transactions on Power Delivery*, **18** (1), 2003, 220–225.

Padiyar, K.R., Sachchidanand stability of converter control for multiterminal HVDC systems. *IEEE Transactions on Power Apparatus and Systems* **PAS 104** (3) 1985, 690–696.

Rahimi, E., Gole, A.M., Davies, J.B., Fernando, I.T., and Kent, K.L. Commutation failure in single- and multi-infeed HVDC systems. AC and DC Power Transmission, 2006. *ACDC 2006. The 8th IEE International Conference on*, pp. 182, 186, 28–31 Mar 2006.

Sadek, K., Pereira, M., Brandt, D.P., Gole, A.M., and Daneshpooy, A. Capacitor commutated converter circuit configurations for DC transmission. *Power Delivery, IEEE Transactions on*, **13** (4), Oct 1998, 1257, 1264.

Sucena-Paiva, J.P., and Freris, L.L. Stability of rectifiers with voltage-controlled oscillator firing systems. *Proceeding of IEE*, **120** (6), 1973, 667–673.

Szechman, M., Wess, T., and Thio, C.V. First Benchmark Model for HVDC Control Studies. CIGRE WG 14.02 Electra No. 135, pp. 54–73, April 1991.

Thio, C.V., Davies, J.B., and Kent, K.L. Commutation failures in HVDC transmission systems. *IEEE Transactions on Power Delivery*, **11** (2), Apr 1996, 946, 957.

Yacamini, R., and Oliveira, J.C. Instability in HVDC schemes at low-order integer harmonics. *IEEE Transactions on Power Apparatus and Systems*, **87** (3), 1980, 179–188.

Zhang, Y. Investigation of reactive power control and compensation for HVDC systems PhD thesis. University of Manitiba, 2011.

Woodfford, D.A. Solving the feroresonnace problem when compensating a DC converter station with a series capacitor. *IEEE Transactions on Power Systems*, **11** (3), 1996, 1325–1331.

Part II

HVDC with Voltage Source Converters

12

VSC HVDC Applications and Topologies, Performance and Cost Comparison with LCC HVDC

12.1 Voltage Source Converters (VSC)

Voltage source converter (VSC) systems are based on self-commutating switches, typically insulated-gate bipolar transistor (IGBT) technology, which have a number of advantages compared with thyristors. The use of self-commutating devices allows the application of high-frequency (over 1 kHz) pulse-width modulation (PWM) techniques, which have been in use in the industrial drives sector since the early 1990s. By the use of PWM, the VSC may synthesize a fully controlled AC voltage, which enables precise control of active and reactive powers. This voltage appears as a fundamental frequency *sine* with harmonics at the switching frequency and its multiples. The control is very fast and, with appropriate feedback, the voltage source inverter may respond as a current source. Because of higher switching frequencies, harmonics are lower and therefore filtering requirements are reduced.

The use of VSCs for DC power transmission (VSC transmission) was introduced with the 3 MW, ±10 kV demonstrator at Hellsjön, Sweden in 1997. Figure 12.1 shows the basic schematic of a two-level VSC high-voltage direct current transmission system.

The main advantages of VSC converters, compared with line-commutated converters) (LCCs) are summarized as:

- Active and reactive power can be controlled independently. The VSC is capable of generating leading or lagging reactive power, independently of the active power level. Each converter station can be used to provide voltage support to the local AC network while transmitting any level of active power, at no additional cost.
- If there is no transmission of active power, both converter stations operate as two independent static synchronous compensators (STATCOMs) to regulate local AC network voltages.
- The use of PWM with a switching frequency in the range of 1–2 kHz is sufficient to separate the fundamental voltage from the sidebands, and suppress the harmonic components around and beyond the switching frequency components. Harmonic filters are at higher frequencies and therefore have low size, losses and costs.

High-Voltage Direct-Current Transmission: Converters, Systems and DC Grids, First Edition.
Dragan Jovcic and Khaled Ahmed.

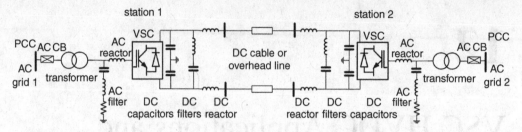

Figure 12.1   A symmetrical monopolar VSC-HVDC transmission system based on a two-level converter.

- Power flow can be reversed instantaneously (50–100 ms) without the need to reverse the DC voltage polarity (only DC current direction reverses). This implies that simpler cross-linked polyethylene (XLPE) DC cables can be used.
- Good response to AC faults. The VSC converter actively controls the AC voltage/current, so the VSC-HVDC contribution to the AC fault current is limited to rated current or controlled to lower levels. The converter can remain in operation to provide voltage support to the AC networks during and after the AC disturbance.
- Black-start capability, which is the ability to start or restore power to a dead AC network (network without generation units). This feature eliminates the need for a startup generator in applications where space is critical or expensive, such as with offshore wind farms.
- VSC-HVDC can be configured to provide faster frequency or damping support to the AC networks through active power modulation.
- Connection to passive AC systems. A VSC converter can generate AC voltage using an internal independent oscillator. This facilitates transition from the grid mode to an intended or unintended islanded mode, as is required in the development of smart grids. Some VSC HVDC are used to drive dead loads (no active power generation).
- It is more suitable for paralleling on the DC side (developing multiterminal HVDC and DC grids) because of constant DC voltage polarity and better control.
- It facilitates modular development using basic constant-voltage cells. In particular, the latest modular multilevel converters (MMCs) use standard low voltage cells with capacitors.

   The limitations and disadvantages of VSC technology are summarized as:

- Improved control is achieved at the expense of increased losses in the power converter. Increased losses are the result of:
  - Application of high frequency switching leads to increased switching loss.
  - Insulated-gate bipolar-transistor devices exhibit significantly higher on-state voltage drop compared to thyristors of similar voltage ratings. For a given power level, this will lead to increased conduction loss.
- Higher costs than LCC converters. A VSC requires a larger number of semiconductors and in particular with modular systems the number of switches is much higher than with an LCC.
- IGBT switches/modules have lower power capability than available thyristor packages, leading to an increased power component count. The semiconductor footprint is larger than with thyristors.
- IGBT devices have lower current overload capability than thyristors.
- High *dv/dt* transitions may be present at the connecting points (both AC and DC). This can cause problems with dominantly capacitive cable systems and also electromagnetic interference.

- DC-side faults are a serious issue because a VSC behaves like an uncontrolled diode bridge during a DC fault. Diode overrating is typically required and the fault is cleared by AC circuit breakers (CBs). The VSC converter reconnection after AC CB tripping may take a long time.
- The development of multiterminal DC and DC grids requires very fast DC CBs.
- Insulated-gate bipolar transistors naturally fail in open circuits. A failure in a short circuit can be achieved at the expense of special packaging, or using much more complex valve design compared with thyristors.

12.2   Comparison with Line-Commutated Converter (LCC) HVDC

High-voltage direct-current systems are currently implemented using both LCC and VSC technologies. Despite evident technological advantages, VSC HVDC did not completely replace LCC HVDC. Line-commutated HVDC systems retain some advantage, in particular related to costs and ratings, but also, importantly, LCC represents a well established technology with many years of experience. LCC HVDC provides a practical solution using robust devices and well proven circuits. At highest power levels (over 2 GW) they are exclusively used. The DC systems with overhead lines are exposed to frequent DC faults and LCC converters are preferred in these applications because of good handling of such faults. Moreover, LCC HVDC generally has lower costs, and so it is still preferred in applications where performance and station space are not critical.

The IGBT converters used in voltage-source HVDC are still in a state of development with relatively limited operation experience. Voltage-source convertor HVDC has evolved in several different technologies, which have shown progressively improving performance, namely: two level, three-level and MMC (half bridge or full bridge). The latest generation of MMC VSC HVDC has converter station losses, which are not significantly larger than those with LCC HVDC.

The LCC HVDC has been studied in part I, while only the main advantages are summarized in this section in order to provide comparison with VSC HVDC:

- Thyristor devices are available in robust high voltage (6–10 kV), high-current (3–4 kA) capacity single wafer capsules.
- Thyristors have excellent overcurrent capability.
- Thyristors naturally fail in a short circuit, which is desirable for series strings in high-voltage valves.
- Thyristors have a lower on-state voltage drop and costs are lower than with IGBT. Line-commutated converters require switches for one current direction while VSC converters have semiconductors for two current directions.
- LCC HVDC have an established track record at transmission voltage and power levels.
- Resilience to DC-side short circuit faults. The system can operate well at low DC voltage and DC fault recovery is well controlled. Direct current faults do not affect the AC system.
- LCC HVDC provides AC system frequency stabilization in many installations but performance is limited because of reactive power issues. The AC voltage control is also implemented in some installations but performance is very limited.

The notable shortcomings of LCC HVDC are:

- Line-commutated thyristor-based systems inject significant low order harmonics, which must be eliminated using large passive filter arrangements. The presence of these filters may lead to circulating harmonic currents and stability issues which must be mitigated by damping networks. Filters and damping networks may have to be designed specifically for each location and may not be optimal for all operating conditions.

- Line-commutated converters are inherently limited in their response time (limited to line-frequency switching). The control response time is in the order of 100–200 ms.
- They require large reactive power compensation. This reactive power is supplied using passive filters and switched capacitors and reactors. The switched banks are required since the amount of reactive power needed varies with the active power magnitude and operating mode (inversion or rectification). As a result, the LCC HVDC station has larger footprint that VSC HVDC station.
- The power reversal requires the DC link voltage polarity to be reversed. This limits the range of DC cables that can be used (XLPE cables cannot be used if voltage reversal is needed). The time required to reverse the power flow is also limited to around 500 ms, which imposes operational constraints.
- There is a practically minimal power flow of around 10%.
- It cannot connect to very weak AC networks (with a low short circuit ratio, $SCR < 2.0$).
- Suffers from commutation failure, which is caused by 5–10% voltage drop on the inverter side. A commutation failure causes transient (0.5–1 s) interruption in power transfer. This can be partially mitigated with a capacitor-commutated converter (CCC) HVDC.
- Vulnerable to DC-side open-circuit conditions. This is very unlikely event and causes overvoltage, which can be limited by surge arresters.
- Multiterminal topologies can be built but with serious performance and cost penalties. Generally only three-terminal systems are accepted in practice. DC grids are very challenging.

Table 12.1 provides a detailed comparison between high-voltage DC transmission system technologies. The comparison focuses on most important aspects, listed in the first column, such as control flexibility, fault ridethrough capability, conversion losses, electromagnetic compatibility (EMC) issues, and provision of auxiliary functionality such as voltage, frequency and damping support.

12.3   Overhead and Subsea/Underground VSC HVDC Transmission

To transmit electrical energy using direct current over a distance, cables and overhead transmission lines can be used. The choice is influenced by environmental constraints as well as an overall optimization that considers total capital cost, performance, losses and transmission system reliability. There is also a small difference between underground and submarine DC cables. The cost of cables is considerably larger than overhead lines – perhaps an order of magnitude larger – but there are also other aspects of cable projects that may influence the choice:

- Cables often have less impact on the environment than overhead transmission lines.
- In the recent years, it has become increasingly difficult to secure right of way for new overhead line corridors, or the permission process takes many years, and this has influenced decisions to use underground cables in some projects.
- Since a VSC allows only one DC voltage polarity, the cable does not need to be designed for voltage polarity reversal. This allows use of less expensive and simpler XLPE DC cables.
- Cables are less likely to experience faults than overhead transmission lines. Since overhead transmission lines are always exposed to lightning strikes and pollution, faults are likely. Most line outages are transient, and transmission recommences once air insulation is restored, which can be readily managed with LCC HVDC. With VSC HVDC (except full-bridge MMC) even transient DC faults may require tripping the whole DC system with significant power-flow disturbance. If overhead line faults are permanent they are typically easier to locate and repair. In the event of a cable fault, the outage would be permanent and repair time may be significant.

Table 12.1 Comparison between different HVDC technologies.

	Current source converter based		Voltage source converter based		
	LCCs (line commutated converters)	CCCs (capacitor commutated converters)	Two/three-level	MMC half bridge	MMC full bridge
Switching device	Thyristor	Thyristor	IGBT	IGBT	IGBT
Switching losses	Negligible	Negligible	High	Low	Low
On-state losses	Low	Low	Moderate	Moderate	Moderate, higher than with half bridge
Station size	Large	Large but reactive power banks and their circuit breakers are lower	Significantly reduced (around 50% of LCC)	Significantly reduced (around 50% of LCC)	Significantly reduced but higher than with half bridge
Active power control	Continuous with fast reversal but cannot operate within ±10%	Continuous with fast reversal but cannot operate within ±10%	Continuous from 0 to ±100%	Continuous From 0 to ±100%	Continuous From 0 to ±100%
Active power reversal	DC voltage polarity must be changed (0.5–1 s)	DC voltage polarity must be changed (0.5–1 s)	Instantaneous (0.1 s) and no change of DC voltage polarity	Instantaneous (0.1 s) and no change of DC voltage polarity	Instantaneous (0.1 s) either DC voltage or current polarity
Independent control of active and reactive power	No	No	Yes	Yes	Yes
Reactive power demand	50–60%	20–40%, but additional series capacitors are required	No	No	No
Reactive power control	Limited (lagging VAr only) and discontinuous using switch shunt capacitors for leading VAr	Limited (lagging VAr only)	Continuous and inherent within the converter control at no additional cost	Continuous and inherent within the converter control at no additional cost	Continuous and inherent within the converter control at no additional cost
Power levels	Up to 6400 MW	Up to 6400 MW	Up to 1200 MW Higher currents by paralleling converters	Up to 1200 MW Higher currents by paralleling converters	Up to 1200 MW Higher currents by paralleling converters
Controllability (response time (s))	Fast (0.1–0.2)	Fast (0.1–0.2)	Very fast (0.03–0.05)	Very fast (0.03–0.05)	Very fast (0.03–0.05)
AC filters	Large	Large	Small	No	No
DC filter	Might be required	Might be required	No (rarely used)	No (rarely used)	No (rarely used)

(continued overleaf)

Table 12.1 (*continued*)

	Current source converter based		Voltage source converter based		
	LCCs (line commutated converters)	CCCs (capacitor commutated converters)	Two/three-level	MMC half bridge	MMC full bridge
Converter transformer	Expensive with high insulation requirement to withstand harmonics and voltage stresses during power reversal	Expensive with high insulation requirement to withstand harmonics and voltage stresses during power reversal, but with reduced MVA rating	Expensive with high insulation requirement to withstand switching of large voltage steps with high frequency. Additional large AC inductor is needed	Standard AC transformer might be used Additional large AC inductor is needed	Standard AC transformer might be used Additional large AC inductor is needed
DC cable	Must withstand fast voltage polarity reversal during power reversal	Must withstand fast voltage polarity reversal during power reversal	Less expensive and light weight extruded cable	Less expensive and light weight extruded cable	Less expensive and light weight extruded cable
HVDC with overhead lines	Yes	Yes	Yes but DC CBs might be needed	Yes but DC CBs might be needed	Yes
Commutation failure	Present for AC disturbances (5–10%)	Present but significantly reduced	No	No	No
Applications with weak AC systems	Connection of strong systems (SCR > 3); connection of weak system is possible but at additional cost by using STATCOM or synchronous condenser	Connection of strong and weak systems with SCR >=2	Very weak and network without generation can be connected	Very weak and network without generation can be connected	Very weak and network without generation can be connected
AC fault ride-through capability	Possible, undergoing commutation failure and 0.5–1 s recovery if inverter	Possible, undergoing commutation failure and recovery 0.5–1 s if inverter	Excellent (three-level has issues with asymmetrical faults)	Excellent	Excellent
DC fault ride-through capability	Excellent	Excellent	Difficult AC CB tripping	Difficult AC CB tripping	Excellent
Multiterminal configuration	Feasible but with performance limitation. Three-terminal topologies are preferred	Feasible but with performance limitation. Three-terminal topologies are preferred	Extension to any number of terminals is feasible assuming single protection zone DC side isolation is very difficult	Extension to any number of terminals is feasible assuming single protection zone Fast DC CB are needed	Extension to any number of terminals is feasible. Mechanical DC CBs can be used
Redundancy at switch level	Yes	Yes	Yes	Yes	Yes

- VSC HVDC converters have difficulties with DC faults and most of the existing links operate with DC cables. It is accepted that if a cable fault occurs, the DC system will be tripped.

Manufacturers have traditionally been reluctant to offer VSC HVDC with overhead lines. The 300 MW, 350 kV, 2010 Caprivi VSC HVDC link is currently the only VSC HVDC with overhead DC lines. The challenge with VSC overhead transmission is to avoid tripping the whole HVDC system (by AC CBs) for transient DC line faults (lightning strikes), which might be frequent on some corridors. A DC arc on overhead lines will not clear until the circuit is interrupted and this is only achieved using AC CBs with VSC converters. Even when AC CBs are tripped, the high-impedance DC circuits will continue to conduct DC fault current through converter anti-parallel diodes. Caprivi HVDC link has an additional DC-side technology for managing DC faults.

With the introduction of full bridge MMC topologies (and other new MMC concepts) the VSC HVDC will be in much better position to operate with overhead lines.

12.4 DC Cable Types with VSC HVDC

The most common cable technologies have been discussed in Chapter 1, and include:

- mass-impregnated cables (MI);
- low-pressure oil-filled (LPOF) cables;
- extruded cross-linked polyethylene (XLPE) cables.

Mass-impregnated and LPOF cables can also be used with VSC HVDC but, so far XLPE cables have been mostly used. Extruded polyethylene cables are less expensive, their construction is simpler, they allow a lower bending radius and have a number of other installation/transport advantages when compared with MI or LPOF cables. The insulation material (extruded polyethylene), conductor screen and insulation shields are extruded and chemically cross linked. Most XLPE DC cable installations are at voltages up to 320 kV, but currently XLPE cables are available at 525 kV.

XLPE DC cable technology has been closely linked with the development of VSC HVDC. Extruded polyethylene cables have influenced VSC technology and VSC technology has opened a market for XLPE cables. Ongoing research and development aimed at increasing the rating of XLPE cables and reducing its cost may also make VSC transmission more attractive for a number of applications. Figure 12.2 shows a typical XLPE DC cable.

12.5 Monopolar and Bipolar VSC HVDC Systems

All the HVDC topologies, including monopolar and bipolar, discussed with LCC HVDC in section 1.4 can also be used with VSC transmission. However, most installed VSC HVDC systems are symmetrical monopoles. Symmetrical monopoles have positive and negative DC cables (as bipoles) but the system is controlled/operated as a single unit (as a monopole), as schematically shown in Figure 12.1. There are a number of reasons for using symmetrical monopole (which is not used with LCC HVDC) with the early VSC HVDC. Most VSC HVDC are cable systems and considering that XLPE cables have been available with limited voltage ratings (320 kV) only symmetrical configuration achieves required pole-pole DC voltage. The power ratings and voltage levels with installed VSC transmission have also been relatively low to justify full bipolar topologies.

The Caprivi overhead link is one of the few monopolar VSC HVDC in operation, which is planned to become a bipole when the second pole is added in the future. The 78 MW, 150 kV, 2011 Valhall HVDC also employs monopolar topology. The Skagerrak 4 is a unique topology, since it is a monopolar VSC HVDC, which will operate with another pole of LCC HVDC type (Skagerrak 3).

Figure 12.2 A polymeric insulated cable for HVDC. Reproduced with permission of ABB.

12.6 VSC HVDC Converter Topologies

The VSC-HVDC transmission system has been implemented using two-level VSCs, three-level VSCs and multilevel VSCs.

12.6.1 HVDC with Two-Level Voltage Source Converter

Two-level Sinusoidal Pulse Width Modulation (SPWM) topology was used in the first generation of VSC HVDC in the period 1996–2006, and Table 12.2 shows some example HVDC installations. It is also currently commercially available in addition to MMC HVDC but mainly for lower power ratings.

Table 12.2 Examples of the VSC-HVDC transmission systems based on the two-level converter.

Project	Rating	PWM strategy	Applications	Distance (km)	Commissioning year
Terranora (Australia)	180 MW, $V_{dc} = \pm 80$ kV	SPWM	Controlled asynchronous connection for trading	59	2000
Tjareborg (Denmark)	8 MW, $V_{dc} = \pm 9$ kV	SPWM	Wind-power connection	4.3	2000
Gotland (Sweden)	50 MW, $V_{dc} = \pm 80$ kV	SPWM	Permission for under-ground cable	70	1999

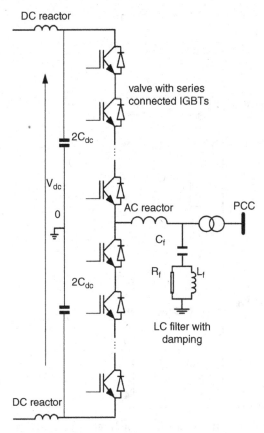

Figure 12.3   One phase of VSC HVDC station based on the two-level converter topology.

Practically, the maximum switching frequency with two-level HVDC converters is around 1.5 kHz (27–33 frequency modulation ratio). Figure 12.3 shows a two-level VSC that uses self-commutated switching devices, mainly IGBTs. The capacitor C_{dc} of the VSC must be sized to maintain a constant DC voltage. The size (footprint) of the two-level VSC-HVDC terminal is around 50% size of the LCC HVDC station.

The main characteristics of this technology are:

- Simple converter construction that requires a simple control strategy to guarantee stable operation over the entire operating range.
- Lowest number of switches.
- High switching losses and relatively high filtering requirements (requires relatively large AC filters with damping, which adds losses). Note that filters are still much smaller than with LCC HVDC.
- High dv/dt because of large voltage difference at each switching, with a relatively high switching frequency and high common mode voltage. These impose high insulation requirements on the interfacing transformers, and also generate high electromagnetic interference.
- Poor DC fault ridethrough capability. This represents a major obstacle to the development of multiterminal HVDC, especially with the absence of reliable and proven DC CBs.

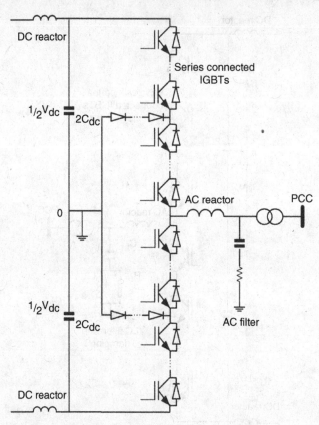

Figure 12.4 One phase of VSC HVDC station based on the neutral-point clamped (NPC) voltage source converter.

12.6.2 HVDC with Neutral Point Clamped Converter

As the second generation VSC HVDC, the three-level diode clamped converter (also known as neutral point clamped (NPC) converter) was developed around 2000, with the following objectives:

- To reduce effective switching frequency per device (consequently, lowering switching losses).
- To lower *dv/dt* enabling the use of a transformer with reduced insulation requirements. It also enables low total harmonic distortion at the point of common coupling (PCC) (achieves a further reduction in filter size).

Figure 12.4 shows one-phase leg of the neutral-point clamped converter, which halves the voltage stress and the effective switching frequency per device compared to the two-level converter in Figure 12.3. As an additional benefit, the converter station loss is reduced significantly. However, with the NPC converter it is difficult to meet some grid code transient requirements, for example that the converter must remain in operation during AC grid faults in order to provide reactive power to support the grid. This is an issue because the NPC converter DC capacitor voltage balancing (at $\frac{1}{2}V_{dc}$) is challenging under asymmetrical AC faults, such as a single-phase open circuit fault, single-phase-to-ground and line-to-line faults. Consequently, HVDC manufacturers eventually abandoned the NPC converters in favour of improved two-level converters. Table 12.3 lists NPC VSC HVDC transmission installations.

Table 12.3 Examples of the VSC-HVDC transmission systems based on the NPC converter.

Project	Rating	PWM strategy	Applications	Distance	Commissioning year
Cross sound (USA) cable	330 MW, $V_{dc} = \pm150$ kV, $f_s = 1.26$ kHz	SPWM	Grid Re-enforcement	40 km	2002
Murray link (Australia)	220 MW, $V_{dc} = \pm150$ kV, $f_s = 1.35$ kHz	SPWM	Grid re-enforcement	180 km	2002
Eagle Pass (USA)	36 MVA, $V_{dc} = \pm16$ kV, $f_s = 1.26$ kHz	SPWM	Power trading and power quality	Back-to-back	2000

12.6.3 Modular Multilevel Converter VSC-HVDC Transmission Systems

This converter concept exploits the benefits of the multilevel structure and PWM. The filtering requirements are greatly reduced because of the generation of high-quality AC voltage (small AC filters might be required). One cell (module) per arm is switched at a time, which results in only 1–2 kV voltage increments at each switching instant, although some topologies use cells of higher voltages. The use of a large number of levels with small voltage steps results in low dv/dt at each switching and reduced voltage stress on the insulation of the interfacing transformers. This allows the use of standard transformers without the need to withstand the DC link voltage or harmonic currents. Furthermore, the effective switching frequency per device is low, resulting in lower switching losses and lower harmonics. On the downside, modular converters require larger number of switches, at least twice the number compared with two-level VSC.

Figure 12.5 shows one-phase of a $n + 1$ level modular converter, consisting of n cells in each arm. This converter relies on the cell capacitors to create a multilevel voltage waveform at the converter terminal. Each cell has low voltage, which is of the order of a single switch rating or it is built up of a small number of series connected switches. Typically, hundreds of cells are required to build a single valve for DC transmission requirements. As the number of levels increases, the quality of AC voltage waveform becomes better and the harmonic content reduces. The DC link capacitors are not required as individual modules have capacitors. However, some small DC capacitors (around a few microfarads) might be installed in a typical symmetrical monopole configuration in order to create ground reference for the AC voltages. As this topology benefits from the redundant combination of module connections for each required AC level, the balancing capability is better than with NPC converters. The MMC performs better than the NPC converter during unbalanced operation and symmetrical/asymmetrical AC faults which reduces the risk of device failure and system collapse.

The ability of the modular converter to ride through different types of AC faults makes it suitable for applications subjected to stringent grid codes requirements. If there is an issue with one phase on the AC system, the remaining two phases of the converter will operate unaffected, potentially at full per-phase power, since there is no common DC capacitor to transfer ripple between phases. The absence of DC capacitors also reduces issues with DC faults, considering that MMC cell capacitors will not discharge into DC fault and this leads to faster post fault recovery. With 2/3 level VSC, post-DC fault recovery requires a period of DC capacitor charging.

The sizing of the cell capacitors requires careful considerations. They dominate the volume requirement for the modules but they are required to store sufficient energy to support the converter DC voltage during transient events. Otherwise the system may fail to meet transient requirements. It will be shown in the modelling section that the cell capacitors behave as series connected AC-side components.

The first commercial MMC based HVDC transmission system project is the 85 km, 400 MW, ±200 kV (±170 MVAr STATCOM functionality) Trans Bay cable project, commissioned in the United States in 2010, and some other example projects are listed in Table 12.4.

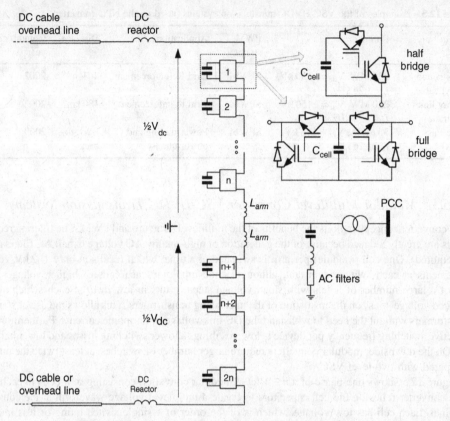

Figure 12.5 One-phase of a $n + 1$ level MMC HVDC converter.

Table 12.4 Examples of the VSC-HVDC transmission systems based on the MMC converter.

Project	Rating	PWM strategy	Applications	Distance (km)	Commissioning year
Trans Bay (USA)	400 MW, $V_{dc} = \pm200\,kV$	MMC	Network upgrade using cable supply	85	2010
INELFE (Spain-France)	$2 \times 1000\,MW$, $V_{dc} = \pm320\,kV$	MMC	Interconnector between two AC systems	65	2014
SylWin1 (Germany)	864 MW, $V_{dc} = \pm320\,kV$	MMC	Offshore windfarm connection	204	2014

12.6.4 MMC HVDC Based on Full Bridge Topology

MMC HVDC with full bridge converters is commercially available, however it has not been implemented at the time of writing. The full bridge MMC HVDC converter has similar overall structure as a half-bridge MMC, but each cell uses four switches in full H connection, as shown in Figure 12.5. The full bridge cell facilitates either positive or negative voltage at the cell terminals while the half bridge gives only positive voltage. Therefore full bridge gives more control flexibility, although

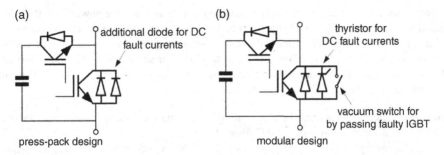

Figure 12.6   MMC cell designs.

at the expense of twice the number of switches. Note that at least 3/4 of the (but not all) cells are required to assume full bridge topology in order to achieve system-level benefits. There are some important operational advantages of full bridge over half bridge MMC:

- Full bridge topology can actively control and interrupt DC fault current. It need not be tripped for DC faults. If it is tripped the full bridge converter becomes open circuit for DC faults.
- It can operate with reduced DC voltage. This implies that full bridge MMC can supply full reactive power to the AC system during DC fault conditions.
- The overrating of antiparallel diodes (which is required with half-bridge cells) may not be needed because a high fault current cannot occur. The cell design is further discussed in Section 12.7.
- Direct current voltage polarity reversal is possible. This is very important to rapidly extinguish the DC fault current path (it can deionize and extinguish arcs on the overhead DC lines as is done with a LCC HVDC in Figure 10.2). The postfault DC voltage ramp up can be controlled.
- It is possible to provide higher (typically by up to 20–30%) AC voltage for the same DC voltage using overmodulation feasible only with full bridge topology. This implies that for a given IGBT current and DC voltage limitation, full bridge topology can transfer more power, at the expense of increased number of cells.

12.7   VSC HVDC Station Components

The basic structure of a VSC-based HVDC system station (terminal) is shown in Figure 12.1. The function and design of the major power components will be summarized in the following sections.

12.7.1   AC Circuit Breaker

The AC CB is employed to connect and disconnect the VSC-HVDC system during normal and fault conditions. There are no special design requirements compared to normal AC CB used in power systems. During a DC fault, VSC converter cannot interrupt fault current, since free-wheeling diodes inside VSC converter will uncontrollably feed the fault from the AC system. Therefore the AC CB is tripped to disconnect the HVDC terminal to prevent feeding the fault from the AC side. In the case of a temporary DC fault, power transmission can be resumed after a normal startup sequence of the HVDC system. This can be achieved in a time-frame of a few seconds.

12.7.2   VSC Converter Transformer

A three-phase 50 Hz/60 Hz converter-grade transformer with tap-changer is used (with MMC HVDC standard transformers can be employed). The converter-side voltage (filter bus voltage) is commonly controlled by the tap changer to achieve the maximum active and reactive power from the VSC, both consumed and generated. The tap changer can be located at any side though. The transformer sometimes has a tertiary winding, which feeds the station auxiliary power system and if delta connected it suppresses any core triplen fluxes. With the use of the AC filter and the VSC PWM, the current in the transformer windings contains minimal harmonics and is not exposed to any DC voltage. The converter transformer can provide the following functions:

- It provides a coupling reactance between the VSC and the AC system, which also reduces fault currents and can decrease size of the AC filter.
- It matches the voltage between the AC system and the VSC converter, which in turn is determined by the DC voltage. This results in optimum use of the switch ratings and DC cable insulation.
- It provides galvanic isolation, enabling optimized grounding of the DC link.
- It prevents a flow of zero sequence current between the AC system and VSC.

During the transformer design, the specific transformer requirements include:

- the fundamental current stresses;
- saturation characteristics of the transformer magnetic field;
- low and medium frequency harmonics stresses;
- dielectric stresses caused by the normal/transient operating voltage especially with PWM VSC, and harmonics.

12.7.3   VSC Converter AC Harmonic Filters

AC filters for VSC HVDC converters have lower ratings than those for LCC HVDC converters and are not required to provide reactive power compensation. Contrary to filtering with LCC HVDC, VSC filters are permanently connected to the converter bus (they are not switched with power loading). A low-pass LC-filter is typically used to suppress high-frequency harmonic components and avoid interaction with fundamental frequency components. An MMC HVDC may not require AC filters.

12.7.4   DC Capacitors

The DC capacitor is the energy storage element in VSC. It provides the VSC with the stiff DC voltage between switching instants, which is an essential presumption with all VSC topologies. The primary functions of the DC-side capacitor are:

- To provide a low-inductance path for switch turnoff current. Because of the stray inductance, the turnoff commutating current results in transient voltage stresses on the switching devices. These stresses can be minimized by reducing the length of the connection between the DC link capacitor and the switching devices.
- Temporary energy storage between the switching instants, stabilizing high frequency dynamics, which allows the VSC closed loop control to adjust the control signals in lower frequencies.
- Reduce the DC voltage harmonic ripple. The HVDC capacitor commonly uses a dry, self-healing, metallized film design, which has advantages over oil-filled technologies. Dry capacitor design is safer, it offers high capacity and low inductance, in noncorrosion, no radiating plastic housing.
- Decrease the harmonic coupling between different VSC substations connected to the same DC bus.

The design requirements for the DC capacitor include:

- continuous operating DC voltage;
- the limits of DC voltage ripple under transient conditions, such as faults to the AC system;

- harmonic currents passed to the DC side;
- peak discharge current for DC faults.

A requirement for a small voltage ripple implies a large capacitor. On the other hand, a small capacitor has advantages considering the control and dynamics of the converter, which results in fast active power control. Selecting the size of the DC capacitor is a tradeoff between voltage ripple, lifetime, costs and the fast control of the DC voltage. Based on the ripple specification, a lower limit can be established for the capacitor value:

$$C_{dc} > \frac{S_{VSC}}{2\omega V_{dc}\Delta V_{dc}}$$ (12.1)

where, C_{dc} is DC capacitance, S_{VSC} is the converter MVA rating, V_{dc} is rated DC voltage, ω is electrical frequency and ΔV_{dc} is the allowed voltage ripple (peak–peak). DC cable manufacturers typically specify DC voltage ripple of around 3–10% but other limits may also apply. Based on the control speed requirements, it is possible to set the upper limit:

$$C_{dc} < \frac{2\tau S_{VSC}}{V_{dc}^2}$$ (12.2)

where, τ is the time constant of DC capacitor charging. This time constant is commonly selected to be less than 10 ms to meet the required performance for speed of transient response.

Commonly, however, the high-power converter capacitor size is determined considering the total energy stored. The energy-to-power ratio E_s [J/VA] is defined using capacitor energy E_c [J] and the converter power S_{VSC} [VA]:

$$E_s = \frac{E_c}{S_{VSC}}$$ (12.3)

The energy to power ratio in practical converters is: 10 (kJ/MVA) $< E_s <$ 50 (kJ/MVA), which is a good tradeoff between harmonic penetration and control performance. Therefore, using the expression for capacitor energy:

$$E_c = \frac{1}{2}C_{dc}V_{dc}^2$$ (12.4)

it is possible to obtain a practical formula for capacitor size:

$$C_{dc} = \frac{2S_{VSC}E_s}{V_{dc}^2}$$ (12.5)

Example 12.1
Calculate pole-neutral DC capacitor for a two-level 800 MW, 640 kV VSC.

Solution
Assuming Es = 20 kJ/MVA:

$$C_{dc} = \frac{2SE_{dc}}{V_{dc}^2}$$

$$C_{dc} = \frac{2 \times 800\,MVA \times 20000\,J/MVA}{640000^2}$$

$$C_{dc} = 78\,\mu F$$

The required Pole-neutral capacitance is 156 μF at 320 kV DC.

12.7.5   DC Filter

Instead of increasing the size of the DC capacitor, a DC filter can be used to eliminate the targeted harmonics, which may be injected into the DC line. It is connected in parallel to the DC capacitor to decrease the total equivalent impedance of the DC circuit. Direct-current cables naturally have a good low-frequency attenuation but they may amplify high-frequency harmonics. In some VSC HVDC systems special high-frequency tuned DC filters are installed to avoid DC-side harmonic resonances. The design criteria of the DC filter for the VSC HVDC are similar to those for LCC HVDC systems.

12.7.6   VSC HVDC Cells and Valves

IGBTs have been used in low- and medium-power applications for many years. However, in developing high-voltage VSC HVDC converter valves, there are two new challenges:

- A valve in VSC HVDC converter requires numerous series connected cells (or individual IGBTs with two-level designs) to achieve the required blocking voltage. Several redundant cells are typically provided in each valve chain which ensure that voltage stress is within acceptable range even when a single (or few) cells fail. When a cell fails, the valve should still be able to commutate ON and OFF. The failed cell can then be replaced at the next scheduled maintenance period. However, an ordinary IGBT will normally fail in an open circuit, which would be a dangerous valve open-circuit fault in an HVDC converter. A valve design must ensure that IGBT failure leads to short circuit across the individual cell/IGBT. During normal operation, all cells (including redundant units) in a valve are operated normally, and the presence of redundant cells means that voltage stress on all cells is sightly lower than rated value.
- The DC fault current is much larger than the rated current. A valve design must ensure that IGBTs are not thermally destroyed under DC fault currents. Diodes take fault current, and this implies that the antiparallel diode must be rated for a much larger current than the IGBT switch.

There are two different MMC cell designs, as shown in Figure 12.6:

- Press-pack design. In this case special press-pack IGBT packaging is used. The IGBTs are physically located in a string and clamped together in a manner that resembles thyristor press-pack packaging. These IGBTs are designed to always weld themselves to a permanent short-circuit fault. The IGBTs also have antiparallel diodes with higher current rating than the active switch, which allows for large DC fault currents. All two-level VSC HVDC valves use this design method.
- Modular design. In this case, standard commercially available IGBT switches are employed. An external thyristor is added in parallel with the diode on one of the switches, which is triggered in case of DC faults. This provides adequate total semiconductor rating for DC fault currents. An external fast vacuum switch is also added, which is closed in case of IGBT failure. This is a one-use mechanical vacuum switch, which will permanently weld its contacts, thus bypassing the IGBT in both directions if a large voltage is detected across the IGBT under a gate pulse.

Some other auxiliary components are also present in the cells (not shown in Figure 12.6) like:

1. Discharge resistors, which ensure that cells are discharged 20–30 minutes after tripping converter to enable safe handling.
2. Overvoltage protection, which may require another small IGBT and a resistor.

12.8 AC Reactors

AC reactors are commonly added in series with VSC converter transformers on the converter side. They increase series reactance where it would be difficult to achieve such large transformer leakage inductance.

The main purpose of AC reactors are:

- They reduce DC fault currents. This is required to prevent overcurrents in diodes for DC faults, in particular with strong AC systems. The alternative is to increase diode ratings, which is costly and has size/volume implications.
- They reduce peak switch currents for AC faults. AC faults are managed by converter controls but control delays require certain minimal AC reactance in order to limit peak transient currents to within IGBT ratings.

At HVDC power levels the inductors are always of air-core design, most often dry type but some are oil immersed. The textbook theory of electromagnetism with concentrated conductors gives good estimates for magnetic circuits with small number of turns, whereas for large inductors with high currents experimental design formulae are usually used.

12.9 DC Reactors

In VSC HVDC systems, a DC reactor may be connected after the DC capacitor, as shown in Figure 12.3 for the following reasons:

- to reduce gradient of DC fault current in order to protect freewheeling diodes and capacitors;
- harmonic current reduction in the DC overhead line or cable;
- critical resonance detuning within the DC circuit.

A DC reactor in a VSC HVDC system is considerably smaller than those used in LCC HVDC schemes and typically it is below 5 mH. However, the design methodology is quite similar.

The following study will present one practical approach in designing a high-power air core inductor, for either AC or DC circuits. Figure 12.7 shows the topology of a solenoid inductor. The inductance can be reasonably well estimated with the following formula:

$$L = 0.2 \times 10^{-6} \pi^2 a \frac{2a}{l_{th}} N^2 K \tag{12.6}$$

where N is the total number of turns and K is Nagaoka's constant

$$K = \frac{1}{1 + 0.9\dfrac{a}{l_{th}} + 0.32\dfrac{t}{a} + 0.84\dfrac{t}{l_{th}}} \tag{12.7}$$

and where all parameters are defined in Figure 12.7. The maximum value of L in Equation (12.6) is obtained for $2a = 3t$, however this design option would lead to a densely wound coil, implying the smallest surface for heat dissipation and likely high temperature of inner turns for high-power designs. With high current designs the current density is usually taken as 1–4 A/mm^2 depending on the cooling arrangements. The cross section will be large and the wire is usually implemented using a large number of stranded wires of smaller radius, or a series of rectangular wires similarly to those used with transformer designs. For high-frequency applications (HF filters or DC/DC converters), Litz wire is used.

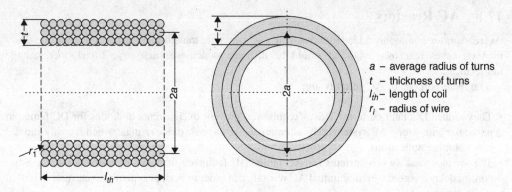

Figure 12.7   Solenoid air-core inductor.

The total length of wire can be determined from the above geometry of the solenoid and knowing the number of turns. Once length is known, the inductor mass can be obtained by multiplying by specific mass and multiplying with specific resistance to determine the total resistance.

Example 12.2
Design a 200 mH inductor for 1200A LCC HVDC converter. Calculate the total inductor mass, resistance and determine inductor loss assuming that copper wire is used. The density of copper is $\rho_m = 8930 \, kg/m^3$ and specific resistance is $\rho_r = 1.73 \times 10^{-8} \, \Omega/m$.

Solution
Assuming that current density is $2 \, A/mm^2$, this gives the conductor a cross section of $600 \, mm^2$, or a radius of $r_1 = 13.8 \, mm$. A 10% enamel insulation thickness is also assumed. Taking the average radius of turns $a = 0.918 \, m$, with 14 layers and 25 turns in each layer ($N = 350$) the length is obtained $l_{th} = 0.759 \, m$, with $t = 0.43 \, m$, giving $L = 0.193 \, H$.

The total length of wire is approximately $2007 \, m$. The total mass is $M_t = 10\,756 \, kg$ and the total resistance is $R_t = 0.0579 \, \Omega$. Therefore, at the rated current the total inductor loss is: $P_{loss} = 83.347 \, kW$.

13

IGBT Switches and VSC Converter Losses

13.1 Introduction to IGBT and IGCT

High-power self-commutated voltage source converter (VSC) converters can be built using switches from one of the two power electronic device families: 1) the insulated gate bipolar transistor (IGBT) or injection-enhanced gate transistor (IEGT)) and 2) integrated gate commutated-thyristor (IGCT), or gate-commutated thyristor (GCT). Although IGBTs and IGCTs are available with broadly similar voltage ratings, they represent significantly different device technologies. Insulated-gate bipolar transistors are voltage-controlled devices characterized by fast switching times and simple gate drive requirements. Individual semiconductor chip size for IGBTs is constrained, with the consequence that high-power switches consist of numerous internal parallel connected units. IGBTs are asymmetrical devices (have no reverse blocking capability) and therefore cannot be used with current source converters.

IGCT devices comprise an improved gate turn-off thyristor with an integral gate drive board. Single-chip solutions for the main power device are available to high current ranges. For a given rating, IGCTs achieve lower conduction loss than IGBTs at the expense of a large and complex gate drive, and high gate current requirements. IGCTs are also available as symmetrical devices (with reverse blocking) but such devices will have an increased ON-state voltage drop. IGCTs do not have active gate control (they are either on or off like all thyristors) and consequently are more difficult for connecting in series compared with IGBTs. Furthermore IGCTs have lower switching speeds, consequently higher switching losses, and for these reasons they have not been used with two-level pulse width modulation (PWM) VSC-HVDC converters.

IGBT and IGCT devices are reported with maximum blocking voltages in the region of 6.5 kV, with recommended DC operating voltages between 50% and 70% of this value. The rated current of the commercially available 6.5 kV IGBT is only 750 A. Because of lower conduction and switching losses, Medium voltage IGBT devices in the 3.3–6.5 kV range, are most often used in HVDC valves and have current ratings of around 1200–1500 A. These current ratings are significantly lower than those with the comparable IGCT devices.

High-Voltage Direct-Current Transmission: Converters, Systems and DC Grids, First Edition.
Dragan Jovcic and Khaled Ahmed.
© 2015 John Wiley & Sons, Ltd. Published 2015 by John Wiley & Sons, Ltd.

13.2   General VSC Converter Switch Requirements

A power semiconductor switch is a component that can either conduct current when it is commanded ON or block a voltage when it is commanded OFF through a control input. This change of conductivity in a semiconductor is made possible by specially arranged device structures that control the carrier transportation. Generally, the following properties are important for a semiconductor switch in power-conversion applications:

- maximum current-carrying capability;
- maximum voltage-blocking capability;
- forward voltage drop in ON state and its temperature dependency;
- leakage current in OFF state;
- thermal capability;
- switching transition times during both turn on and turnoff;
- capability to sustain dv/dt when the switch is OFF or during turnoff;
- capability to sustain di/dt when the switch is ON or during turn on;
- controllable di/dt or dv/dt capability during switching transition;
- ability to withstand both high current and voltage simultaneously (operating area);
- ability to conduct in reverse direction;
- low switching losses;
- control power requirement and control circuit complexity.

13.3   IGBT Technology

The history of IGBTs started in the early 1980s but a major technological advance was made in the early 1990s when several generations of IGBT devices were developed by a number of companies. The IGBT is a voltage-controlled device from the transistor family. It can be switched on with a +15 V gate voltage and turned off when the gate voltage is zero. In practice, a negative gate voltage of a few volts may be applied during the device-off transient to increase its noise immunity. The IGBT does not require any gate current when it is fully turned on or off. However, it does need a peak gate current of a few amperes during switching transients because of the gate-emitter capacitance. Insulated-gate bipolar transistors combine the on-state voltage characteristics of bipolar transistors with the low power gate characteristics of (metal-oxide semiconductor field-effect transistor) MOSFET devices. Fast switching speeds and a near rectangular safe operating area (SOA) allow the use of IGBTs without external snubber circuits. A circuit symbol for the IGBT is shown in Figure 13.1, where the symbols are: C-collector, E-emitter, G-gate. Figure 13.2 shows commercially available modular type IGBTs.

An IGBT is developed by exploiting best properties of a Power MOSFET and a bipolar junction transistor (BJT), with the following advantages:

- It has a low ON-state voltage drop and has superior on-state current density. A smaller chip size is therefore possible and the cost can be reduced.
- Low driving power and a simple drive circuit because of the input metal-oxide gate structure (MOS). It can be controlled easily in comparison with current-controlled devices (thyristor, BJT) in high-voltage and high-current applications.
- Wide SOA. It has superior current-conduction capability compared with the bipolar transistor. It also has excellent forward-blocking capabilities.
- The IGBT is suitable for scaling up the blocking voltage capability. In case of Power MOSFET, the on-resistance increases sharply with the breakdown voltage because of an increase in the resistivity

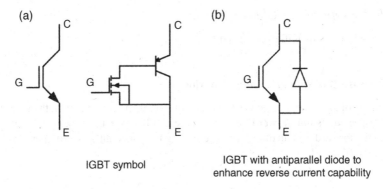

Figure 13.1   IGBT circuit symbol.

Figure 13.2   IGBT switch modules for high-power converters. Reproduced with permission of ABB.

and thickness of the drift region required to support the high operating voltage. For this reason, the development of high-current power MOSFET with high-blocking voltage rating has been abandoned. In contrast, with the IGBTs, the drift region resistance is drastically reduced by the high concentration of injected minority carriers during on-state current conduction.

The drawbacks of high power IGBTs are:

- Switching speed is lower than power MOSFET but still better than that of a BJT.
- There is a possibility of latch-up because of the internal PNPN thyristor structure.
- IGBTs normally fail in an open circuit, which is a significant problem in HVDC applications where numerous switches are series connected in each valve. Manufacturers have now developed IGBTs (cell assemblies) that fail in short circuits, but complexity and cost increase.

13.3.1 IGBT Operating Characteristics

The main characteristics of IGBTs are listed below:

13.3.1.1 Forward Blocking or Conduction Modes

When a positive voltage is applied across the collector to emitter terminals with the gate shorted to the emitter, the device enters into forward-blocking mode. An IGBT in the forward-blocking state can be transferred to the forward-conducting state by removing the gate emitter shorting and applying a positive voltage of sufficient level.

13.3.1.2 Reverse Blocking Mode

When a negative voltage is applied across the collector to emitter terminals, the junction becomes reverse biased and its depletion layer extends. The breakdown voltage for the reverse blocking is determined by an open-base BJT. The desired reverse voltage capability can be obtained by optimizing the resistivity and thickness of the N-drift region. However reverse blocking is not required in typical VSC topologies, and reverse blocking voltage with high-power IGBT is very low (around 20 V). The full output characteristic for a range of gate voltages is shown in Figure 13.3. In most commercial switches, a fast recovery antiparallel diode is integrated inside the switch module as shown in Figure 13.1. This diode is essential in all VSC applications in order to prevent destruction of IGBTs under reverse voltage.

13.3.1.3 Transfer Characteristics

The transfer characteristic is defined as the variation of I_C with V_{GE} values at different temperatures, namely, 25, 125 and −40 °C. A typical transfer characteristic is shown in Figure 13.4a. The gradient of transfer characteristic at a given temperature is a measure of the transconductance (g_{fs}) of the device at that temperature. A large g_{FS} is desirable to obtain a high current handling capability with low gate drive voltage. The channel and gate structures dictate the g_{FS} value. Both g_{FS} and $R_{DS}(on)$ (on resistance of IGBT) are controlled by the channel length, which is determined by the difference in diffusion depths of the P base and N+ emitter. The point of intersection of the tangent to the transfer characteristic determines the threshold voltage (V_{GEth}) of the device (gate voltage for switching on).

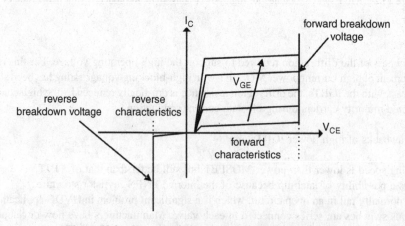

Figure 13.3   Shape of full output characteristic of the IGBT.

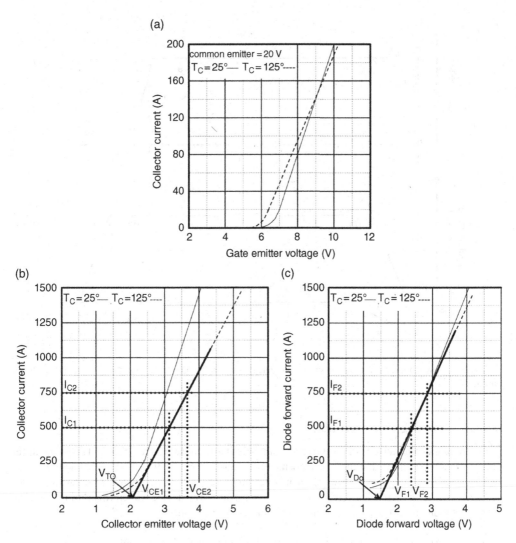

Figure 13.4 Characteristics for a typical high-power (6.5 kV/750 A) IGBT: (a) transconductance, (b) IGBT output characteristics and (c) antiparallel freewheeling diode ON characteristics.

13.3.1.4 Output Characteristics

The plot of a typical forward output characteristics of an IGBT is shown in Figure 13.3. It has a family of curves, each of which corresponds to a different gate-to-emitter voltage (V_{GE}). The collector current (I_C) is measured as a function of the collector-emitter voltage (V_{CE}) with the gate-emitter voltage (V_{GE}) constant. In high-power applications, the IGBT is always operated in a saturated region (operates as a switch in either ON or OFF state). Although current control is possible using the gate signal, this would lead to excessive losses. The output characteristic in saturated region is shown in Figure 13.4b. Note that gate control also enables variation in switching speed, which is very useful in balancing turn-on stress in series connected switches.

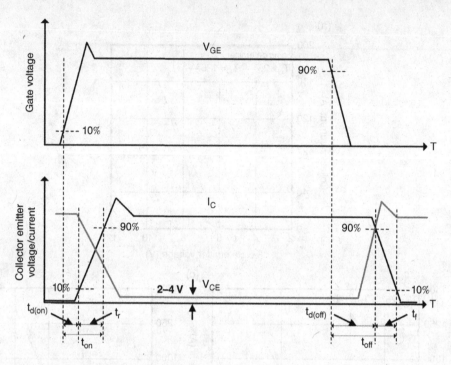

Figure 13.5 Switching waveforms of the IGBT gate-emitter voltage, collector-emitter voltage and collector current.

13.3.1.5 Switching Characteristics

The switching characteristics of an IGBT are quite similar to those of a MOSFET. The major differentiating characteristic is that the IGBT characteristic has a tailing collector current. The tail current increases the turnoff loss and requires an increase in the dead time between the conduction of the two devices in a converter leg. Figure 13.5 shows typical IGBT device waveforms during one cycle of switching on and off, where the following parameters are defined:

- *Turn-on delay time ($t_{d(on)}$):* this is defined as the time between 10% of gate voltage to 10% of the final collector current.
- *Rise time (t_r):* this is the time required for the collector current to increase to 90% of its final value from 10% of its final value.
- *Turn-off delay time ($t_{d(off)}$):* this is the time between 90% of gate voltage to 10% of final collector voltage.
- *Fall time (t_f):* this is the time required for the collector current to drop from 90% of its initial value to 10% of its initial value.

13.3.1.6 IGBT Parameters

The following general parameters for an IGBT are introduced:

Maximum power dissipation (PD). This parameter represents the power dissipation required to raise the junction temperature to its maximum value of 150 °C, at a case temperature of 25 °C. Normally, a curve is provided to show the variation of this power with temperature.

Junction temperature (Tj). This specifies the allowable range of the IGBT junction temperature during its operation.

Collector-emitter leakage current (I_{CEL}). This parameter determines the leakage current at the rated voltage and specific temperature when the gate is shorted to the emitter.

Gate-emitter threshold voltage (V_{GEth}). This parameter specifies the gate-emitter voltage range when the IGBT is turned on to conduct the collector current. The threshold voltage has a negative temperature coefficient.

Total gate charge (Q_G). This parameter is important in designing a suitably sized gate-drive circuit and to approximately calculate its losses. It varies as a function of the gate-emitter voltage.

Safe operating area (SOA). The SOA determines the current and voltage boundary within which the IGBT can be operated without destructive failure. At low currents the maximum IGBT voltage is limited by the open-base transistor breakdown. A forward biased SOA is defined during the turn on transient of the inductive load switching when both electron and hole current flow in the IGBT in the presence of high voltage across the device. The reverse biased SOA is defined during the turn off transient, where only a hole current flows in the IGBT with high voltage across it. If the time duration of simultaneous high voltage and high current is long enough, the IGBT failure will occur because of thermal breakdown. However, if this time duration is short, the temperature rise caused by the power dissipation may not be sufficient to cause thermal breakdown.

13.3.2 Fast Recovery Antiparallel Diode

In addition to the power switch, all self-commutating converter systems (based on voltage source principles) require a reverse current path at each switch. Commonly with asymmetrical devices, an antiparallel diode of similar rating is provided to conduct the reverse current. These devices differ from rectifier diodes in that they must be compatible with the operating frequency (in the order of kilohertz) and switching speed of the switches in the converter.

These devices must have suitably fast turn-off behaviour (good reverse recovery characteristic). Of primary interest is the reverse current that is passed by the diode and circulates in the switch, and also the transient over voltage during diode turn-off, because this same voltage is experienced on the main switch. Fast recovery diodes satisfy the above requirements but on the downside they display significantly higher conduction loss and reduced transient overload capability when compared to rectifier diodes. In HVDC applications, transient diode ratings are commonly a limiting factor, considering for instance DC side faults.

Recently, there have been significant developments in silicon carbide diodes which are now commercially available with voltage ratings around 1.7 kV, 10 A. Although these devices offer much superior characteristics to silicon diode devices they have yet to achieve the ratings required for transmission applications. The diode reverse recovery characteristic is analysed in a similar way to the thyristor, which is studied in Part I.

13.4 High Power IGBT Devices

Table 13.1 shows technical details of some commercially available IGBTs and IGCTs for high-voltage, high-current modules. This table indicates the advantage of IGCT in terms of the current rating and on-state losses. As a further comparison, a 8 kV, 2 kA phase-control thyristor (ABB 5STP 20 N8500) has only 2.2 V on state voltage.

The high current IGBTs use multiple *p-n* junctions in parallel inside a device to achieve high current ratings. The positive temperature coefficient and the use of single driver enable internal paralleling. Note, however that paralleling at device level is not normally used, and the maximum current ratings are not likely to increase much beyond 1.5 kA. If higher current is required for an HVDC terminal, then complete VSC converters can be connected in parallel.

Table 13.1   IGBT/IGCT devices ratings for high-voltage, high-current applications.

Device	Rated voltage (kV)	Rated current (kA)	On-state voltage
IGBT – MG1200FXF1US53 (Toshiba)	**3.3**	1.2	3.7 V
IGBT (ABB 5SNA 1200G450300)	**4.5**	1.2	3.5 V
IGBT (ABB 5SNA 0750G650300)	**6.5**	0.75	3.8 V
IGBT (Infineon FZ750R65KE3)	**6.5**	0.75	3.7 V
IGCT (ABB 5SHY 35L4522)	**4.5**	4.0	1.8 V

IGBT technology requires significantly more silicon surface area for a given current rating than IGCT technology because of the parallel connection of multiple devices in press-pack IGBTs. This brings some advantage since there is more surface area available for cooling though this has to be balanced against the increased conduction loss.

High-power IGBTs are available in press-pack and single-sided plastic modules, as discussed with cell design in Figure 12.6. Insulated-gate bipolar transistor press-pack technology brings the same advantages as press pack thyristors as it provides an opportunity to cool the IGBT devices on both sides thereby improving heat dissipation and reducing junction temperatures. Press pack IGBTs can be mechanically configured to fail in a short circuit. The design requires a paste, foil or component that melts and fuses with the silicon wafer and the pressure plate contact. Under fault current conditions, this material creates a low-resistance conducting path (a short circuit) between the external collector and emitter terminals of the device. This device design is employed by some HVDC manufacturers.

A conventional module package IGBT technology is also employed with some VSC HVDC. In case of device failure, the fault current breaks the bond wires and the device will fail to an open circuit. When this switch package is employed in HVDC valves, a specially developed parallel bypass vacuum switch is used to provide current continuity in case of a switch failure. This is discussed with cell analysis in Section 12.7.

13.5   IEGT Technology

Some manufacturers have made IEGT available. It represents improved IGBT technology for medium-voltage applications. By optimizing the gate structure it is claimed that the IEGT achieves a reduction in both on-state voltage and switching loss. A 42 chip, 4.5 kV, 5.5 kA press pack IEGT has been reported. The IEGT is manufactured in both press pack and single sided modular versions.

13.6   Losses Calculation

This section will review the different loss components occurring in VSC semiconductor devices.

13.6.1   Conduction Loss Modelling

Conduction losses occur because of the voltage drop across switching devices during conduction, similarly as discussed with thyristors. The device voltage V_{FT} is a function of the current, and can be approximated by:

$$V_{FT} = V_{To} + r_{FT} i_{ce} \tag{13.1}$$

where V_{To} is the forward voltage drop across the device at no load, r_{FT} is the on resistance of the switch and i_{ce} is collector-emitter current flow through the device during conduction.

A typical output characteristic for the high-power IGBT is shown in Figure 13.4b. The values of V_{To} and r_{FT} can be obtained from this output characteristic as follows:

$$r_{FT} = \frac{\Delta V_{CE}}{\Delta I_{CE}} = \frac{V_{CE2} - V_{CE1}}{I_{CE2} - I_{CE1}} \tag{13.2}$$

Using Eqs (13.1) and (13.2), the conduction loss is calculated as:

$$P_{cT} = \frac{1}{T} \int_0^T V_{FT} i_{ce} dt = \frac{1}{T} \int_0^T [V_{To} + r_{FT} i_{ce}] i_{ce} dt = \frac{1}{T} \int_0^T V_{To} i_{ce} dt + \frac{1}{T} \int_0^T r_{FT} i_{ce}^2 dt \tag{13.3}$$

$$P_{cT} = V_{To} I_{avT} + r_{FT} I_T^2 \tag{13.4}$$

where T is fundamental period, I_{avT} and I_T are the average and RMS currents in the switching device over one fundamental period.

Similarly, the antiparallel diode conduction losses can be calculated as:

$$V_{DT} = V_{Do} + r_{FD} i_d \tag{13.5}$$

where V_{Do} is the forward voltage drop across the device at no load, r_{FD} is the on resistance of the diode and i_d is current flow through the device during conduction. Therefore the diode conduction loss is:

$$P_{cD} = \frac{1}{T} \int_0^T V_{FD} i_d dt = \frac{1}{T} \int_0^T [V_{D0} + r_{FD} i_d] i_d dt = \frac{1}{T} \int_0^T V_{FD0} i_d dt + \frac{1}{T} \int_0^T r_{FD} i_d^2 dt \tag{13.6}$$

$$P_{cD} = V_{Do} I_{avD} + r_{FD} I_D^2 \tag{13.7}$$

The values of V_{Do} and r_{FD} can be obtained from the diode specific output characteristics (Figure 13.4c) with the same calculation method in Eq. (13.2). Note that diodes will have ON-state characteristics different from IGBTs.

13.6.2   Switching Loss Modelling

The switching losses in semiconductors such as the IGBT can be divided into two components:

- *Turn-on switching loss (E_{on})* is the total energy lost during turn on of an inductive load. In practice, it is measured from when the collector current begins to flow to when the collector-to-emitter voltage completely falls to zero (ON-state level) in order to exclude any conduction loss.
- *Turn-off switching loss (E_{off})* is the total energy lost during turn off of an inductive load. In practice, it is measured from when the collector-emitter voltage begins to rise from zero (ON-state level) to when the collector current falls to zero.

The total switching loss in one cycle is the sum of the E_{on} and E_{off}. In all VSC HVDC converters, the switching losses can be substantial and must be considered in the thermal design. The switch and antiparallel diode should be considered separately but note that diode turn-on loss is typically neglected.

The average switching losses in the IGBT can be computed using:

$$P_{swT} = f_s \left(E_{on} + E_{off} \right) \tag{13.8}$$

where f_s is the switching frequency. E_{on} and E_{off} represent energy loss for a single event at voltage V_{CE} and current I_C. They can be calculated as:

$$E_{on} = E_{on\_test} \frac{V_{CE}}{V_{CEtest}} \frac{I_C}{I_{Ctest}} \tag{13.9}$$

$$E_{off} = E_{off\_test} \frac{V_{CE}}{V_{CEtest}} \frac{I_C}{I_{Ctest}} \tag{13.10}$$

where E_{on_test} and E_{off_test} are the switching losses from manufacturers' sheets at I_{Ctest} and V_{CEtest}. Figure 13.6 shows the test switch (6.5 kV/750 A) values for E_{on_test} and E_{off_test} at a specific test voltage $V_{CEtest} = 3.6$ kV, and current I_{Ctest} values used for manufacturer's tests.

The switching loss in the antiparallel diode is primarily the reverse recovery loss:

$$P_{swD} = f_s E_{rec} \tag{13.11}$$

The reverse recovery switching energy, E_{rec} can be calculated as

$$E_{rec} = E_{rec\_test} \frac{V_{DC}}{V_{DCtest}} \frac{I_C}{I_{Ctest}} \tag{13.12}$$

where E_{rec_test} is the loss at test conditions which is shown in Figure 13.7 for the test diode.

Example 13.1
Figure 13.8 shows a 500 A, 300 kV VSC converter, which employs the test IGBT from Figure 13.6. Insulated-gate bipolar transistors have a 6.5 kV, and 750 A rating and 100 devices are used in each valve. The operating blocking voltage across each IGBT is therefore 3 kV. The IGBT and freewheeling diode characteristics are shown in Figures 13.4, 13.6 and 13.7. Neglecting turn-on loss of the free-wheeling diode, calculate total losses in this converter. Assume that the converter operates in square

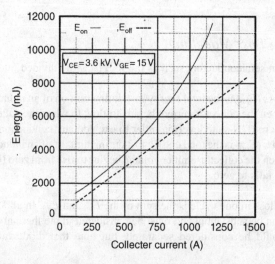

Figure 13.6   Switching energy characteristics for the test IGBT.

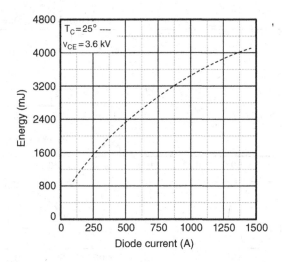

Figure 13.7 Reverse recovery energy characteristics for the antiparallel diode in the test switch.

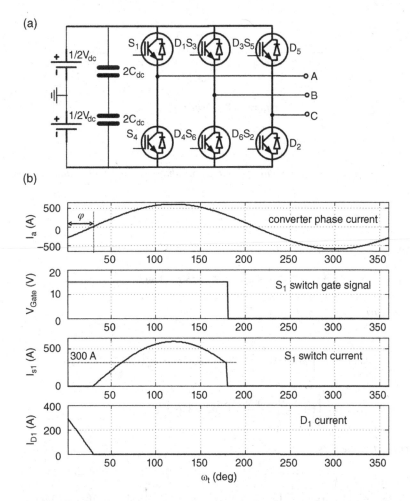

Figure 13.8 VSC converter variables in Example 13.1; (a) two-level three-phase converter, (b) Valve 1 variables and (c) Valve 4 variables.

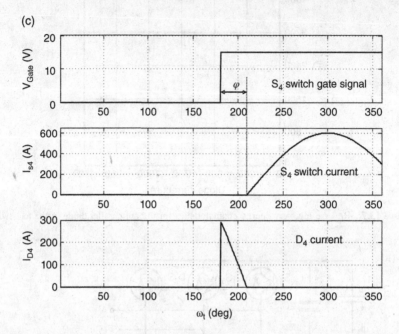

Figure 13.8 (*Continued*)

wave control (single pulse per cycle) with AC current as show in Figure 13.8, and the power factor is 0.866.

Solution

a. Conduction losses
 For the IGBT
 From the above figures, $V_{T0} = 2\ V$, $r_{FT} = (3.7 – 3.2)/250 = 2\ m\Omega$
 The AC current is $i_a = I\sin(\omega t - \varphi_i)$ and $\varphi_i = \cos^{-1}(0.866) = 30°$
 The average current of the S_1 switch is:

$$I_{S1,av} = \frac{1}{2\pi}\int_{\varphi}^{\pi} i_a d\omega t = \frac{1}{2\pi}\int_{\varphi}^{\pi} I\sin(\omega t - \varphi)d\omega t = \frac{I}{2\pi}[-\cos(\omega t - \varphi_i)]_{\varphi}^{\pi}$$

$$I_{S1,av} = \frac{I}{\pi}\left[1 + \frac{\sqrt{3}}{2}\right] = \frac{600}{\pi} \times 1.866 = 356.38 A$$

The rms current of the S_1 switch:

$$I_{S_1} = \sqrt{\frac{1}{2\pi}\int_{\varphi}^{\pi} i_a^2 d\omega} = \sqrt{\frac{1}{2\pi}\int_{\varphi}^{\pi} I^2\sin^2(\omega t - \varphi_i)d\omega t} = \sqrt{\frac{I^2}{2\pi}\int_{\varphi}^{\pi}\frac{1}{2}[1 - \cos 2(\omega t - \varphi_i)]d\omega t}$$

$$I_{S_1} = \sqrt{\frac{I^2}{4\pi}\left[\omega t - \frac{1}{2}\sin 2(\omega t - \varphi_i)\right]_{\varphi}^{\pi}} = \sqrt{\frac{I^2}{4\pi} \times 3.05} = 295.64 A$$

IGBT conduction loss is:

$$P_{cT} = 6n\left(V_{To}\bar{I}_{avT} + r_{FT}I_T^2\right) = 6 \times 100 \times \left(2 \times 356.38 + 0.002 \times 295.64^2\right) = 1064.53\,kW$$

The average current of the D_1 diode:

$$I_{D1,av} = \frac{1}{2\pi}\int_0^{\varphi} -i_a d\omega t = \frac{I}{2\pi}[1 - \cos\varphi_i] = 12.79\,A$$

The RMS current of the D_1 diode:

$$I_{D_1} = \sqrt{\frac{1}{2\pi}\int_0^{\varphi_i} i_a^2 d\omega} = \sqrt{\frac{1}{2\pi}\int_0^{\varphi_i} I^2 \sin^2(\omega t - \varphi_i) d\omega t} = \sqrt{\frac{I^2}{2\pi}\int_0^{\varphi}\frac{1}{2}[1 - \cos 2(\omega t - \varphi_i)]d\omega t}$$

$$I_{D_1} = \sqrt{\frac{I^2}{4\pi}\left[\omega t - \frac{1}{2}\sin 2(\omega t - \varphi_i)\right]_0^{\varphi_i}} = 50.9\,A$$

The freewheeling diode D_1 conduction loss:
From the above figures, $V_{D0} = 1.5\,V$, $r_{FD} = (3 \text{ to } 2.4)/250 = 2.4\,m\Omega$

$$P_{cD} = 6n\left(V_{Do}\bar{I}_{avD} + r_{FD}I_D^2\right) = 6 \times 100 \times \left(1.5 \times 12.79 + 0.0024 \times 50.9^2\right) = 15.24\,kW$$

In case of a pure unity power factor, the full current will follow through only the IGBT, and the diode losses would be negligible.

b. *Switching losses*
   The IGBT loss:
   The current at turn on is Icon = 0, (IGBTs always turn on at zero current when current is lagging).
   The current at turn off Icoff = 300 A.
   From the device characteristics curves and above figures, $E_{on\_test} = 0\,mJ/pulse$, $E_{off\_test} = 1800\,mJ/$
   pulse at 300 A, assuming $V_{CE} = 3.6\,kV$
   $E_{on} = 0$, current is zero during turn on

$$E_{off} = E_{off\_test}\frac{V_{DC}}{V_{DCtest}} = 1800 \times \frac{3000}{3600} = 1500\,mJ$$

$$P_{swT} = 6 \times n \times f_s\left(E_{on} + E_{off}\right) = 6 \times 100 \times 50 \times (0 + 1.5) = 45\,kW$$

   The diode loss:
   The turn on energy is always zero because of zero voltage. Diodes always turn on at zero voltage.
   From the current curves it is seen that at diode turn off Icoff = 0.
   $E_{swD} = 0$, therefore

$$P_{swD} = f_s E_{swD} = 0$$

Total losses for IGBT = 1064.53 + 45 = 1109.53 kW
Total losses for diode = 15.24 + 0 = 15.24 kW
Total losses = 1.1248 MW

13.7 Balancing Challenges in Series IGBT Chains

At present, the voltage ratings of IGBT devices are limited to 3.5–6.5 kV, depending on required current and there is little scope for significant increase in these figures. For HVDC converter applications, a series connection of devices or cells is required in order to achieve the necessary operating voltage. In case of two-level pulse width modulation converters, each valve should be rated for full DC voltage. Up to 200 series connected IGBTs have already been used in HVDC installations.

It is required to balance the switch voltage in a valve in conduction (static) and during switching (dynamic balancing). Static balancing is resolved using large resistors in parallel R_g (grading resistors similarly as with thyristor chains) with each switch in the chain as shown in Figure 13.9. This in turn

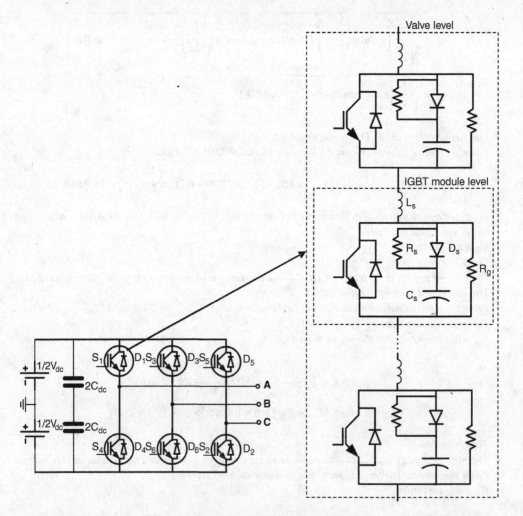

Figure 13.9   VSC with RCD voltage clamp snubbers for *dv/dt* protection, and L_s for *di/dt* protection.

Figure 13.10   Testing VSC HVDC valves on BorWin1 project. Reproduced with permission of ABB.

increases the power loss. Dynamic balancing is a more difficult issue because of very short time frames for the switching transients. If dynamic balancing is not used then the switch that turns off first (or turns on last) would have to sustain full valve voltage. The problem is normally reduced by the use of snubbers, which slow down the switching process at the expense of additional costs and reduced efficiency. Another more commonly used solution is to control actively the switching speed in IGBTs. As transistors have capability of active control of emitter current, the gate voltage can be regulated in fast local (driver-level) feedback loop to make minor adjustments during the switching interval in response to on-line voltage measurements. This method is also called asynchronous firing control. Figure 13.10 show practical design of VSC converter valves.

The emergence of multilevel topologies eliminates/reduces the need for series connection of switches and can provide more cost-effective solutions because cells in a valve are switched one at a time.

13.8 Snubbers Circuits

Snubbers are used to protect the semiconductors from the voltage and current transient stresses that occur during the turn-on and turn-off switching events. They limit the voltage and current magnitudes as well as their rates of rise. The reduction of dv/dt and di/dt also lowers the electromagnetic interference (EMI) levels. Snubber circuits can also help in reducing the static and dynamic balancing problem but they add complexity and may increase the power losses.

Snubbers will shift the switching losses from the silicon to a low-cost passive component where it can be dissipated further away from the main switching device or recovered back into the supply or load. The reduction in losses in turn results in lower junction temperatures. Thus, for a given rating, snubbers may allow the use of higher switching frequencies.

IGBT snubbers are required to limit the switching overvoltage that occurs because of the wire inductance between the IGBT and the DC capacitor. Different snubber configurations can be used in VSC-converters. The resistor-capacitor-diode (RCD) voltage clamp snubber, as shown in Figure 13.9, is preferred over resistor-capacitor (RC) snubbers for medium- and high-current VSC applications because it gives reduced losses in the snubber resistor. Also, a small inductor (L_s) may be added in series with the switch in order to limit the di/dt for transient turn-on voltage protection.

14

Single-Phase and Three-Phase Two-Level VSC Converters

14.1 Introduction

This chapter will study the principles of voltage source converters (VSCs), which have two-level AC waveform and which operate with single switching per cycle. Such topology underpins the operating principles of pulse-width modulation (PWM) VSC high-voltage direct current (HVDC) but it also applies to cell-level operation of multilevel VSC HVDC converters.

The main function of static power VSC inverters is to produce an AC waveform from a DC power supply. The AC waveform should ideally be controllable in terms of magnitude, frequency and phase. A VSC relies on a stiff DC voltage source, which implies approximately constant DC voltage between consecutive switchings. In practice, a large capacitor is required at the DC side of the VSC converter.

14.2 Single-Phase Voltage Source Converter

A single-phase VSC converter is shown in Figure 14.1. The converter consists of four switches (IGBT) $(S_1–S_4)$) and four antiparallel diodes $(D_1–D_4)$, where each diode is connected in antiparallel with an IGBT switch to protect the switch and return the current to the supply when current is not in phase with the voltage.

If S_1 and S_2 are turned on simultaneously, the DC voltage V_{dc} appears across the AC load. While turning on S_3 and S_4 simultaneously, the negative voltage $-V_{dc}$ appears across the load. If the AC current has different sign from the AC voltage, the current will be conducted through freewheeling diodes D_1 to D_4. These four diodes clamp the load voltage to within the DC supply voltage rails ($-V_{dc}$ to V_{dc}). Figure 14.2 shows the phase-to-zero, phase-to-phase and AC current waveforms assuming $V_{dc} = 200$ V. An inductive load is assumed and therefore AC current is lagging the AC voltage.

High-Voltage Direct-Current Transmission: Converters, Systems and DC Grids, First Edition.
Dragan Jovcic and Khaled Ahmed.

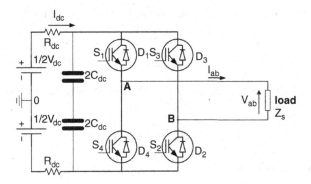

Figure 14.1   A single-phase VSC converter.

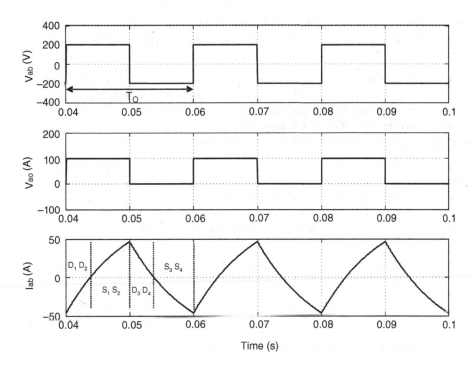

Figure 14.2   Single-phase VSC converter waveforms with inductive AC load.

The RMS AC voltage is

$$V_{ab} = \sqrt{\frac{2}{T_o} \int_{0}^{T_o/2} V_{dc}^2 dt} = V_{dc} \tag{14.1}$$

The instantaneous AC voltage can be expresses as a sum of fundamental and odd harmonic voltages by using the Fourier series:

$$v_{ab} = \sum_{n=1,3,5}^{\infty} \frac{4V_{dc}}{n\pi} \sin n\omega t \tag{14.2}$$

where $\omega = 2\pi f$ is the frequency of the AC voltage in *rad/s*. For $n = 1$, the fundamental RMS voltage is:

$$V_{ab1} = \frac{4V_{dc}}{\sqrt{2}\pi} = 0.9\, V_{dc} \qquad (14.3)$$

The current is typically determined using the AC load impedance, or from the power-balance equation.

$$V_{ab}I_{ab}\cos\varphi = V_{dc}I_{dc} \qquad (14.4)$$

where φ is the power factor angle.

Example 14.1
A single-phase VSC inverter has an RLC load with R = 8 Ω, L = 21.5 mH and C = 100 μF. The inverter operates in a square-wave fashion with an AC frequency of 50 Hz and DC voltage 200 V.

- *Express the instantaneous converter voltage and load current using the Fourier series.*
- *Calculate the RMS AC voltage and load current at the fundamental frequency.*
- *Calculate the power absorbed by the load.*
- *Determine average current of the DC supply.*
- *Determine the peak and RMS current for each transistor.*

Solution
The inductive reactance for the n-harmonic is:

$$X_{Ln} = 2n \times \pi \times 50 \times 0.0215 = 6.75n\,\Omega$$

The capacitive reactance for the n-harmonic is:

$$X_{cn} = \frac{10^6}{2n\pi \times 50 \times 100} = \frac{31.83}{n}\,\Omega$$

The impedance magnitude for the n-harmonic:

$$z_n = \sqrt{8^2 + (6.75n - 31.83/n)^2}$$

The power-factor angle for the n-harmonic is:

$$\theta_n = \tan^{-1}\frac{6.75n - 31.83/n}{8} = \tan^{-1}\left(0.843n - \frac{3.98}{n}\right)$$

The instantaneous AC voltage can be calculated using the Fourier series:

$$v_{ab} = 254.6\sin(314.1t) + 84.9\sin(3 \times 314.1t) + 45.47\sin(5 \times 314.1t) + 36.38\sin(7 \times 314.1t)$$
$$+ 28.29\sin(9 \times 314.1t) + \dots$$

By using the load impedance, the instantaneous AC current can be calculated:

$$i_{ab} = 9.67 \sin(314.1t + 72.31°) + 6.77 \sin(3 \times 314.1t - 50.25°) + 1.6 \sin(5 \times 314.1t - 73.7°)$$
$$+ 0.83 \sin(7 \times 314.1t - 79.37°) + 0.49 \sin(9 \times 314.1t - 82.03°) + \ldots$$

The fundamental RMS load current is $I_{ab1} = 9.67/\sqrt{2} = 6.83\,A$
Considering up to the ninth harmonics, the peak load current is:

$$I_{ab} = \sqrt{9.67^2 + 6.77^2 + 1.6^2 + 0.83^2 + 0.49^2} = 11.95\,A$$

The load current RMS

$$I_{ab} = 11.95/\sqrt{2} = 8.45\,A$$

The load power is $P_{ab} = 8.45^2 \times 8 = 571.22\,W$
The average DC current is (neglecting converter losses):

$$I_{dc} = 571.22/200 = 2.85\,A$$

The peak transistor current is 11.95 A. The RMS current of each transistor is:

$$I_T = 0.5 I_{ab}/\sqrt{2} = 4.2\,A$$

14.3 Three-Phase Voltage Source Converter

Normally, three-phase converters are used for high-power applications and the VSC topology is shown in Figure 14.3a. If isolation is required, as is frequently the case in transmission applications, the converter is connected through a delta-star transformer, as illustrated in Figure 14.3b. The transformer leakage inductance could then function as the interface inductance. In this topology, the three-wire system on the delta side has only two independent dimensions and zero current cannot flow, which makes control simpler. The drawback is the requirement for the costly, heavy and bulky transformer.

14.4 Square-Wave, Six-Pulse Operation

The simplest method of generating AC voltage is using a square-wave, six-pulse operation. This modulation method involves one switching per half cycle for each switch.

14.4.1 180° Conduction

For the three-legged conventional converter, shown in Figure 14.3a, the conduction angle of each switch is 180°, which is modulated in such a way that the two switches on the same converter leg do not conduct simultaneously (to prevent a DC short circuit). The firing pulses are shown in Figure 14.4, and the phase voltages with respect to DC central point '0' have the same waveform. Six pulse/steps exist for one AC fundamental cycle, giving a dominant harmonic of the order of 6*f*, where *f* is the fundamental frequency. The frequency of AC voltage can be controlled by changing the period between

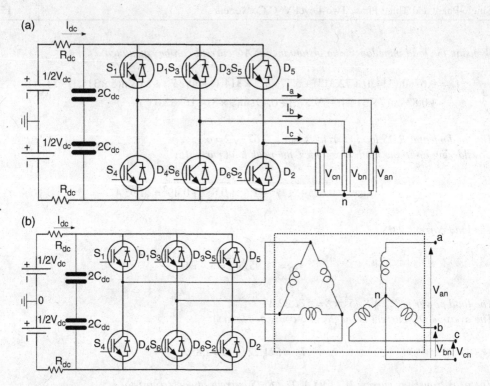

Figure 14.3 Two-level converter topologies: (a) three legged conventional converter and (b) converter interfaced with delta-star transformer.

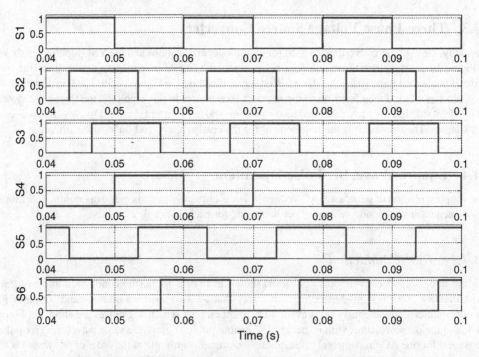

Figure 14.4 Switch firing signals for 180° conduction modulation method.

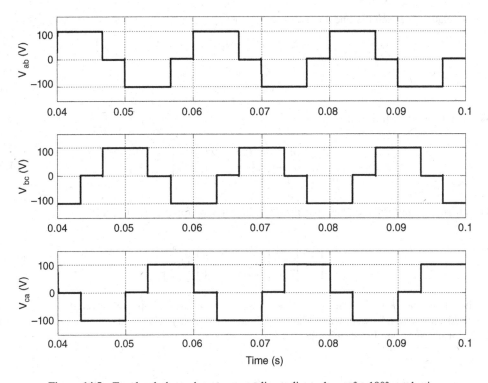

Figure 14.5 Two-level, three-phase converter line-to-line voltages for 180° conduction.

pulses, whereas the phase shift of AC voltage can be controlled directly by varying the phase displacement of the pulses. A phase-locked loop (PLL) is used to provide the reference phase. The main drawback of this modulation technique is that the amplitude of the AC voltage can be changed only by varying the DC voltage magnitude.

The line-to-line voltages are shown in Figure 14.5. The instantaneous line-to-line voltage can be expressed using the Fourier series as follows:

$$v_{ab} = \sum_{n=1,5,7,11}^{\infty} \frac{4V_{DC}}{n\pi} \cos\frac{n\pi}{6} \sin n\left(\omega t + \frac{\pi}{6}\right) \tag{14.5}$$

The other two phases can be represented with the same equation except that the phase is shifted by 120° or 240°. It should be noticed that the triplen harmonics (n = 3, 9, 15, …) will be zero in the line-to-line voltages. The line-to-line RMS voltage is:

$$V_{ab} = \sqrt{\frac{2}{2\pi} \int_{0}^{2\pi/3} V_{DC}^2 d\omega t} = \sqrt{\frac{2}{3}} V_{DC} = 0.8165\, V_{DC} \tag{14.6}$$

For n = 1, the line to line RMS fundamental voltage is:

$$V_{ab1} = \frac{4V_{DC}\cos 30°}{\sqrt{2}\pi} = 0.7797\, V_{DC} \tag{14.7}$$

The relation between the line and phase voltage is:

$$v_{ab} = v_{an} - v_{bn}$$
$$v_{bc} = v_{bn} - v_{cn} \tag{14.8}$$
$$v_{ca} = v_{cn} - v_{an}$$

This can be expressed in matrix form:

$$\begin{bmatrix} v_{ab} \\ v_{bc} \\ v_{ca} \end{bmatrix} = \begin{bmatrix} 1 & 1 & 0 \\ 0 & 1 & -1 \\ -1 & 0 & -1 \end{bmatrix} \begin{bmatrix} v_{an} \\ v_{bn} \\ v_{cn} \end{bmatrix} \tag{14.9}$$

The phase voltage for three phase balanced system can therefore be calculated as:

$$\begin{bmatrix} v_{an} \\ v_{bn} \\ v_{cn} \end{bmatrix} = \frac{1}{3} \begin{bmatrix} 1 & 0 & -1 \\ -1 & 1 & 0 \\ 0 & -1 & 1 \end{bmatrix} \begin{bmatrix} v_{ab} \\ v_{bc} \\ v_{ca} \end{bmatrix} \tag{14.10}$$

The line-to-load neutral voltages are shown in Figure 14.6. The three-phase converter currents with a balanced resistive load are shown in Figure 14.7, and conducting switches at different intervals in one cycle for phase '*a*' are listed in Table 14.1.

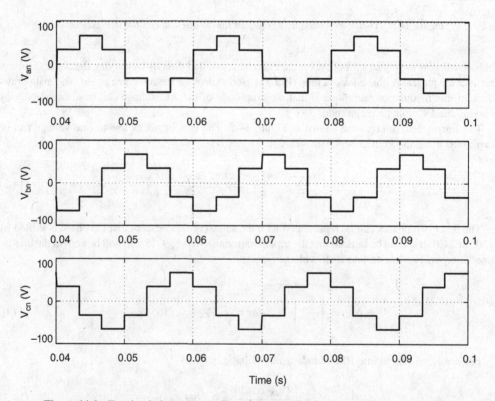

Figure 14.6 Two-level, three-phase converter line-to-neutral voltages for 180° conduction.

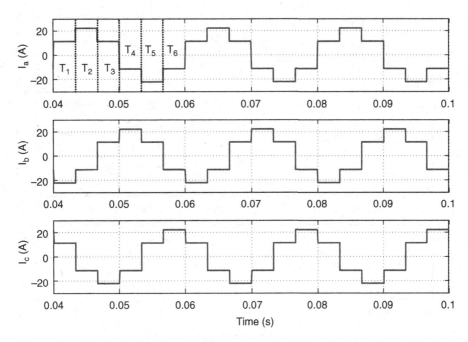

Figure 14.7   Two-level, three-phase converter currents for 180° conduction with a resistive load.

Table 14.1   Conducting switches in case of resistive load, for 180° conduction.

Intervals	Conducting switches
T_1	$S_1\ S_5\ S_6$
T_2	$S_1\ S_2\ S_6$
T_3	$S_1\ S_2\ S_3$
T_4	$S_4\ S_2\ S_3$
T_5	$S_4\ S_5\ S_3$
T_6	$S_4\ S_5\ S_6$

The three-phase converter currents with inductive load are shown in Figure 14.8, and conducting switches at different intervals in one cycle for phase 'a' are listed in Table 14.2.

14.4.2   120° Conduction

With this modulation technique the switches are fired for 120° (instead of 180°). As a result, at any instant, only two switches conduct. The switch firing signals are shown in Figure 14.9. A dead time of 60° exists between two switches on the same converter leg (phase), resulting in a lower device use and lower RMS AC voltage than with 180° conduction. The line-to-line and line-to-load neutral voltages are shown in Figure 14.10.

A full comparison between 120° and 180° conduction is shown in Table 14.3. Both techniques give the same harmonic spectrum but 180° conduction gives 15% increase in AC voltage magnitude compared to 120° conduction.

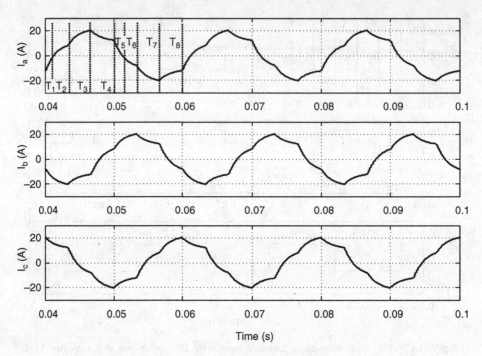

Figure 14.8   Two-level, three-phase converter currents assuming 180° conduction and inductive load.

Table 14.2   Conducting switches in case of inductive load with 180° conduction.

Intervals	Conducting switches
T_1	$D_1\ D_6$
T_2	$S_1\ S_6$
T_3	$S_1\ S_2\ S_6$
T_4	$S_1\ S_2$
T_5	$D_3\ D_4$
T_6	$S_3\ S_4$
T_7	$S_3\ S_4\ S_5$
T_8	$S_4\ S_5$

Example 14.2

A three-phase converter operates with 180° conduction and it is connected to star resistive load of $R = 200\,\Omega$. The converter frequency is $f = 50\,Hz$ and the DC voltage is 180 kV.

a. *Express the instantaneous line-to-line voltage and line current.*
b. *Determine the RMS line and phase voltage.*
c. *Determine the RMS line and phase voltage at the fundamental frequency.*
d. *Determine load power and total harmonic distortion (THD) of the AC voltage.*

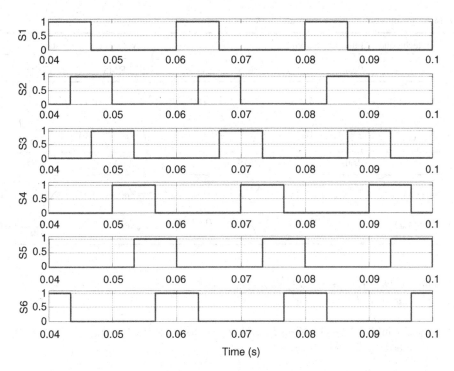

Figure 14.9   Switch firing signals for 120° conduction modulation.

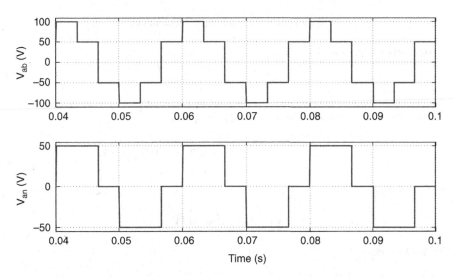

Figure 14.10   Line-to-line and line-to-load neutral voltages for one phase with 120° conduction.

Solution

a. The instantaneous line to line voltage can be calculated by using the Fourier series:

$$v_{ab} = 198.47\,\text{kV} \sin(314.1t + 30°) - 39.7\,\text{kV} \sin 5(314.1t + 30°) - 28.35\,\text{kV} \sin 7(314.1t + 30°)$$
$$+ 18\,\text{kV} \sin 11(314.1t + 30°) + 15.26\,\text{kV} \sin 13(314.1t + 30°) + \ldots$$

Table 14.3 Comparison between 120° and 180° modulation.

Conduction technique	Fundamental voltage		Waveform characteristic	
180°	Peak	RMS	Total RMS	THD
Phase voltage (V_{an})	$\frac{2}{\pi}V_{DC}=0.637V_{DC}$	$\frac{\sqrt{2}}{\pi}V_{DC}=0.45V_{DC}$	$\frac{\sqrt{2}}{3}V_{DC}=0.471V_{DC}$	$\sqrt{\frac{\pi^2}{9}-1}=0.311$
Line voltage (V_{ab})	$\frac{2\sqrt{3}}{\pi}V_{DC}=1.1V_{DC}$	$\frac{\sqrt{6}}{\pi}V_{DC}=0.78V_{DC}$	$\sqrt{\frac{2}{3}}V_{DC}=0.816V_{DC}$	$\sqrt{\frac{\pi^2}{9}-1}=0.311$
120°	Peak	RMS	Total RMS	THD
Phase voltage (V_{an})	$\frac{\sqrt{3}}{\pi}V_{DC}=0.551V_{DC}$	$\frac{\sqrt{6}}{2\pi}V_{DC}=0.39V_{DC}$	$\frac{1}{\sqrt{6}}V_{DC}=0.408V_{DC}$	$\sqrt{\frac{\pi^2}{9}-1}=0.311$
Line voltage (V_{ab})	$\frac{3}{\pi}V_{DC}=0.955V_{DC}$	$\frac{3}{\sqrt{2}\pi}V_{DC}=0.673V_{DC}$	$\frac{1}{\sqrt{2}}V_{DC}=0.707V_{DC}$	$\sqrt{\frac{\pi^2}{9}-1}=0.311$

The phase voltage is $v_{an}=v_{ab}/\sqrt{3}$ with a delay of 30°.
The instantaneous phase and line current is:

$$i_{ab}=573\sin(314.1t)-114.46\sin(5\times314.1t)-81.8\sin(7\times314.1t)+52\sin(11\times314.1t)$$
$$+44.1\sin(13\times314.1t)+\ldots$$

b. The line to line RMS voltage is:

$$V_{LL}=0.8165V_{dc}=0.8165\times180=147\,kV$$

The phase RMS voltage is:

$$V=V_{LL}/\sqrt{3}=84.8\,kV$$

c. The line-to-line RMS voltage at the fundamental frequency is:

$$V_{LL1}=\frac{4V_{dc}\cos30°}{\sqrt{2}\pi}=0.7797V_{dc}=0.7797\times180=140.34\,kV$$

The phase RMS voltage is:

$$V_1=V_{LL1}/\sqrt{3}=81\,kV$$

d. THD and load power:

$$THD=\frac{1}{V_1}\sqrt{\sum_{n=2,3,\ldots}^{\infty}V_n^2}=\frac{1}{V_1}\sqrt{V_L^2-V_1^2}=0.24236V_{DC}/0.7797V_{DC}=31.08\%$$

$$I_L=V/R=424\,A$$

The load power is $P_l=3VI_l=3\times84.8\times424=107.8\,MW$.

15

Two-Level PWM VSC Converters

15.1 Introduction

The simple two-level converter topologies described in the previous chapter have very little control capability. AC voltage magnitude control using converter modulation techniques has been the subject of intensive research during the last few decades. A large variety of methods have been developed and implemented. Their implementation in converters depends on the application type, the power level, the semiconductor devices and the harmonic level requirements. It is commonly performance and cost criteria that determine the choice of a modulation method in a specific application.

Pulse-width modulated (PWM) voltage source inverters represent the dominant technology in use by industry today and they were used exclusively in voltage source converter (VSC) high-voltage direct current (HVDC) systems in the early period 1996–2010. As shown in Figure 15.1 the modulation techniques can be classified as:

- *Six-pulse mode*, as studied in Chapter 14. A similar method is used at cell level with some modular multilevel HVDC.
- *Pulse-width modulation*: Used with a two- or three-level HVDC and with some modular multilevel converter (MMC) HVDC.
- *Stepping*. Used with some low-power, variable speed drives but not with HVDC.

15.2 PWM Modulation

A general hierarchal consensus have emerged from substantial research/development work which ranks space vector modulation techniques, regular sampled modulation and *sine*-triangle modulation strategies in decreasing order of merit based on harmonic performance. The following are primary modulation aims: wide linear modulation range; low switching loss; low total harmonic distortion (THD); easy implementation and low computation time.

High-Voltage Direct-Current Transmission: Converters, Systems and DC Grids, First Edition.
Dragan Jovcic and Khaled Ahmed.
© 2015 John Wiley & Sons, Ltd. Published 2015 by John Wiley & Sons, Ltd.

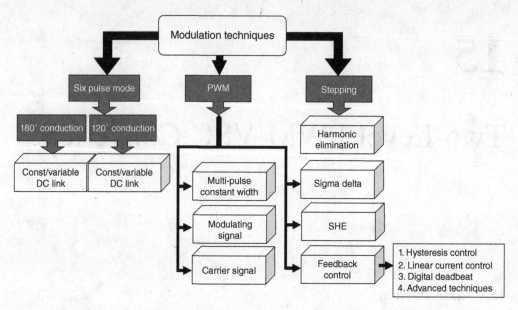

Figure 15.1   Converter modulation techniques.

15.2.1   Multipulse with Constant Pulse Width

This method allows the fundamental component of the AC voltage to be linearly controlled only by changing the DC voltage magnitude. A further disadvantage is that the low-frequency harmonic content is high. The maximum modulation index (ratio between the AC voltage to the DC voltage) for constant pulse-width modulators is $(2V_{dc}/\pi)$ in the linear modulation range as shown in Figure 15.2.

15.2.2   Modulating Signal

The modulating signal can be used as a criterion for the classification of PWM techniques, as shown in Figure 15.3, where the first two methods (sinusoidal pulse-width modulation (SPWM) and third harmonic injection (THI)) have been implemented with HVDC converters. Sinusoidal PWM refers to the generation of PWM AC voltage with a *sine* wave as the modulating signal. The ON and OFF instants of a PWM signal, in this case, can be determined by comparing a reference *sine* wave (the modulating wave) with a high-frequency wave (the carrier wave). The frequency of the modulating wave determines the frequency of the AC voltage. Converter switching frequency is determined from the carrier wave, which is kept constant. The peak amplitude of the modulating wave determines the modulation index, which in turn controls the *rms* value of AC voltage. This technique improves distortion significantly compared to previous modulation techniques. It eliminates all harmonics less than or equal to $2p - 1$, where p is the number of pulses per half cycle of the *sine* wave. The AC voltage contains harmonics, which are, however, pushed to the range around the carrier frequency and its multiples.

15.3   Sinusoidal Pulse-Width Modulation (SPWM)

The most frequently used modulation method is the SPWM, where the modulating signal is a pure sinusoidal waveform.

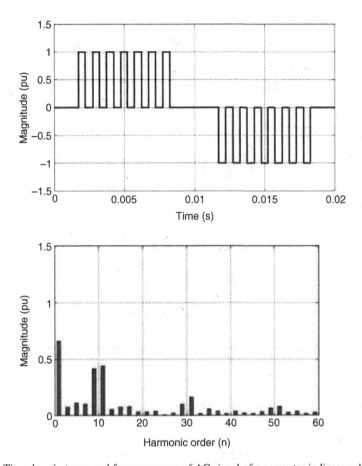

Figure 15.2   Time domain traces and frequency scan of AC signal of a converter in linear pulse modulation.

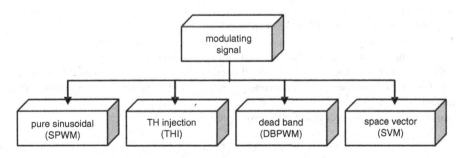

Figure 15.3   Modulating signal classification.

Consider a three-phase two-level converter as shown in Figure 15.4. The gate signals δ_1–δ_6 are generated by comparing the two control signals, for phase A: reference signal M_a and carrier signal C_a. The reference signal is a *sine* signal of variable magnitude and phase which is derived from the control system according to higher control loops. The carrier signal is a triangular wave with fixed parameters.

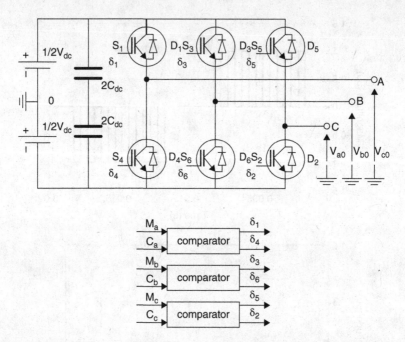

Figure 15.4   Three-phase two-level converter with gate signal generators for SPWM control.

With bipolar modulation, as shown in Figure 15.5, gate signals are generated at the intersection of the two control signals, resulting in two complementary signals δ_1 and δ_4 on phase A. The two switches cannot be ON simultaneously (this would result in a DC short circuit); this is prevented by a short dead time during commutation. The resulting phase to zero AC voltage can be either $\frac{1}{2}V_{dc}$ or $-\frac{1}{2}V_{dc}$, as shown in Figure 15.5, assuming $V_{dc} = 640\,\text{kV}$. The control is achieved by changing the width of the pulses.

The intersection of the negative slope of the triangular signal with the reference *sine* signal triggers switches to the ON state. The intersection with the positive slope sends an OFF signal. If the reference signal is higher, therefore, the ON pulses will be wider.

Using Fourier series analysis it is possible to determine the fundamental component and all harmonics of the AC voltage. The fundamental component is shown in Figure 15.5 as V_{a01}, which resembles the input reference signal M_a and this property is a great advantage of the modulation method. This property implies that the converter operates almost as an ideal linear amplifier (neglecting harmonics).

Assuming that the converter phase A control signal is

$$m_a = M_a \cos(\omega t + \varphi_m),\tag{15.1}$$

and therefore the three independent control variables are used: M_a, ω and θ_a. Developing Fourier coefficients, it can be shown that the fundamental component of phase A line-neutral AC voltage is:

$$
\begin{aligned}
v_{a01} &= \frac{1}{2}V_{dc}m_a \\
v_{a01} &= V_{a01}\cos(\omega t + \varphi_m) = \frac{1}{2}V_{dc}M_a\cos(\omega t + \varphi_m)
\end{aligned}\tag{15.2}
$$

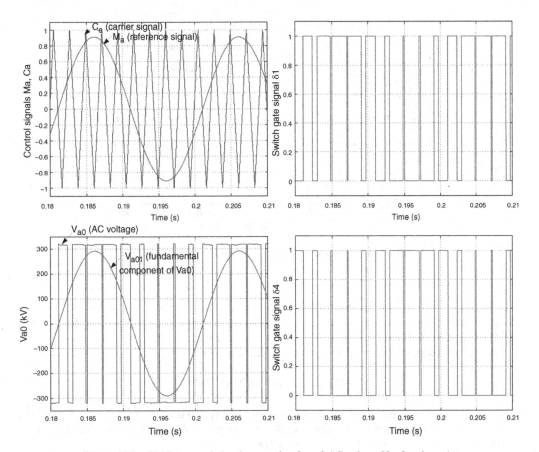

Figure 15.5 SPWM control signals, gate signals and AC voltage V_{a0} for phase A.

Although the implementation of this technique on microcontrollers is relatively simple and control is good, there are two drawbacks when compared with the six-step inverter as follows:

- Attenuation of the fundamental component of the AC waveform. The maximum line to neutral voltage amplitude is $\frac{1}{2}V_{dc}$, which is lower than with square-wave six-pulse modulation ($0.637\,V_{dc}$).
- High switching frequency. The switching frequency is higher than with the six-step inverter, which leads to increased stresses on converter semiconductor elements and switching losses.

15.4 Third Harmonic Injection (THI)

This technique has been derived as an improvement on the original *sine* PWM technique by injecting a 17% third harmonic component to the reference modulating signal. The hardware implementation of this additional signal is quite simple. It provides a higher fundamental amplitude and low distortion of the AC voltage. The linear modulation range of line-to-line voltage is extended as shown in Figure 15.6.

The carrier frequency index (the ratio between the carrier and fundamental frequency) should be an odd multiple of three as with all PWM methods. The generated AC voltages are symmetrical and balanced (all phase voltages are identical, but 120° out of phase without even harmonics). On the downside, this technique also generates a substantial AC third harmonic component on line to neutral voltage

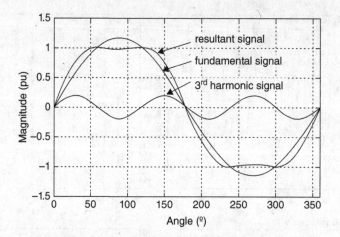

Figure 15.6 Modulating signal with third harmonic injection.

but, for a balanced load with floating neutral point, the third harmonic current cannot flow and therefore third harmonic voltages are not present on the line-to-line waveforms. It can be concluded that the overall PWM gain is increased by 15% over the conventional SPWM.

The modulated signal is commonly used in the following form:

$$M = 1.15\sin(\omega t) + 0.19\sin(3\omega t) \tag{15.3}$$

From this equation, it is evident that the linear modulation region is now extended to 1.15 compared with 1, which is the maximum magnitude with conventional SPWM modulation. This modulation method was widely used with two-level VSC HVDC.

15.5 Selective Harmonic Elimination Modulation (SHE)

A generalized selective harmonic elimination technique (SHE) was proposed in the 1970s to selectively eliminate harmonics generated by switching voltage waveforms. Since then it has been widely used in power converters and inverters. This method calculates individual firing angles in multipulse waveform in such way that it eliminates specific harmonics (typically lowest order harmonics).

The number of pulses per half cycle defines the degrees of freedom and the number of independent control angles. For example, there are nine angles per cycle in PWM AC signal in Figure 15.5. Each control angle is used to eliminate one harmonic. Fourier series modelling is adopted to develop a set of equations for magnitudes of the lowest order harmonics. The number of equations equals the number of independent control angles. The values for control angles are then calculated offline and the calculations are repeated for each magnitude of modulation signal. These angles are stored in memory in tables and the actual angles are calculated for each particular operating point using interpolation.

The technique, however, suffers from the following disadvantages:

• Calculation of angles is complex. An initial guess is extremely important to reduce the number of iterations and to assure a faster convergence.
• The switching angles have to be stored with adequate resolutions. The storage (memory) space depends upon the number of harmonics to be eliminated, carrier to modulating frequency ratio and operating range of the fundamental magnitude.

Because of the above reasons, the technique is restricted to applications with a low number of pulses. This modulation method has been used with some VSC HVDC in 2006–8 where frequency modulation index was around 23.

15.6 Converter Losses for Two-Level SPWM VSC

The conduction and switching losses formulae in this section are based on SPWM. Table 15.1 shows the possible conduction paths in one leg of the power converter in Figure 15.4.

The switching signals for phase a of the inverter are shown in Figure 15.5. If the load current and modulation waveform of the phase a are defined as:

$$i_a = I \sin(\omega t - \varphi_i) \quad m = M \sin \omega t \tag{15.4}$$

where m is the control signal and φ_i is the load angle. Note that the AC voltage will have the same angle as the modulating signal.

Observing Figure 15.5, the duty cycle of switch S_1 is defined by $d_1 = {}^1\!/_2[1 + M \sin \omega t]$

Therefore, the average current of S_1 switch is:

$$I_{S1av} = \frac{1}{2\pi} \int_{\varphi_i}^{\pi + \varphi_i} d_1 i_a d\omega t = \frac{1}{2\pi} \int_{\varphi_i}^{\pi + \varphi_i} \frac{I}{2} \sin(\omega t - \varphi_i)[1 + M \sin \omega t] d\omega t \tag{15.5}$$

$$I_{S1av} = \frac{I}{8\pi}[4 + \pi M \cos \varphi_i] \tag{15.6}$$

The *rms* current of S_1 switch:

$$I_{S1} = \sqrt{\frac{1}{2\pi} \int_{\varphi_i}^{\pi + \varphi_i} d_1 i_a^2 d\omega t} = \sqrt{\frac{I^2}{24\pi}[3\pi + 8M \cos \varphi_i]} \tag{15.7}$$

The average current of D_1 diode:

$$I_{D1av} = \frac{1}{2\pi} \int_{\pi + \varphi_i}^{2\pi + \varphi_i} d_1 i_a d\omega t = \frac{I}{8\pi}[4 - \pi M \cos \varphi_i] \tag{15.8}$$

Table 15.1 Conduction paths in one leg of a conventional two-level inverter.

Switching states		V_{a0}	Conducting devices in phase 'a'	
S_1	S_4		$i_a > 0$	$i_a < 0$
1	0	$0.5V_{DC}$	S_1 or D_4	D_1
0	1	$-0.5V_{DC}$	D_4	S_4 or D_1

The rms current of the D_1 diode:

$$I_{D1} = \sqrt{\frac{1}{2\pi} \int_{\pi+\varphi_i}^{2\pi+\varphi_i} d_1 i_a^2 d\omega t} = \sqrt{\frac{I^2}{24\pi}[3\pi - 8M\cos\varphi_i]} \qquad (15.9)$$

Considering that S_1 and S_4 switches operate in a complementary way, the average and *rms* S_4 switch currents are:

$$I_{S4av} = \frac{1}{2\pi} \int_{\pi+\varphi_i}^{2\pi+\varphi_i} (1-d_1) i_a d\omega t = -\frac{I}{8\pi}[4 + \pi M \cos\varphi_i] \qquad (15.10)$$

$$I_{S4} = \sqrt{\frac{1}{2\pi} \int_{\varphi_i}^{\pi} (1-d_1) i_a^2 d\omega t} = \sqrt{\frac{I^2}{24\pi}[3\pi + 8M\cos\varphi_i]} \qquad (15.11)$$

while the average and *rms* currents of D_4 diode are:

$$I_{D4av} = -I_{D1av} \quad I_{D4}^2 = I_{D1}^2 \qquad (15.12)$$

The negative sign in the average current indicates the current direction and is not considered when calculating power loss. It will be noticed that diodes D_1 and D_4 carry the same current and similarly for switches S_1 and S_4, as they supply the same phase. Therefore using Eqs (15.6)–(15.9), the total conduction losses for one device (diode and IGBT) of the three-phase inverter can be estimated as:

$$P_{c1} = \frac{V_{D0}I}{8\pi}[4 - \pi M\cos\varphi_i] + \frac{r_{FD}I^2}{24\pi}[3\pi - 8M\cos\varphi_i] + \frac{V_{T0}I}{8\pi}[4 + \pi M\cos\varphi_i] + \frac{r_{FT}I^2}{24\pi}[3\pi + 8M\cos\varphi_i]$$

$$(15.13)$$

The diode is considered as an ideal switch at turn on because it turns on rapidly at zero voltage. The turn-on switching energy of the diode is therefore neglected. Therefore the total switching losses for one switching device of the three-phase inverter is:

$$P_{swT} = f_s \left(E_{on\_test} + E_{off\_test}\right) \frac{V_{dc}}{V_{dctest}} \frac{I_C}{I_{Ctest}} \qquad (15.14)$$

$$P_{sD} = f_s E_{rec\_test} \frac{V_{dc}}{V_{dctest}} \frac{I_C}{I_{Ctest}} \qquad (15.15)$$

Example 15.1

A 500 A, 300 kV, SPWM VCS HVDC converter with switching frequency of 1050 Hz employs 6.5 kV, 750 A IGBTs and 100 devices are used in each valve. The blocking voltage across each IGBT is 3 kV. The IGBT and freewheeling diode characteristics are shown in Figure 13.4, Figure 13.6 and Figure 13.7. Assume that the modulation index is 0.9, and the AC current angle is zero. Neglecting turn-on loss of the freewheeling diode, calculate total losses in this converter.

Solution

a. Conduction losses

For the IGBT

From the switch figures, $V_{T0} = 2$ V, $r_{FT} = (3.7-3.2)/250 = 2$ $m\Omega$

For the freewheeling diode

From the diode figures, $V_{D0} = 1.5$ V, $r_{FD} = (3-2.4)/250 = 2.4$ $m\Omega$

Total conduction loss (for all six valves):

$$P_c = 6 \times n \times \left\{ \frac{V_{D0}I}{8\pi}[4 - \pi M \cos\varphi_i] + \frac{r_{FD}I^2}{24\pi}[3\pi - 8M\cos\varphi_i] + \frac{V_{T0}I}{8\pi}[4 + \pi M \cos\varphi_i] + \frac{r_{FT}I^2}{24\pi}[3\pi + 8M\cos\varphi_i] \right\}$$

$$P_c = 6 \times 100 \times \left\{ \frac{1.5 \times 500}{8\pi}[4 - \pi \times 0.9 \times 1] + \frac{0.0024 \times 500^2}{24\pi}[3\pi - 8 \times 0.9 \times 1] + \frac{2 \times 500}{8\pi}[4 + \pi \times 0.9 \times 1] + \frac{0.002 \times 500^2}{24\pi}[3\pi + 8 \times 0.9 \times 1] \right\}$$

$$P_c = 6 \times 100 \times [34.974 + 24.068 + 271.654 + 110.246] = 264.6\,kW$$

b. Switching losses

For the IGBT

From the above figures, $E_{on\_test} = 4000$ mJ/pulse, $E_{off\_test} = 2800$ mJ/pulse,

The switch current is normally taken as average current

$$I_c = \frac{I}{8\pi}[4 + \pi M \cos\varphi_i]\frac{500}{8\pi}[4 + \pi M \cos\varphi_i] = 136\,A$$

$$P_{swT} = 6nf_s\left(E_{on\_test} + E_{on\_test}\right)\frac{V_{DC}}{V_{DCtest}}\frac{I_C}{I_{Ctest}}$$

$$P_{swT} = 6 \times 100 \times 1050(4 + 2.8)\frac{3000}{3600}\frac{136}{500} = 97\,kW$$

For the diode

From the above figures, $E_{rec\_test} = 2400$ mJ/pulse

$$P_{sD} = 6nf_sE_{rec\_test}\frac{V_{DC}}{V_{DCtest}}\frac{I_C}{I_{Ctest}}.$$

$$P_{sD} = 600 \times 1050 \times 2.4\frac{3000}{3600}\frac{136}{500} = 342\,kW$$

Total switching losses = 1313 kW

Total losses = 1577.6 kW

15.7   Harmonics with Pulse-Width Modulation (PWM)

Voltage-sourced converters will generate harmonics on both the AC and DC sides because of the switching actions of the converter. For VSC-HVDC integration, the amplitude of the harmonics entering the AC network and the DC line should be limited according to the harmonics standards. Different methods are used to reduce the harmonics to required limits such as:

- PWM techniques;
- multipulse techniques;

- multi-level techniques;
- harmonic filters.

In general, the harmonic spectrum of the AC inverter voltage is affected by the used modulation technique. In the case of six step modulation, the harmonic spectrum can be obtained by using the Fourier analysis. The situation will be different in the case of PWM modulation. The non periodic nature of a PWM switched waveform makes determination of harmonic components difficult. The waveform can be described as two variables function that is periodic across both the carrier and the *sine* reference waveforms.

Figure 15.7 shows the SPWM VSC control signals (the triangular carrier signal and the sinusoidal reference voltages), and also the AC voltage of phase *a* with respect to the midpoint of the DC link capacitor. In this example the frequency of the carrier is nine times the fundamental AC frequency ($f_m = 9$). The switching angles α_0, α_1, α_2, α_M, π are shown in the figure, which are used by the controller to send the firing signals to the IGBT switches. There are $2f_s + 1$ switching times per cycle. The amplitudes of the harmonics of the AC waveform are given by:

$$V_n = \frac{2V_{DC}}{n\pi}\left(1 + \sum_{k=1}^{2f_s}(-1)^k \cos n\alpha_k\right) \tag{15.16}$$

If triangular carrier signal frequency is an odd integer multiple of the fundamental frequency, the waveform does not contain even order harmonics.

In a three-phase bridge circuit, all the triplen harmonics, that is, third, ninth ... are eliminated in the line voltages, provided that the AC system voltages are balanced and symmetrical. If the triangular carrier signal frequency is a multiple of three, the harmonics of the order of the triangular carrier frequency are cancelled in the line and phase to floating neutral voltages.

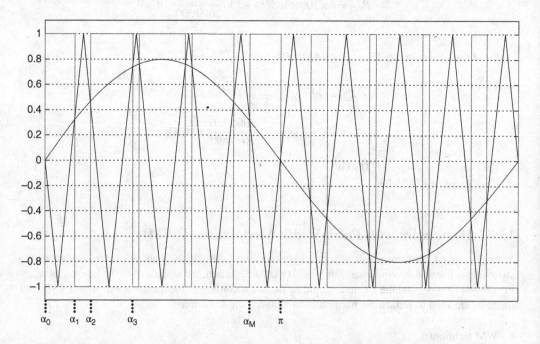

Figure 15.7   SPWM VSC modulation and AC voltage, assuming frequency modulation ratio of $f_m = 9$.

Table 15.2 Magnitude of harmonics on line-line voltage (%) relative to fundamental for a three-phase SPWM converter.

Harmonic order (n)	Modulation index M				
	0.2	0.4	0.6	0.8	1
fm ± 2	8	15	22	28	32
fm ± 4	0	0	0	1	2
2fm ± 1	95	82	62	39	18
2fm ± 5	0	0	0	2	3
3fm ± 2	22	35	34	22	6
3fm ± 4	0	3	8	13	16

The harmonics produced from the switching inverter AC voltage can be classified as:

- low-frequency harmonics (baseband harmonics);
- switching-frequency harmonics (carrier harmonics); and
- high-frequency harmonics (sideband harmonics).

Table 15.2 shows the order of the dominant harmonics and their magnitude in percentages relative to fundamental, for line-line voltage of three-phase SPWM converter. It is seen that harmonic magnitude depends on the modulation index. It is assumed that the frequency ratio is a multiple of 3 and that it is large (at least 9).

The AC phase to floating neutral and line voltages spectra for SPWM VSC converter are shown in Figure 15.8, for the test system, which has carrier frequency of 1050 Hz (21 times fundamental

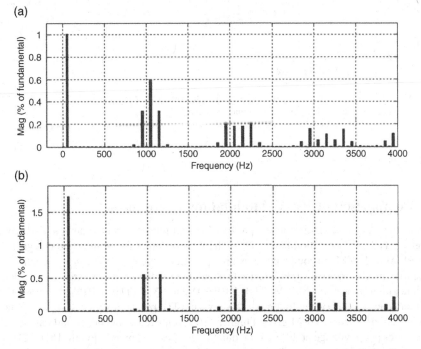

Figure 15.8 Spectra analysis (a) phase to floating neutral voltage spectrum and (b) line voltage spectrum, assuming modulation ratio of 21 (fs = fmf = 1050 Hz).

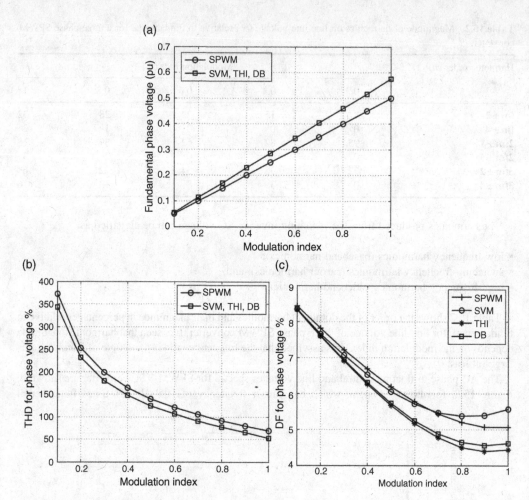

Figure 15.9 Performance comparison for different modulation techniques: (a) fundamental phase voltage; (b) THD for the phase voltage; and (c) DF for the phase voltage.

frequency). It can be shown that THD of phase to floating neutral voltage is 68%, while THD of line-line voltage is 99%.

15.8   Comparison of PWM Modulation Techniques

The harmonic performance indicators are presented in Part I in section 11.1, and include magnitude of fundamental, THD and distortion factor (DF). The three different PWM modulating signals: SPWM, THI, DBPWM and SVM are compared for a three-phase converter at a switching frequency of 450 Hz using the triangular carrier signal. Figure 15.9a shows the fundamental converter phase voltage at different modulation indices for SPWM, THI, DBPWM and SVM. THI, DBPWM and SVM give the highest fundamental component. Figure 15.9b shows the THD of the inverter AC phase voltage at different modulation indices. SPWM gives a higher THD compared to the other modulation techniques. DF for inverter phase voltage at different modulation indices is shown in Figure 15.9c. THI gives the lowest DF at higher modulation indices (which are most commonly used).

16

Multilevel VSC Converters

16.1 Introduction

Nowadays, multilevel converters are considered to be overall preferable topology at high voltages and have drawn considerable research/development in the high-voltage direct current (HVDC) professional community. To obtain high-quality AC voltage or current with a two-level converter requires high switching frequencies (1–2 kHz). Such high frequencies, combined with high voltages in HVDC applications, cause significant switching losses, device/valve voltage rating constraints and high electromagnetic interference (EMI).

Multilevel converters use an array of modules to build the required AC voltage level from a number of individual DC power supplies, as shown in Figure 16.1. The DC voltage sources are typically formed using capacitors and a charge-balancing scheme is used to maintain the capacitor voltages constant.

A two-level inverter generates an AC-phase voltage with either of the two voltages ($\frac{1}{2}V_{dc}$ or $-\frac{1}{2}V_{dc}$) as shown in Figure 16.1a. The three-level inverter generates three voltage levels ($\frac{1}{2}V_{dc}$, $-\frac{1}{2}V_{dc}$ or 0), as shown in Figure 16.1b, while the five-level inverter generates five voltage levels as shown in Figure 16.1c. As may be appreciated from this figure, as the number of levels increases, the AC voltage appears closer to the sinusoidal waveform and the total harmonic distortion (THD) reduces.

The most utilized multilevel topologies in a wide range of power electronics applications are:

- diode clamped converter;
- flying capacitor converter;
- H-bridge cascade converter;
- modular multilevel converter.

Only the diode clamped three-level converter and the modular multilevel converter (MMC) have been used in HVDC converters. An H-bridge cascade converter has been used with other transmission system converters, like STATCOM (static synchronous compensator) for reactive power-control applications only.

High-Voltage Direct-Current Transmission: Converters, Systems and DC Grids, First Edition.
Dragan Jovcic and Khaled Ahmed.
© 2015 John Wiley & Sons, Ltd. Published 2015 by John Wiley & Sons, Ltd.

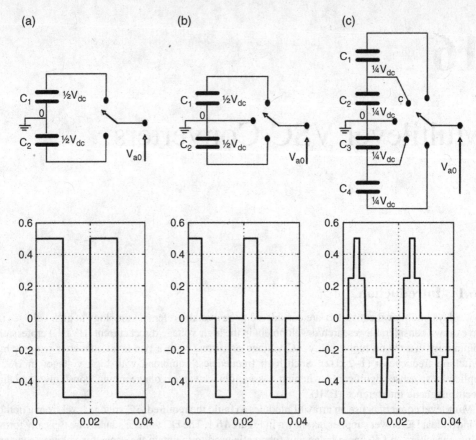

Figure 16.1   One phase leg of a voltage-source converter with: (a) two levels; (b) three levels; and (c) five-levels.

The main purpose of multilevel circuits is to generate a high-voltage waveform using low-voltage basic modules. Typically, series-connected modules are switched sequentially, producing an AC voltage pattern that contains defined discrete steps.

The attractive features of multilevel converters are:

- Low THD distortion and lower *dv/dt* compared to conventional two-level converters at the same effective switching frequency. This implies a reduced footprint because of reduced filtering.
- Blocking voltage of each module/switch is clamped to a cell capacitor voltage level. This also reduces switching losses.
- Smaller common mode (CM) voltage, thus reducing insulation stresses. In addition, by using sophisticated modulation methods, CM voltage can be eliminated.
- Modularity and ability to scale converters to high voltage and power levels.
- Inherent reliability. If a multilevel converter module fails, the desired output voltage can still be produced because of module redundancy.

Multilevel converter disadvantages involve:

- Control becomes complex with increased number of levels.
- Multiple DC voltages are required, which are usually provided by capacitors.

- Balancing the capacitor voltage is a challenge, because of the requirement for very fast simultaneous feedback control of large number of cells.
- Conduction losses increase with an increase in the number of switches.
- Typically two times (or more) switches are required compared with two-level converters. Beside costs, this also implies increased valve-hall footprint.
- Different current ratings for switching devices because of their different conduction duty cycle.

In HVDC applications, multilevel converters have inherent advantages because an HVDC valve always consists of a large number of series connected switches. Instead of firing simultaneously a string of series switches, MMC technology allows building AC voltage in steps using individually controlled modules.

16.2 Modulation Techniques for Multilevel Converters

All modulation schemes aim to create a train of pulses, which have the same fundamental voltage-second average as a target reference waveform at any instant. The major difficulty with these pulse trains is that they also contain unwanted harmonic components, which should be minimized. Hence, for any pulse-width modulation (PWM) scheme, a primary objective is to calculate the duty cycles for the converter switches that create the desired low frequency target voltage or current output. The secondary objective is to determine the most effective way of arranging the switching processes to minimize unwanted harmonic distortion, switching losses, and so on.

The modulation methods with multilevel converters include multipulse PWM and fundamental frequency (single pulse per module per half cycle), as shown in Figure 16.2. The common two multilevel PWM techniques are multilevel carrier based and space vector PWM which are extensions of conventional two-level PWM strategies, discussed in section 15.2.

With MMCs in high-voltage applications, however, fundamental frequency switching (a module switches once per cycle) has become attractive because of the low switching frequency, which reduces switching losses. Staircase (fundamental frequency) multilevel modulation methods can be classified as selective harmonic elimination (SHE) and nearest level modulation (NLM), as shown in Figure 16.2. In SHE methods, the switching angles for all modules are computed offline to eliminate harmonics at each value of the modulation index and stored in lookup tables, which are then interpolated according to the operating condition. This requires a large memory and with a high number of cells this method is not practical.

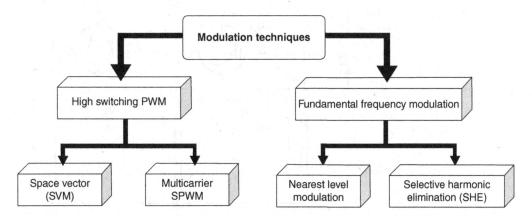

Figure 16.2 Multilevel converter modulation techniques.

The NLM approximates the desired AC voltage to the closest available voltage level of the converter. Considering that there are numerous modules with converters at HVDC voltage ranges, the number of available voltage levels becomes large, and NLM approximates the sine waveform very well. The computation is not complex because at each sampling instant the controller makes only a YES/NO decision if another module should be switched.

16.3   Neutral Point Clamped Multilevel Converter

Since the 1980s, a three-level neutral point-clamped (NPC) or diode-clamped PWM converter has been a very practical and widely adopted multilevel topology in drives and other industries. It was implemented in the second generation (voltage source converter) (VSC) HVDC in the period 2002–11, as shown in Table 12.3.

Figure 16.3 shows a three-level diode-clamped converter. The DC bus voltage is split into three voltage levels by two series-connected capacitors, C_1 and C_2. The connection point between the two capacitors is defined as the zero voltage point (label 0). The AC voltage V_{a0} has three states/levels: 0, $\frac{1}{2}V_{DC}$ and $-\frac{1}{2}V_{DC}$. If leg a is considered as an example, for voltage level $\frac{1}{2}V_{DC}$, switches S_{a1} and S_{a2} are turned on; for 0 switches S_{a2} and S_{a3} are turned on; and for a $-\frac{1}{2}V_{DC}$ level, S_{a3} and S_{a4} are turned on. The components that distinguish this circuit from a conventional two-level converter are clamping diodes D_{a5} and D_{a6}. They provide AC voltage connection to zero potential, which is not available

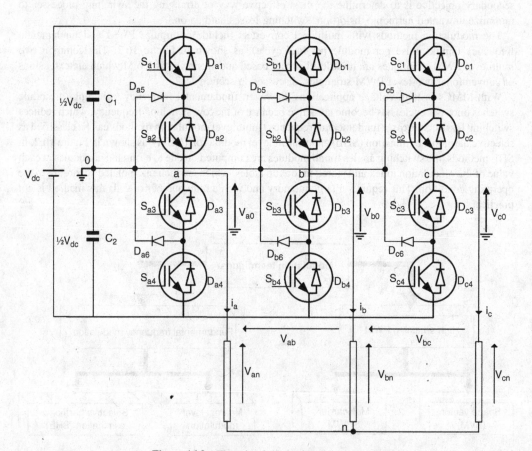

Figure 16.3   Three-level diode clamped converter.

in two-level converters. Another purpose of these two diodes is to clamp specific switch voltages to half the DC bus voltage. For example, when S_{a1} and S_{a2} turn on, $V_{a0} = \frac{1}{2}V_{DC}$; and diode D_{a6} balances the voltage sharing between S_{a3} and S_{a4} with S_{a3} blocking the voltage across C_2 and S_{a4} blocking the voltage across C_1.

A three-level diode-clamped converter is particularly attractive because only these diodes (D_{a5} and D_{a6}) are required as an extra switches over a two-level converter, while the quality of AC voltage improves substantially. Figure 16.4 shows the phase to zero voltage, line-to-line and line-to-load neutral voltages for a thee-level diode-clamped converter.

For a general $(n+1)$ level diode clamped multilevel converter, n storage capacitors are required, the voltage stress across each switching device is clamped to V_{dc}/n (one capacitor voltage), n consecutive switching devices in each leg are turned on, and the blocking voltage of each clamping diode is dependent on its position (k) in the structure according to: $V_{diode} = (k/n)V_{dc}$ where $1 \le k \le n-1$.

Some practical difficulties with the basic diode clamped multilevel topology can be summarized as:

- It requires high-speed clamping diodes that must be able to carry the full load current and are subjected to severe reverse recovery stresses.
- If the number of levels is more than three, the clamping diodes are subjected to different voltage stress. Therefore a series connection of diodes is required. This complicates the design and raises reliability and cost concerns.
- Maintaining the charge balance of the DC capacitors with more than three levels needs careful attention for some operating conditions.

Although a three-level NPC converter was used successfully with some HVDC projects, this topology was not used at higher number of levels because of the above issues.

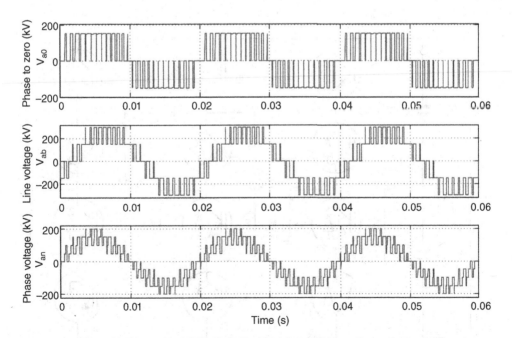

Figure 16.4 The pole, line to line and line to load neutral voltages of phase 'a' for three-level diode clamped multilevel converter at 1.35 kHz switching frequency.

16.4 Flying Capacitor Multilevel Converter

In the 1990s, the flying capacitor (or capacitor clamped) converter was introduced in low-power industry applications. It is used in some high-power drives and it was seriously studied for HVDC application. Figure 16.5 illustrates a three-level flying capacitor converter. Assuming that each capacitor has the same voltage rating, the series connection of capacitors indicates that the voltage sharing between these capacitors should be equal. The converter leg provides a three-level phase to zero AC voltage, which is $V_{a0} = \{0, \frac{1}{2}V_{dc}\ or\ -\frac{1}{2}V_{dc}\}$. The operating principle is illustrated for leg a: for voltage level $\frac{1}{2}V_{dc}$, switches S_{a1} and S_{a2} are ON; for a 0 level, switches S_{a1} and S_{a3} or S_{a2} and S_{a4} are ON; and for a $-\frac{1}{2}V_{dc}$ level, switches S_{a3} and S_{a4} are ON. Capacitor C_a therefore introduces negative voltage in series with either C_1 or C_2.

The AC voltage levels of the flying capacitor converter are similar to those defined for the diode clamped converter. The voltage between the phase and DC-link neutral point 0 of an $n+1$-level converter has $n+1$ levels including the reference level while the line-to-line voltage has $2n$ levels, which is achieved with n DC capacitors.

Flying capacitor converter has the following advantages:

- large storage capacitance provides extra ridethrough capability during power outage;
- switch combination redundancy for balancing capacitor voltages;
- both real and reactive power flow can be controlled, making it a suitable candidate for high-voltage DC transmission.

The disadvantages are:

- an excessive number of storage capacitors are required when the number of AC levels is high;

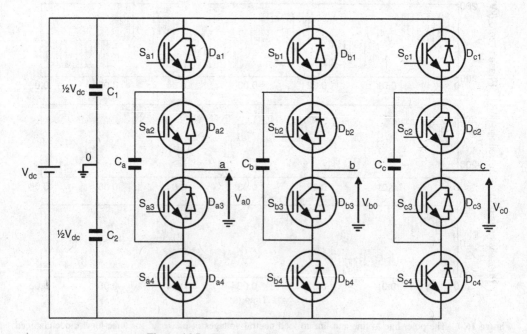

Figure 16.5 Three-level flying capacitor converter.

- inverter control is complicated, while the switching frequency and switching losses will be high with real power transmission;
- lack of modularity: high-level systems become difficult to package and are expensive.

This topology has never been used in practical HVDC transmission systems.

16.5   H-Bridge Cascaded Converter

The cascaded multileyel converters use series connection of single-phase H-bridge modules each with a separate DC source (capacitors) as shown for a five-level topology in Figure 16.6. Each module is capable of producing an output voltage of $+V_{cell}$, 0 and $-V_{cell}$, by connecting the capacitors to the AC terminals by different combinations of the four switches, S_{a1}, S_{a2}, S_{a3} and S_{a4}. The resulting AC voltage swings from $+2V_{cell}$ to $-2V_{cell}$ with five-levels.

In a chain of x V_{cell} cells and $4x$ switches, each of which is individually controlled, the AC voltage can vary between $+xV_{cell}$ to $-xV_{cell}$ with $2x + 1$ levels ($n + 1$ levels require $0.5n$ cells).

The main advantages of this topology are:

- It requires fewer components than the diode-clamped and capacitor-clamped circuits for the same number of levels.
- Optimized circuit layout and packing are possible because each level has the same modular structure.
- The charge balance of the separate DC capacitors can be readily achieved. Each module can be used in a cyclic way throughout each semicycle of a fundamental period.

The major disadvantage is:

- Each DC capacitor must be isolated. If there is no active power transfer, this is not a limitation and transmission-level converters (chain-circuit STATCOMs) have been successfully implemented in

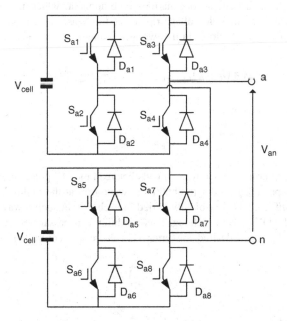

Figure 16.6   Five-level cascaded inverter.

many projects in the late 1990s. However in an HVDC application there would be requirement for a separate DC cable for each capacitor (to connect to corresponding capacitor on the remote terminal) and this is not cost effective. The topology might be feasible with back-to-back HVDC, but chain circuit topology has not been implemented with VSC transmission.

16.6 Half Bridge Modular Multilevel Converter (MMC)

16.6.1 MMC Topology and Operation

An MMC is an attractive alternative to conventional diode-clamped multilevel converters in medium/high voltage applications. HVDC experience is demonstrating that this topology it provides a viable and cost-effective approach to the construction of a multilevel DC/AC converter with very high number of levels. It can potentially operate continuously under unbalanced conditions including surviving symmetrical and asymmetrical AC faults without increasing the risk of system collapse and some topologies have DC fault management capability. As with all VSC converters, it can control active and reactive AC power independently as it readily controls AC voltage magnitude, phase and frequency. The most attractive feature is the modularity as the same basic modules are used in all projects whereas only the number of modules differs. The cells are independently controlled; however, they also need to be controlled jointly in very fast manner in order to maintain balanced voltage on all cells. This demand for simultaneous balancing of a large number of cells using a central signal processor is a significant new technical challenge. Currently there are two main configurations, namely half-bridge and full-bridge modular converters.

Figure 16.7 shows the topology of one phase MMC with $n + 1$ levels. The converter leg consists of positive and negative arms (in power electronics two 'arms' make one 'leg'). Each arm comprises a series chain of n-cells (and each of these can be half or full bridge). Each cell (module) includes one capacitor, which will store energy and several switches. Also there are two interfacing inductors L_{arm}, which are necessary to interface the two arms (which have different AC waveforms) and also to limit the circulating current harmonics caused by the switching of the cells. This circulating current includes a large second harmonic component, which not only deforms the arm currents but also increases the module voltage ripples and increases losses.

The circulating current i_{diff} that flows in the two arm inductors of one phase is expressed as:

$$i_{diff} = (i_P - i_N)/2, \quad \left\{ I_{dc} = i_{diff\_a} + i_{diff\_b} + i_{diff\_c} \right\} \tag{16.1}$$

The grid current is by inspection:

$$i_g = i_P + i_N \tag{16.2}$$

There are n cells in each arm and therefore $2n$ cells between the DC poles (in each leg). Each cell is either connected or bypassed. In order to obtain AC waveform at the middle point on each leg, the number of ON and OFF cells is sinusoidally varied in each arm in such a way that there are always n OFF (bypassed) and n ON cells in each leg. This condition is necessary in order to provide constant pole-pole DC voltage, V_{dc}. For a desired converter AC voltage v_c (a controlled sine waveform) the arm voltages are:

$$v_N = v_c + \frac{V_{dc}}{2}$$

$$v_P = -v_c + \frac{V_{dc}}{2} \tag{16.3}$$

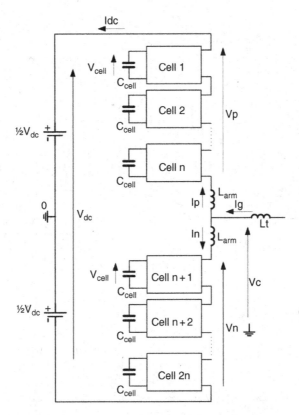

Figure 16.7 $n + 1$ level, single-phase modular multilevel converter.

As there are always n-cells ON in each leg, the voltage of each MMC cell is:

$$V_{cell} = \frac{1}{n}V_{dc},\qquad(16.4)$$

The DC voltage equals:

$$V_{dc} = v_P + v_N\qquad(16.5)$$

while converter AC voltage is obtained from the circuit diagram as:

$$v_c = \frac{v_P - v_N}{2},\quad\left\{-\frac{1}{2}V_{dc} < v_c < \frac{1}{2}V_{dc}\right\}\qquad(16.6)$$

Figure 16.8 shows the basic variables for one phase of the MMC converter in Figure 16.7 for a typical 1 GW, 640 kV MMC converter (harmonic suppression control is not used). It is seen that the arm currents (i_p and i_n) are distorted because of a significant second harmonic. However, the line current (i_g) has no distortion as second harmonics will cancel in Eq. (16.2). The circulating current one each leg i_{diff} contains significant second harmonic but when the differential currents on the three phases are summed on the DC side; the DC current I_{dc} has no harmonics. The arm voltages (v_p and v_n) are 50 Hz sine waveforms phase shifted by 180° and their sum is DC voltage V_{dc}. The circulating current

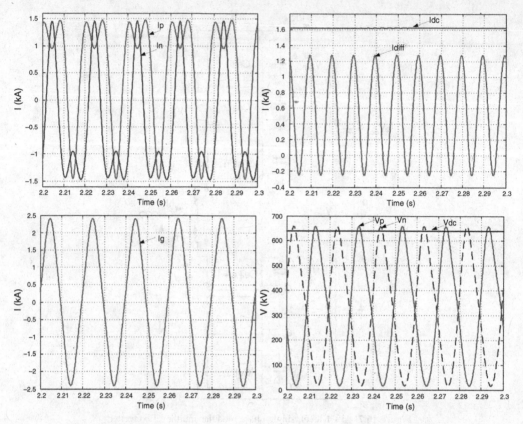

Figure 16.8 MMC converter basic variables. Circulating current control is not used.

harmonics have little impact on the connecting AC and DC systems; however, they are very undesirable because they create cell overvoltages and increase losses in the switches. Special controls are used to eliminate the circulating current, which are discussed later.

16.6.2 Operating Principles of Half-Bridge MMC Cells

Each half-bridge cell consists of two insulated-gate bipolar transistor (IGBT) packs (IGBT plus freewheeling diode) and a capacitor. The two switches are operated exactly like one leg of a two-level half-bridge converter. When one of the switches is fired ON the other switch will be OFF and vice versa. The operation of a half-bridge module is shown in Table 16.1 for the two cell states: ON (cell inserted, represented by state 1) and OFF (cell bypassed, represented by state 0), under two different current directions. The third column shows the current flow when the cell is blocked (both switches are OFF but converter is still online). This case is important in two instances: (i) cell passive charging for positive cell current and (ii) converter blocked under DC fault for negative cell current. The capacitor voltage is kept constant at V_{cell} on all cells. The cell voltage is either $V_0 = V_{cell}$, or $V_0 = 0$, depending on switch states and current direction.

Figure 16.9 shows phase a of a three level half-bridge modular converter, which is studied first. There are two groups of switches: main switches S_{a1}, S_{a2}, S_{a3} and S_{a4}, and auxiliary switches S_{xa1}, S_{xa2}, S_{xa3} and S_{xa4}. There are four modules per phase and therefore four complementary switch pairs per phase

Table 16.1 Operation of half-bridge module.

	Cell ON state, (symbol 1) $S_x = 1$ (ON), $S = 0$ (OFF)	Cell OFF state (symbol 0) $S_x = 0$ (OFF), $S = 1$ (ON)	Cell blocked $S_x = 0$ (OFF), $S = 0$ (OFF)
Positive current	V_{cell} S_x S $V_0 = V_{cell}$	V_{cell} S_x S $V_0 = 0$	V_{cell} S_x S $V_0 = V_{cell}$
Negative current	V_{cell} S_x S $V_0 = 0$	V_{cell} S_x S $V_0 = 0$	V_{cell} S_x S $V_0 = 0$

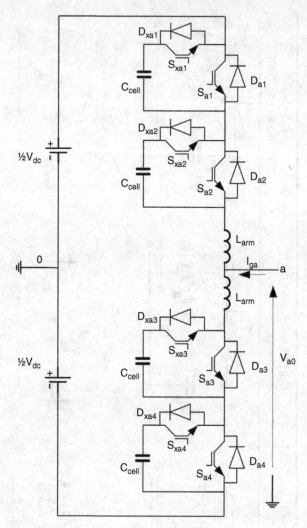

Figure 16.9 Phase A of a three-level, half-bridge modular converter.

$(S_{a1}, S_{xa1}), (S_{a2}, S_{xa2}), (S_{a3}, S_{xa3})$ and (S_{a4}, S_{xa4}), such that $S_{ai} + S_{xai} = 1$, where $i = 1–4$, and 1 represents the ON state while 0 represents the OFF state of the switch.

At any time, two cells will be ON in each leg, and voltage of each capacitor should be maintained at $\frac{1}{2}V_{DC}$. In each instant, four switches will be in the ON state, two from the main switches and the remaining two from the auxiliary switches. The voltage stress on each switch is limited to one capacitor voltage. If the supply midpoint (0 in Figure 16.9) is the AC voltage reference, then the three-level waveform of phase *a* voltage can be synthesized as follows:

For voltage level $V_{a0} = \frac{1}{2}V_{DC}$, turn on all the upper main switches (S_{a1} and S_{a2}) and all lower auxiliary switches (S_{xa3} and S_{xa4}).

For voltage level $V_{a0} = 0$, there are four different switching combinations:

- Turn on S_{a1}, S_{a3}, S_{xa2} and S_{xa4}.
- Turn on S_{a2}, S_{a3}, S_{xa1} and S_{xa4}.

- Turn on S_{a2}, S_{a4}, S_{xa1} and S_{xa3}.
- Turn on S_{a1}, S_{a4}, S_{xa2} and S_{xa3}.

For voltage level $V_{a0} = -\frac{1}{2}V_{DC}$, turn on all upper auxiliary switches (S_{x1} and S_{x2}) and all lower main switches (S_{a3} and S_{a4}).

In a general case of $(n + 1)$ level MMC, the voltage across each capacitor and switching device is limited to V_{dc}/n. At any time n cells will be ON in each leg. By controlling the number of cells from upper and lower arms in the ON state, it is possible to control the magnitude of phase voltage. The total number of capacitors required is $6n$ ($2n$ per phase) and the number of switches (IGBT plus freewheeling diode) is $4n$, which is twice the number required for a diode clamped or flying capacitor multilevel converter. The arm voltage range for half bridge MMC is $0 < v_p < V_{dc}$, and $0 < v_N < V_{dc}$.

16.6.3 Capacitor Voltage Balancing

The MMC cell voltages will deviate rapidly under normal operation because of unequal conduction intervals for each cell. Active balancing is required at each switching instant, to charge or discharge particular cells in order that the capacitor voltage always converges towards the desired steady-state value. The arm current should also be measured as the sign of capacitor voltage gradient depends on the current direction. This balancing technique must be embedded at the fastest control level, within the modulator, to facilitate the selection of the correct switching combinations that will lead to balanced cell voltages. If a modulation controller requires x cells of an n-cell arm to be switched ON at a particular instant, then the balancing controller will determine which particular cells are switched ON. There are several known methods to achieve capacitor balancing. One possible method is discussed below, which is based on switching ON a cell with the highest (or lowest) voltage at each switching instant.

The case of a three-level modular converter is analysed first. There is an opportunity for DC voltage balancing only when the AC phase is attached to a zero voltage level. The zero voltage level can be synthesized by any of the following four states: 1010/0101, 0110/1001, 0101/1010 and 1001/0110 (1010 stands for $S_{x1} = 1$, $S_1 = 0$, $S_{x2} = 1$, $S_2 = 0$). Consider the switching state (1010 for the upper arm and 0101 for the lower arm) as an example and assume the current direction in Figure 16.9 as positive. For $i_a > 0$ the upper capacitor C_2 charges and lower capacitor C_4 discharges while the upper capacitor C_2 discharges and lower capacitor C_4 charges for $i_a < 0$. Table 16.2 summarizes all switching states and their effect on the individual capacitors voltage. Based on this table, the capacitor voltage balancing method for a three-level converter is readily developed.

Table 16.2 Redundant switching states effect on capacitor voltages for a three-level modular converter.

Switching states for the upper arm	Load current		Effect on capacitors voltage
	Direction	Path	
1010	$i_a > 0$	$S_{a1}D_{xa2}$ and $S_{xa4}D_{a3}$	$C_2\uparrow$ and $C_4\downarrow$
	$i_a < 0$	$S_{xa2}D_{a1}$ and $S_{a3}D_{xa4}$	$C_4\uparrow$ and $C_2\downarrow$
0110	$i_a > 0$	$S_{a2}D_{xa1}$ and $S_{xa4}D_{a3}$	$C_1\uparrow$ and $C_4\downarrow$
	$i_a < 0$	$S_{xa2}D_{a1}$ and $S_{a3}D_{xa4}$	$C_4\uparrow$ and $C_1\downarrow$
0101	$i_a > 0$	$S_{a2}D_{xa1}$ and $S_{xa3}D_{a4}$	$C_1\uparrow$ and $C_3\downarrow$
	$i_a < 0$	$S_{xa1}D_{a2}$ and $S_{a4}D_{xa3}$	$C_3\uparrow$ and $C_1\downarrow$
1001	$i_a > 0$	$S_{a1}D_{xa2}$ and $S_{xa3}D_{a4}$	$C_2\uparrow$ and $C_3\downarrow$
	$i_a < 0$	$S_{xa2}D_{a1}$ and $S_{a4}D_{xa3}$	$C_3\uparrow$ and $C_2\downarrow$

$\downarrow$ Represents discharging state and $\uparrow$ represents charging state.

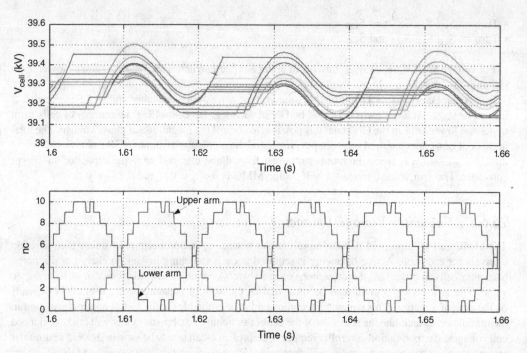

Figure 16.10   Capacitor voltage balancing for a 11-level half-bridge MMC. Top graph: voltages across 10 upper arm capacitors (n=10). Lower graph: number of on-state cells in upper and lower arms.

A generalized balancing method for an arm of n + 1 level converter can be similarly derived. At each sampling instant (instant of switching):

1. Determine all available cells for switching ON or switching OFF.
2. Rank the available cells according to their voltages and identify the cells with maximum and minimum voltages.
3. At the next request for switching ON of an additional cell, if the current is positive, switch in the cell with the lowest voltage,
4. If the current is negative, switch in the cell with the lowest voltage.

The above method is efficient since only one cell is switched at any instant, however it may not provide adequate cell ripple suppression in case of converters with large number of cells.

Some converters with high cell number use balancing algorithm where all the capacitor voltages are measured and ranked (comparing each capacitor voltage with voltage of each other) at each switching instant. Depending on the particular algorithm, a number of cells are re-arranged (switched in or out) according to their voltage magnitudes in order to optimise switching losses and maintain satisfactory cell voltage ripple.

Figure 16.10 illustrates voltages across 10 capacitors on one arm of a 11-level half-bridge 400 kV MMC. The voltage balancing controller maintains cell voltage to 400 kV/10 (around 40 kV). It is seen that the cell voltage ripple is low, at around 10%.

16.6.4   MMC Cell Capacitance

The cell capacitor design is very important as the capacitor in the MMC is carrying the load current and it represents the storage element for the converter instead of the DC-link capacitors. There is always a tradeoff between the cell voltage ripple requirements and capacitor size. The cell capacitor is typically selected to enable that the cell ripple voltage is kept within a range of below 10%.

In practical converters, the capacitor size is commonly determined using the energy-to-power ratio E_s [J/VA], which is defined in Eq. (12.3). With MMC converters the energy-to-power ratio should be in the range of $E_s = 30$–40 kJ/MVA (where MVA refers to the total converter rating). The cell capacitance can be calculated following similar derivation used for two-level VSC in Eq. (12.5), as:

$$C_{cell} = \frac{2 S_{MMC} E_s}{N_{MMC} V_{cell}^2}$$ (16.7)

where, S_{MMC} is the apparent power of the converter (MVA) and N_{MMC} is the total number of cells per converter ($N_{MMC} = 2np$ p-number of phases, n-number of cells in each arm). The total instantaneous arm capacitance is the series connection of cell capacitances where the number of series capacitors depends on the control modulation index $0 < n_c < n$. When the maximum number of cells is ON $n_c = n$, the arm capacitance is given by (equals to leg capacitance):

$$C_{arm} = \frac{C_{cell}}{n}$$ (16.8)

16.6.5 MMC Arm Inductance

The arm inductance L_{arm} is designed considering the following three criteria:

- Magnitude of circulating current second harmonic, which is analysed in depth in the MMC modelling chapter. Although L_{arm} significantly affects I_{diff} all modern MMC converters use active circulating current suppression control, which completely eliminates the second harmonic even with small arm inductance.
- Resonance with the arm capacitance.
- Current derivative in the converter arms for DC faults.

Considering Kirchhoff's voltage law for the MMC-leg circuit, it is possible to determine the value of arm inductance L_{arm_res} that will cause resonance with the arm capacitance C_{arm} at h harmonic under the modulation index magnitude of M, with a converter operating at frequency ω:

$$L_{arm\_res} = \frac{1}{C_{arm} \omega^2} \frac{2(h^2 - 1) + M^2 h^2}{8 h^2 (h^2 - 1)}$$ (16.9)

It is highly desired to avoid resonances at even integer harmonics as these harmonics are naturally excited inside an MMC. Of most concern are the lower order harmonics – that is. second and fourth. Figure 16.11 shows the inductance value that causes resonance at the second and fourth harmonics for a range of arm capacitances and considering two extreme modulation index values. The arm capacitances for two typical DC voltage values are also indicated. For a given arm capacitance, the arm inductance should be selected above the second harmonic resonance curve. This condition gives common values for L_{arm} in the range 30–100 mH.

The critical value for the arm inductor, considering the current derivative in the switches during DC faults, is given by the following formula:

$$L_{arm\_crit} = \frac{V_{dc}}{(dI_{dc}/dt)_{cr}}$$ (16.10)

The critical current derivative $(dI_{dc}/dt)_{cr}$ for diodes and IGBTs is in the range of 100–1000 A/μs. When replaced in Eq. (16.10), the minimal value for arm inductance of below mH is calculated, which is a less

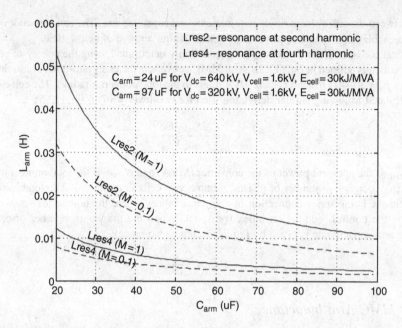

Figure 16.11   Arm inductance value that causes resonance at second and fourth harmonics.

demanding criterion than the resonance condition in Eq. (16.9). The practical rule is to use arm induct-ance of 0.1–0.15 pu, relative to the converter rating. Together with the transformer leakage reactance 0.05–0.15 pu, the total series reactance is in the range of 0.2 pu. Figure 16.12 shows a practical MMC HVDC arm inductor.

Example 16.1
Consider a 1000 MVA, 640 kV, 401 level MMC HVDC converter. Determine:

a. *Cell capacitor value.*
b. *Arm capacitance.*
c. *Arm inductance.*

Solution

a. *Number of modules per arm is 400. Therefore cell voltage is:*

$$V_{cell} = 640/400 = 1.6\,kV$$

Total number of cells in the converter N = 400 × 6 = 2400. Assuming Es = 30 kJ/MVA,

$$C_{cell} = \frac{2SE_s}{NV_{cell}^2} = \frac{2 \times 1000 \times 30000}{2400 \times 1600^2}$$

$$C_{cell} = 9.76\,mF \approx 10\,mF$$

b. *Arm capacitance is:*

$$C_{arm} = \frac{C_{cell}}{n_c} = \frac{0.1F}{400} = 24.4\,\mu F$$

Figure 16.12   Arm reactors at Trans Bay MMC HVDC station. Reproduced with permission of Siemens.

c. *The critical arm inductance for resonance at second harmonic (h = 2, M = 1) is:* $L_{arm\_res} = 43\,mH$. *Therefore the suggested arm inductance is* $L_{arm} = 50{-}60\,mH$.

16.6.5.1   MMC with Fundamental Frequency Modulation

This control method is also called square-wave control (or staircase modulation), and it is characterized by a single switching of each module per cycle. However, it should be noted that because of possible current sign changes and depending on balancing algorithm, a module may actually be switched a few times per cycle. By connecting a sufficient number of modules and using proper selection of conducting intervals, a nearly sinusoidal output voltage waveform can be synthesized.

Figure 16.13 shows a nine-level phase to 0 voltage waveform and the main switches (S_1–S_{16}) conducting periods assuming typical nine levels MMC circuit. The firing angles α_1 to α_4 can be calculated according to the adopted modulation method, which is commonly either SHE or nearest level control (NLC), as presented in Figure 16.2.

If SHE is employed, then the goal is to eliminate 5th, 7th and 11th harmonics by developing a set of equations for voltage magnitude and then using an iterative method such as the Newton–Raphson method to evaluate the angles for each operating point. With a high number of voltage levels, the complexity involved makes this method impractical and unnecessary.

The NLM method is considerably simpler to implement as it compares a desired reference *sine* waveform with the available staircase waveform. When the two signals intersect, the controller sends a request to switch ON (or OFF) one cell. Figure 16.14 shows the NLM for a single phase of the test nine-level converter, where the crossings between the reference *sine* and the available converter staircase voltage are used to determine the switching instant for the next module. As the number of modules is very high in practical HVDC converters, the NLM has proven to be a very attractive modulation technique.

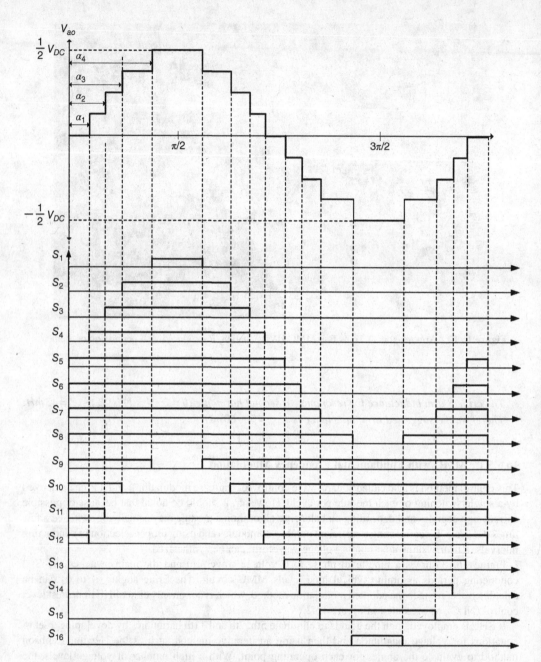

Figure 16.13   The phase-0 voltage of phase 'A' for a nine-level modular multilevel converter.

Staircase modulation methods are suitable with a high number of voltage levels because the high module number achieves a good AC voltage quality, equivalent to using high switching frequency. As an example a stair wave controlled 401 level converter has the equivalent of $800 \times 50\,\text{Hz} = 40$ kHz switchings per cycle and it gives better AC waveform than a two-level converter operating at 40 kHz. Furthermore, in an MMC, each semiconductor is switching only once (or few times) per cycle implying reduction in the switching frequency and the switching losses.

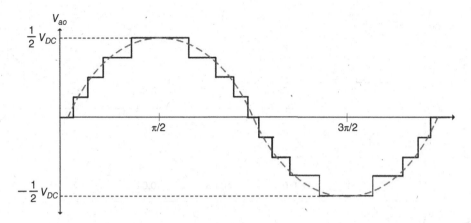

Figure 16.14 A typical AC reference and actual converter AC voltage of a nine-level converter, with nearest level control.

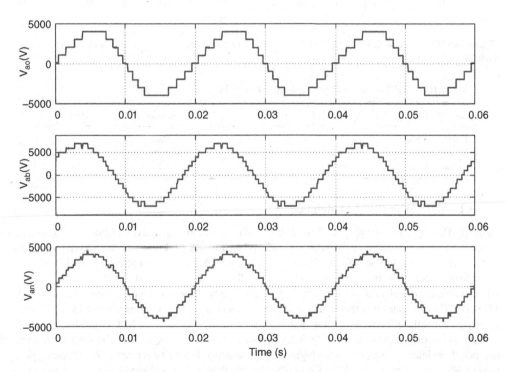

Figure 16.15 The phase to 0, line-to-line and line-to-load neutral voltages of phase A for a nine-level half-bridge modular multilevel converter.

Figure 16.15 shows line-0, line-line and line-to-load neutral voltages for the same nine-level MMC. With the increased number of levels the AC waveforms become closer to the ideal *sine,* with a lower THD as shown in Figure 16.16 for a 101-level MMC. Table 16.3 compares THD of phase-0, line and phase voltages for 3-level, 9-level, 21-level and 101-level MMC.

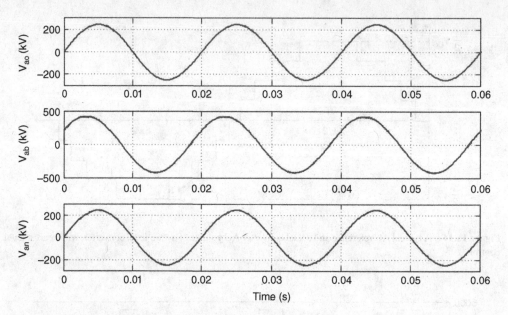

Figure 16.16 The phase-0, line-to-line and line-to-load neutral voltages of phase 'A' for 101-level half-bridge modular multilevel converter.

Table 16.3 THD of AC voltages for MMC of different levels.

	3-level (%)	9-level (%)	21-level (%)	101-level (%)
Phase 0 voltage	55.44	14.05	5.66	1.44
Line voltage	31.30	7.70	3.20	1.16
Phase voltage	30.77	7.55	3.17	1.06

16.6.5.2 MMC with PWM Modulation

Another MMC modulation approach is to adopt PWM modulation with individual modules. This method is also used with some commercial MMC HVDC and the topology is known as cascaded two-level converter. Typically the number of modules is lower than with NLC method but each module is switched multiple times per cycle. In the initial projects, each cell is switched at around 150 Hz, and since for highest DC voltages typically 38 modules are used per arm, the effective switching frequency per leg will be 11.4 kHz. This indicates that the dynamic response of such a converter is much better than a two-level converter, while the effective switching frequency is similar to square-wave NLC controlled topologies.

With this topology, each module is the same half-bridge converter, nevertheless because of the smaller number of modules, each module has a higher voltage and therefore valve is a string of multiple (typically around 10) series-connected IGBTs. Consequently, the challenge of voltage sharing exists in this topology but it is less severe compared with the two-level converter case. On the positive side, cell balancing is simpler and the total capacitance and converter footprint will be lower than with NLC MMC.

Example 16.2
Consider a 1000 MVA, 640 kV, half-bridge MMC VSC-HVDC converter.

a. If a square-wave control is used and the cell voltage is 2 kV, determine the number of cells, number of switches and the effective switching frequency per arm.

b. Assume that a PWM control is used, and there are 38 cells per arm. If each cell is fired at 165 Hz and the operating switch voltage is 2 kV, determine the number of switches in a cell, the total number of the switches and effective switching frequency per leg.

Solution

a. The number of cells per arm is 640/2 = 320. Since a cell will have two switches the total number of switches per arm is 640. The number of cells per leg is 1280. Since the square wave control is used, the frequency per cell is 50 Hz.
   The effective switching frequency per arm is $f_{eff} = 2 \times 320 \times 50 = 32\ kHz$.
b. The cell voltage is 640/38 = 16.8 kV. The number of switches per cell = $2 \times 16.8/2 = 16.8 \approx 17$
   Total number of switches per arm = $38 \times 17 = 646$
   The effective switching frequency per arm is $f_{eff} = 2 \times 38 \times 165 = 12.54\ kHz$.

16.7 MMC Based on Full Bridge Topology

16.7.1 Operating Principles

In this topology, each module has full-bridge structure, which consists of four IGBT packs (IGBT plus freewheeling diode) and a capacitor. Here, a module has more AC voltage states, as the cell voltage can be 0, $+V_{cell}$ and $-V_{cell}$. This increases number of redundant states for each AC voltage level and also opens the possibility for operating with a negative DC voltage. In normal HVDC operation this increased redundancy is not normally required, but it might be advantageous during abnormal operating conditions, like DC faults. Therefore, using double the number of semiconductors, increasing electronics footprint and almost doubling the semiconductor losses (two switches always conduct) in normal operation, should be weighed against performance enhancements.

The operation of a full-bridge module is shown in Table 16.4 for the cell ON, OFF and blocked states.

The cell can be in ON state either with positive voltage V_{cell} or with negative voltage $-V_{cell}$. There are also two possible configurations for a bypassed cell, but in each case current runs through two switches. It should be observed that when cell is blocked there is no direct short circuit current path, and a blocked full bridge cell becomes an open circuit for DC current.

Similarly as with half bridge cells, the voltage across each switching device is limited to one cell capacitor voltage V_{cell}. Modulation and capacitor voltage balancing of the FB (full-bridge) MMC is similar to that of the HB (half-bridge) MMC, except that the use of full bridge introduces new redundant switch states. This allows wider range for arm voltages, eliminates dependency on current direction and cell voltage balancing can be faster.

Figure 16.17 shows one-phase leg of the full bridge MMC, with 2 cells per arm. The converter can generate 3 voltage levels per phase. Similarly to half-bridge topology, the DC voltage equals sum of arm voltages in (16.5). The converter AC voltage is similar as in (16.6) but the range for arm voltages becomes wider if all cells use full-bridge topology.

If all cells in an arm assume FB topology then the arm voltage range is:

$$-V_{dc} < v_P < V_{dc}, \quad -V_{dc} < v_N < V_{dc} \tag{16.11}$$

Since each arm can generate either positive or negative voltage, there is very wide range of possible AC voltage magnitudes and it can be shown that normal AC voltage can be generated even with low DC voltage.

Also, it is not always necessary that all the arm cells assume FB topology. Depending on the required minimal operating DC voltage, a required AC voltage can be generated with only a fraction of the arm cells of FB type (the remaining cells can assume HB topology).

Table 16.4 Operation of a full-bridge cell.

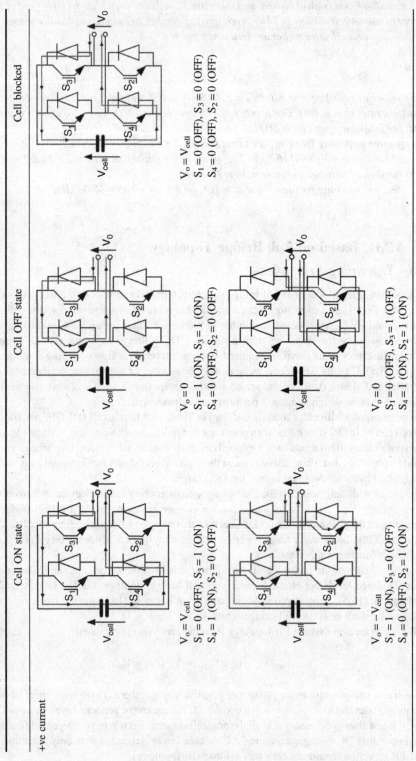

Table 16.4 *(continued)*

−ve current

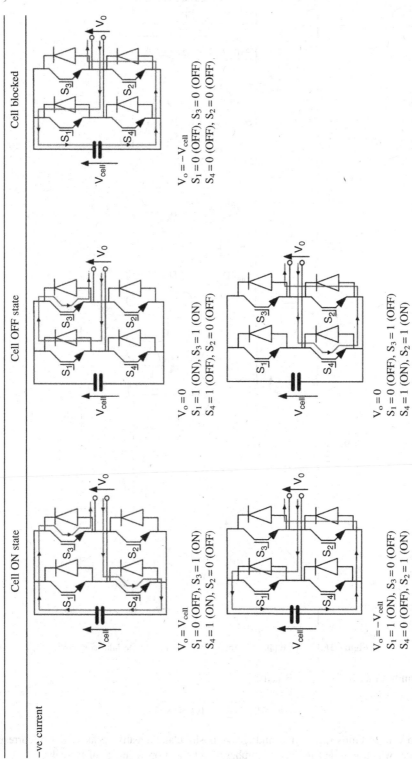

Cell ON state

$V_o = V_{cell}$
$S_1 = 0$ (OFF), $S_3 = 1$ (ON)
$S_4 = 1$ (ON), $S_2 = 0$ (OFF)

$V_o = -V_{cell}$
$S_1 = 1$ (ON), $S_3 = 0$ (OFF)
$S_4 = 0$ (OFF), $S_2 = 1$ (ON)

Cell OFF state

$V_o = 0$
$S_1 = 1$ (ON), $S_3 = 1$ (ON)
$S_4 = 1$ (OFF), $S_2 = 0$ (OFF)

$V_o = 0$
$S_1 = 0$ (OFF), $S_3 = 1$ (OFF)
$S_4 = 1$ (ON), $S_2 = 1$ (ON)

Cell blocked

$V_o = -V_{cell}$
$S_1 = 0$ (OFF), $S_3 = 0$ (OFF)
$S_4 = 0$ (OFF), $S_2 = 0$ (OFF)

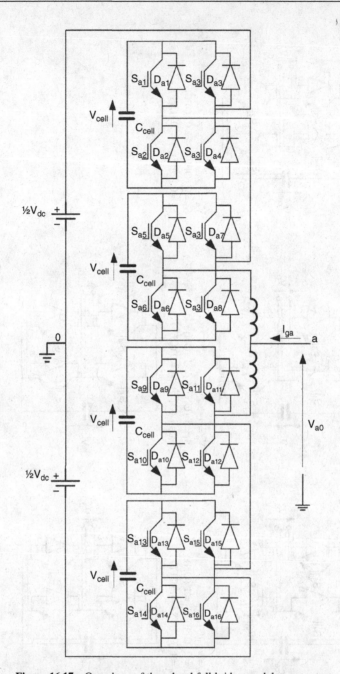

Figure 16.17 One phase of three-level full-bridge modular converter.

If an arm has mix of FB and HB cells:

$$v_{P-} < v_P < v_{P+}, \quad v_{N-} < v_N < v_{N+} \tag{16.12}$$

where arm voltage limits v_{p+}, v_{p-}, v_{N+} and v_{N-} are positive and negative peak values of corresponding arm voltages which will depend on the number of cells and opercentage of FB cells.

16.7.2   Operation Under Low DC Voltage

In practical terms, the most significant benefit of FB MMC is the possibility to operate normally under low DC voltage and DC fault conditions. It will be assumed that the converter can operate normally for any DC voltage V_{dc} in the range:

$$V_{dcmin} \leq V_{dc} \leq V_{dc\_rated} \tag{16.13}$$

where V_{dc_rated} is the nominal/rated DC voltage. The minimal DC voltage V_{dcmin}, is a design requirement which will be achieved by proper selection of arm cells in the design stage. The minimal DC voltage can be specified in the range:

$$-V_{dc\_rated} \leq V_{dcmin} \leq V_{dc\_rated} \tag{16.14}$$

Commonly it will be required that converter operates normally for severe DC faults and therefore $V_{dcmin} = 0$. However, under arcing conditions with overhead lines it is desired that converter generates some negative DC voltage in order to rapidly extinguish the arc current. In such case $V_{dcmin} < 0$. In some other applications it might be benefical to fully reverse DC voltage ($V_{dcmin} = -V_{dc\_rated}$) like in the case when LCC converter is connected to the remote end of the DC line.

16.7.3   Overmodulation Requirements

FB cells enable over-modulation, which produces higher AC voltage for a given DC voltage, as compared with HB MMC. This might be beneficial where DC voltage is constrained, like for example with DC cable insulation limits. At the expense of more arm cells, FB MMC can generate higher power for a given DC voltage and given cell (IGBT) current. Such method of increasing converter power is not as cost effective as power increase by increasing DC voltage, but might be justified where DC voltage is limited.

The over-modulation is defined by the design parameter $k_{MMC} \geq 1$, where the case $k_{MMC} = 1$ corresponds to the AC voltage generated by HB MMC. The maximal converter AC voltage is defined as (assuming maximal modulation index $M = 1$):

$$v_c = k_{MMC} \frac{V_{dc\_rated}}{2} \cos(\omega t + \theta_{vc}) \tag{16.15}$$

Considering the reactive power exchange capability with the AC system, it is essential that the converter maintains full control range for AC voltage magnitude under all operatring conditions on DC side (with any DC voltage magnitude within the design range). Therefore the following capability of minimal v_{c-} and maximal converter AC voltage v_{c+} ($v_{c-} < v_c < v_{c+}$) should be maintained for any DC voltage level:

$$v_{c+} = k_{MMC} \frac{V_{dc\_rated}}{2}, \quad v_{c-} = -k_{MMC} \frac{V_{dc\_rated}}{2} \tag{16.16}$$

16.7.4   Cell Voltage Balancing Under Low DC Voltage

The converter positive arm current can be derived using (16.1):

$$i_P = \frac{I_g}{2} \cos(\omega t + \theta_{lg}) + \frac{1}{3} I_{dc} \tag{16.17}$$

which indicates that the arm current has an oscillating and a DC component. In order to facilitate balancing of HB cell charge, the arm current should have both: positive and negative values in each cycle. This is required since HB cells have only one voltage polarity. With FB cells however, charge balance is possible even with unipolar arm current considering that FB cells can change voltage polarity.

Therefore, considering (16.17) the requirement for sucesfull HB cell voltage balancing gives the grid current peak value I_{g+}:

$$i_P < 0, \quad I_{g+} > \frac{2}{3} I_{dc} \tag{16.18}$$

Considering further the power balance between converter AC and DC sides under a low DC voltage V_{dc}:

$$V_{dc} I_{dc} = \frac{3}{4} I_g k_{MMC} V_{dc\_rated} \tag{16.19}$$

and replacing (16.18) in (16.19), the condition for successful HB cell voltage ballancing is:

$$V_{dc} > \frac{1}{2} k_{MMC} V_{dc\_rated} \tag{16.20}$$

Therefore if DC voltage drops below half the value of $k_{MMC} V_{dc\_rated}$ it will not be possible to balance any HB cell in the converter arms. This means that all the arm cells should be of FB type if DC voltage is expected to be below the value in (16.20). In such case the required peak arm voltage gives required FB voltage capacity and consequently the required number of FB cells can be determined:

$$v_{P+} \left(V_{dc} = k_{MMC} V_{dc\_rated}/2 \right) = \frac{k_{MMC}}{4} V_{dc\_rated} + \frac{k_{MMC}}{2} V_{dc\_rated} \tag{16.21}$$

16.7.5   Optimal Design of Full Bridge MMC

Starting with the operating requirement for V_{dcmin} and k_{MMC}, the FB MMC design determines the number of required cells in each arm and the percentage of cells that should be of FB type in each arm.

Rearranging the MMC voltage expressions from (16.5) and (16.6) it is possible to determine instantaneous arm voltages v_p and v_N if the desired AC and DC voltages are given:

$$v_P = v_c + \frac{V_{dc}}{2}$$
$$\tag{16.22}$$
$$v_N = -v_c + \frac{V_{dc}}{2}$$

Replacing required AC voltage given in (16.16) in the expression for positive arm voltage in (16.22) and considering highest possible DC voltage $V_{dc} = V_{dc\_rated}$, it is possible to obtain range for positive arm voltage $v_{P-} < v_P < v_{P+}$:

$$v_{P+} \left(V_{dc} = V_{dc\_rated} \right) = k_{MMC} \frac{V_{dc\_rated}}{2} + \frac{V_{dc\_rated}}{2} = (k_{MMC} + 1) \frac{V_{dc\_rated}}{2} \tag{16.23}$$

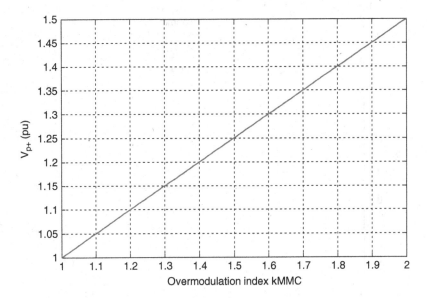

Figure 16.18 Required arm voltage capacity (number of arm cells in pu) for given overmodulation.

$$v_{P-}(V_{dc}=V_{dc\_rated}) = -k_{MMC}\frac{V_{dc\_rated}}{2} + \frac{V_{dc\_rated}}{2} = (-k_{MMC}+1)\frac{V_{dc\_rated}}{2} \tag{16.24}$$

The above equation shows that negative arm voltage is required (since $v_{P-}<0$) for $k_{MMC}>1$. Since only FB cells can generate negative voltage, some arm cells must assume FB topology if overmodulation is required.

Replacing further AC voltage given in (16.16) in the expression for positive arm voltage in (16.22) but considering lowest possible DC voltage $V_{dc}=V_{dcmin}$:

$$v_{P+}(V_{dc}=V_{dcmin}) = k_{MMC}\frac{V_{dc\_rated}}{2} + \frac{V_{dcmin}}{2} \tag{16.25}$$

$$v_{P-}(V_{dc}=V_{dcmin}) = -k_{MMC}\frac{V_{dc\_rated}}{2} + \frac{V_{dcmin}}{2} \tag{16.26}$$

Equation (16.26) shows that FB cells are required (since $v_{P-}<0$) for $V_{dcmin}<k_{MMC}V_{dc\_rated}$. The above equations for extreme conditions will define FB MMC total number of cells and the number of FB cells. Equation (16.23) determines the peak positive arm voltage and gives the total number of arm cells, which is illustrated in Figure 16.18.

The number of FB cells in an arm is larger of the following two requirements: the requirement for FB cells to achieve the peak negative voltage in (16.26) and the required number of FB cells to achieve voltage ballancing in (16.21). The minimal number of required FB cells in *pu* (relative to the total number of cells in arm) is illustrated in Figure 16.19, for given $V_{dcminpu}=V_{dcmin}/(k_{MMC}V_{dcrated})$ and for a range of k_{MMC}. As an example, if minimal expected DC voltage is *0.8pu* and no overmodulation is required, then only *10%* of the cells shoud be of FB type. For any DC voltage below *0.5pu* the number of FB MMC cells should be at least *75%*.

Example 16.3

Consider a 1000 MVA, 640 kV, full bridge MMC VSC-HVDC converter, employing cells of 2.0 kV voltage. The converter is required to generate 400kV peak phase-neutral AC voltage. It is also required to

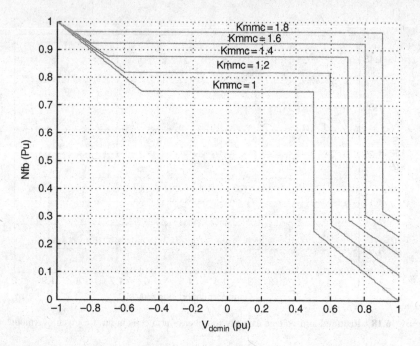

Figure 16.19 Required number of full bridge cells in pu for given k_{MMC}, and for a range of V_{dcmin}.

generate negative DC voltage of $V_{dcmin} = -64\,kV$ under DC faults and converter should be able to exchange rated reactive power under such DC faults.

a. Determine required overmodulation.
b. Determine the total number of cells per arm.
c. Determine number of full bridge cells per arm.
d. Determine number of full bridge cells under normal operation.
e. Sketch waveforms for AC voltage and converter arm voltages assuming nominal DC voltage.

Solution

a. The overmodulation index is $k = 400/320 = 1.25$.
b. Using (16.23) the maximal required arm voltage is:

$$v_{P+} = k_{MMC}\frac{V_{dc\_rated}}{2} + \frac{V_{dc\_rated}}{2} = 1.25\frac{640}{2} + \frac{640}{2} = 720\,kV$$

Therefore, the number of cells in each arm is $n = 720\,kV/2\,kV = 360\,cells$.
c. Using (16.21) the required FB voltage capacity in an arm is:

$$v_{P+}\left(V_{dc} = k_{MMC}V_{dc\_rated}/2\right) = \frac{3k_{MMC}}{4}V_{dc\_rated} = \frac{3 \times 1.25}{4}640 = 600\,kV$$

Therefore number of full bridge cells is:

$$n_{fb} = 600\,kV/2\,kV = 300\,cells$$

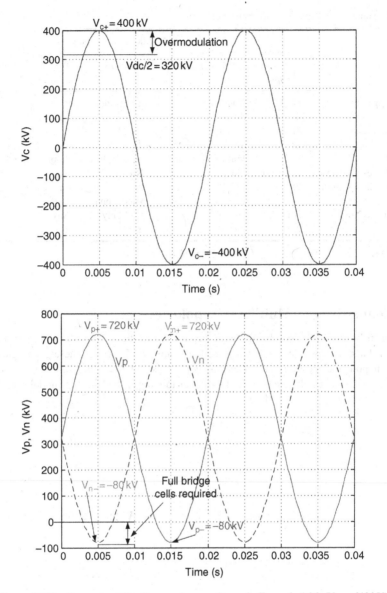

Figure 16.20 Converter AC voltage and arm voltages in Example 16.3 ($V_{dc} = 640\,kV$).

d. *Under normal operation, the required negative voltage is from (16.24):*

$$v_{P-} = \left(-k_{MMC} + 1\right)\frac{V_{dc\_rated}}{2} = \left(-1.25 + 1\right)\frac{640}{2} = -80\,kV$$

Therefore under normal operation ($V_{dc} = V_{dc\_rated}$) only 40 cells will be required to operate with negative voltage.

e. *Figure 16.20 shows the converter AC voltage and arm voltages. It is observed that peak AC voltage is higher than $V_{dcrated}/2 = 320\,kV$ and also that the arm voltages have negative minimal values $v_{p-} = -80\,kV$, $v_{n-} = -80\,kV$.*

Table 16.5 Summary of multilevel inverter topologies.

For ($n + 1$) level	Multilevel topologies				
	Diode clamped	Flying capacitor	Cascaded	Modular (half bridge)	Modular (full bridge)
Phase-0 voltage levels	$n + 1$	$n + 1$	$n + 1$ (odd)	$n + 1$	$n + 1$
Phase voltage levels	$4n + 1$	$4n + 1$	$4n + 1$	$4n + 1$	$4n + 1$
Line voltage levels	$2n + 1$	$2n + 1$	$2n + 1$	$2n + 1$	$2n + 1$
Switches per phase	$2n$	$2n$	$2n$	$4n$	$8n$
Cell capacitors	n	$\frac{1}{2}n(n + 1)$ per phase	$\frac{1}{2}n$ per phase	$2n$ per phase	$2n$ per phase
Clamping diodes per phase	$2(n - 1)$	No	No	No	No
Freewheeling diodes/phase	$2n$	$2n$	$2n$	$4n$	$8n$
DC supplies	1	1	$\frac{1}{2}n$ per phase	1	1
Maximum number of levels	Any	Any	Odd only	Any	Any
Modularity	No	No	Yes	Yes	Yes

Note: ($n + 1$) represents number of phase-0 AC voltage levels.

16.8   Comparison of Multilevel Topologies

A comparison between the four basic multilevel converters that are studied in this chapter for the same AC voltage of $n + 1$ levels is illustrated in Table 16.5. All of these topologies have been employed in power electronics at various power levels but only the following three have been used with HVDC: diode clamped (only three-level topology), half-bridge modular and full bridge modular converter.

17

Two-Level PWM VSC HVDC Modelling, Control and Dynamics

17.1 PWM Two-Level Converter Average Model

17.1.1 Converter Model in ABC Frame

This section uses an average modelling approach to represent converters (high-power circuits only) like the one given in Figure 15.4, using equivalent equations. In this way modelling of switching is avoided, and therefore simulations can be executed at much higher speeds, perhaps in the range of 50–100 µs, as discussed with HVDC (high voltage direct current) modelling methods in section 7.1. Outside the converter, the model retains the full dynamics of analogue components (cables, inductors), and also full control system dynamics.

Considering the converter modelling methods presented in Eq. (15.1), assuming an ideal AC *sine* waveform, the ABC frame equations for a two-level voltage source converter (VSC) converter model are:

$$v_{ca} = \frac{1}{2}V_{dc}m_a = \frac{1}{2}V_{dc}M_a\cos(\omega t + \varphi_m)$$

$$v_{cb} = \frac{1}{2}V_{dc}m_b = \frac{1}{2}V_{dc}M_b\cos\left(\omega t - \frac{2\pi}{3} + \varphi_m\right) \tag{17.1}$$

$$v_{cc} = \frac{1}{2}V_{dc}m_c = \frac{1}{2}V_{dc}M_c\cos\left(\omega t + \frac{2\pi}{3} + \varphi_m\right)$$

It is assumed that the system is symmetrical and balanced. The power conservation equation for the converter is:

$$P_{dc} = P_{ac} \tag{17.2}$$

$$V_{dc}I_{dcc} = \left(i_{ga}v_{ca} + i_{gb}v_{cb} + i_{gc}v_{cc}\right) \tag{17.3}$$

High-Voltage Direct-Current Transmission: Converters, Systems and DC Grids, First Edition.
Dragan Jovcic and Khaled Ahmed.
© 2015 John Wiley & Sons, Ltd. Published 2015 by John Wiley & Sons, Ltd.

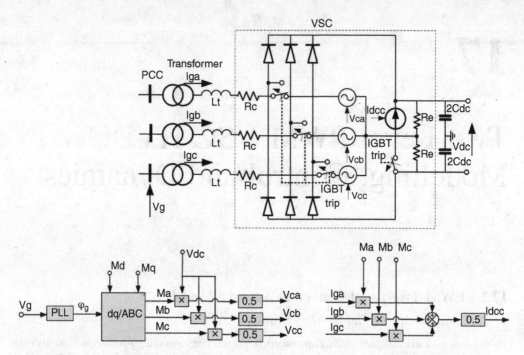

Figure 17.1 Implementation of the average two-level VSC converter model in ABC frame.

Replacing Eq. (17.1) in the above equation, the DC current is obtained:

$$I_{dcc} = \frac{1}{2}\left(i_{ga}m_a + i_{gb}m_b + i_{gc}m_c\right) \tag{17.4}$$

The two-level VSC converter behaves as a controllable voltage source on the AC side according to Eq. (17.1), and a controllable current source on DC side according to Eq. (17.4). Figure 17.1 shows a typical graphical representation of a symmetrical monopole VSC converter simulator model. In this figure, R_c represents all current-dependent losses in the converter and transformer (conduction and switching loss), while R_e represents voltage dependent loss (leakage and DC capacitors). The diode bridge is disconnected during normal operation by the switch 'IGBT trip'. During DC faults, IGBTs are blocked and the diode bridge is connected.

Figures 17.2 and 17.3 show the simulation results for a step input on active power. It is seen that the average model represents well the dominant system dynamics. It has been demonstrated in literature that the averaged models are sufficiently accurate for designing main HVDC control loops, and in general in the frequency range below around 50–100 Hz.

17.2   Two-Level PWM Converter Model in DQ Frame

The control signal in the ABC frame from Eqs (15.1) and (17.1) can be readily converted to DQ rotating frame using methods presented in Appendix B. In the rotating DQ frame the converter will have two control inputs (M_d and M_q), which are related to magnitude and angle (M, θ_m) as shown in Figure 17.4.

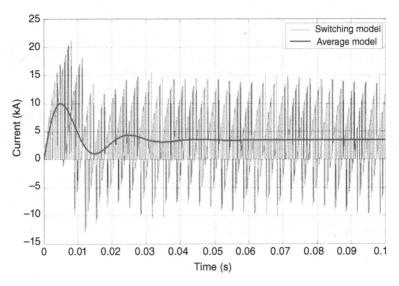

Figure 17.2   DC current comparison between detailed model and the average VSC converter model.

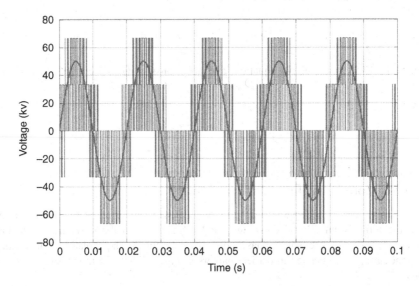

Figure 17.3   AC voltage V$_{aN}$ comparison of the detailed model and the average VSC converter model.

These two signals can be used to control the two components of the generated AC voltage (V_d and V_q). Note that the DC voltage (V_{DC}) can be assumed at rated value for most operating conditions as it is usually controlled at a constant level by one of the HVDC converters. The high-power output voltages V_d and V_q are therefore directly proportional to the low-power control signals M_d and M_q, and VSC converter responds like a linear amplifier.

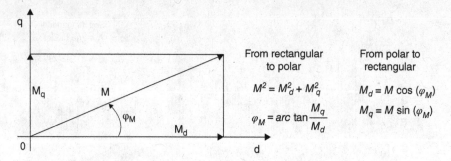

Figure 17.4 Converter control-signal vector representation.

If the converter model Eq. (17.1) is transferred to the synchronously rotating DQ frame:

$$V_{cd} = \frac{1}{2}V_{dc}M_d$$
$$V_{cq} = \frac{1}{2}V_{dc}M_q \tag{17.5}$$

The power balance equation in DQ frame is:

$$V_{dc}I_{dcc} = \frac{3}{4}\left(V_{cd}I_{gd} + V_{cq}I_{gq}\right) \tag{17.6}$$

Replacing Eq. (17.5) in the above equation, and also considering DQ currents, the DC current is obtained:

$$I_{dcc} = \frac{3}{4}\left(I_{gd}M_d + I_{gq}M_q\right) \tag{17.7}$$

17.3 VSC Converter Transformer Model

An HVDC converter is connected to the grid or local network through an isolating transformer. Although theoretically an HVDC converter can be connected to the grid directly, a transformer is used in all systems. The single-phase equivalent circuit of the isolating transformer in the low-frequency range is presented in earlier chapters in Figure 7.8c. The transformer model dynamic equation is:

$$v_1 = R_T i_1 + L_T \frac{di_1}{dt} + nv_2 \tag{17.8}$$

$$i_2 = \frac{1}{n}i_1 \tag{17.9}$$

where n is the transformer ratio and L_T is the transformer (and additional reactor) impedance. In many VSC systems the transformer leakage inductance is smaller than the required interface reactance and an additional series reactor is introduced.

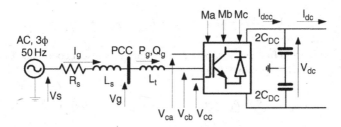

Figure 17.5   VSC converter with local AC grid.

17.4   Two-Level VSC Converter and AC Grid Model in ABC Frame

Figure 17.5 shows the basic schematic of a VSC converter connected to a local AC grid. In this figure, V_s is the voltage of equivalent remote source (fixed magnitude), V_g is the voltage of local point of connection (voltage for phase-locked loop (PLL) synchronization), V_c is the converter AC voltage, V_{dc} is the converter DC voltage, I_g is the grid-side line current, I_{dcc} is the converter DC current and I_{dc} is the current from the DC cable/grid. The converter control signals (in static frame) are M_a, M_b and M_c. The parameters R_s and L_s represent the equivalent grid impedance. L_T is transformer impedance and, for simplicity, unity stepping ratio is assumed, that is $n = 1$.The dynamic equation for each of the three phases (a, b and c) in static coordinate frame is:

$$v_{Sa} = v_{Ca} + R_s \cdot i_{ga} + (L_s + L_T) \cdot \frac{di_{ga}}{dt}$$

$$v_{Sb} = v_{Ca} + R_s \cdot i_{gb} + (L_s + L_T) \cdot \frac{di_{gb}}{dt} \qquad (17.10)$$

$$v_{Sc} = v_{Cc} + R_s \cdot i_{gc} + (L_s + L_T) \cdot \frac{di_{gc}}{dt}$$

Assuming lossless converter, the AC power equation at the point of coupling is:

$$S_g = v_{ga}i_{ga} + v_{gb}i_{gb} + v_{gc}i_{gc} \qquad (17.11)$$

The DC circuit equation is:

$$I_{dc} = C_{dc} \cdot \frac{dV_{dc}}{dt} + I_{dcc} \qquad (17.12)$$

17.5   Two-Level VSC Converter and AC Grid Model in DQ Rotating Coordinate Frame

The Park's transformation is used to convert all the above AC system equations into a synchronously rotating DQ reference frame, as it is introduced in Appendix B. Assuming symmetrical and balanced system, Eq. (17.10) is transferred as:

$$V_{Sd} = V_{Cd} + R_s I_{gd} - \omega (L_s + L_T) I_{gq} + (L_s + L_T) \frac{dI_{gd}}{dt} \qquad (17.13)$$

$$V_{Sq} = V_{Cq} + R_s I_{gq} + \omega(L_s + L_T)I_{gd} + (L_s + L_T)\frac{dI_{gq}}{dt} \qquad (17.14)$$

where ω is electrical system frequency (*rad/s*) and d, q suffix represent coordinate frame axes. The AC power Eq. (17.11) in a rotating frame becomes:

$$S_g = P_g + jQ_g = \frac{3}{4}\left(V_{gd} + jV_{gq}\right)\left(I_{gd} - jI_{gq}\right) \qquad (17.15)$$

$$P_g = \frac{3}{4}\left(V_{gd}I_{gd} + V_{gq}I_{gq}\right) \qquad (17.16)$$

$$Q_g = \frac{3}{4}\left(-V_{gd}I_{gq} + V_{gq}I_{gd}\right) \qquad (17.17)$$

Note that all variables are expressed as instantaneous values.

17.6   VSC Converter Control Principles

At control system level, all the VSC converters have similar general input-output structure and can be viewed as two-input, two-output, nonlinear dynamic amplifiers. The two converter control inputs (Md and Mq) are employed to develop feedback regulators to achieve various control functions depending on the converter application.

A controller for the VSC converter should meet the following goals:

- *Regulation of system variables of importance:* Typically some of the following variables are regulated at reference values:
  - *DC voltage*: to ensure minimum losses and to prevent insulation damage.
  - *Power transfer*: according to scheduling demands.
  - Reactive power exchange.
  - *AC voltage level*: this control may be of importance with very weak (high impedance) AC grids.
- Protect converter from damage caused by currents or voltages exceeding rated values.
- *Ensure system stability and good speed of responses*. This requirement implies *bounded responses* and good transient performance of system variables under all foreseen operating conditions and all disturbances.
- *Local AC grid support*. Typically, with low-inertia AC grids, the VSC converter is required to achieve frequency stabilization using frequency droop feedback. With high-impedance systems the voltage stabilization may be required.

The above control functions are commonly achieved with two-level cascaded converter control where the inner control ensures protection and stability whereas the outer control meets performance goals. The inner loops normally use fast decoupled current control. The outer control achieves various regulation and stabilization functions by sending references for the inner current control loops.

The above control structure has evolved from traditional controllers used with VSC converters in variable speed drives. The primary difference is the size of VSC HVDC, which means that it can have a significant impact on the AC grid. In addition, the VSC with HVDC may connect to long DC cables and therefore DC system dynamics play important role.

The firing circuit level control will depend on the type of converter in use. Without loss of generality it is assumed in this document that sinusoidal pulse width modulation (SPWM) is used with two-level converters and it is not described in detail. The firing synchronization will also be achieved using some form of PLL, which is described in Section 4.2 Part I.

The VSC converter control is based on a good understanding of the converter to AC system dynamic interactions, and consequently detailed converter modelling is an essential first step. The VSC converter modelling is typically accomplished by writing a basic per-phase dynamic model and converting this model into a rotating coordinate frame under the assumptions of a symmetrical and balanced three-phase system.

17.7 The Inner Current Controller Design

17.7.1 Control Strategy

It is assumed that the coordinate frame is aligned with the AC terminal voltage – that is, the voltage vector V_g is located on the d axis. In practice this can be achieved with the use of PLL. Under this assumption $v_{gq} = 0$, Eqs (17.16) and (17.17) become:

$$P_g = \frac{3}{4} V_{gd} I_{gd}$$

$$Q_g = -\frac{3}{4} V_{gd} I_{gq}$$

(17.18)

It can also be accepted that the terminal voltage is maintained at close to rated value, and therefore the influence of terminal voltage can be neglected at this stage, $V_{gd} = const$. From the above Eq. (17.18) it is possible to control active power by controlling i_{gd} and reactive power by controlling i_{gq}. The simplicity of Eq. (17.18) is one of the main reasons for using DQ current control as the fastest inner control. The second, and perhaps more important reason, is that direct current control is important for preventing converter overheating, in particular during fault conditions.

17.7.2 Decoupling Control

The current in Eq. (17.18) is controlled using converter AC voltage components but this relationship is more complex. Replacing Eq. (17.5), in Eqs (17.13) and (17.14) the current equations are:

$$V_{Sd} = 0.5 M_d V_{DC} + R_s I_{gd} - \omega (L_s + L_T) I_{gq} + (L_s + L_T) \frac{dI_{gd}}{dt}$$

(17.19)

$$V_{Sq} = 0.5 M_q V_{DC} + R_s I_{gq} + \omega (L_s + L_T) I_{gd} + (L_s + L_T) \frac{dI_{gq}}{dt}$$

(17.20)

It is seen in the above equations that control signals M_d and M_q can manipulate converter currents I_{gd} and I_{gq}, respectively. However, there are also unwanted crosscoupling terms, because, for example, changing M_d will change both I_{gd} and I_{gq}. Therefore, at the first design stage, decoupling control loops are introduced in an attempt to create two independent control channels. The control signals M_d and M_q consist of two terms: M^C_d and M^C_q are control signals from the main feedback current loops (described below in the next design stage) and the decoupling terms as described by the following expressions:

$$M_d = 2 \frac{M_d^c + (L_s + L_T) \omega I_{gq}}{V_{DC}}$$

(17.21)

$$M_q = 2 \frac{M_q^c - (L_s + L_T) \omega I_{gd}}{V_{DC}}$$

(17.22)

Note that current signals (i_{gd} and i_{gq}) in the above decoupling loops are normally measured in wide bandwidths (filter constants of over 1000 Hz) considering that their dynamics are fast and magnitudes vary significantly during normal operation. These signals are also commonly scaled by dividing by the rated current value. The signal V_{DC} is not a concern as DC voltage is typically tightly controlled $V_{DC} = const$ (as discussed below) and therefore it need not be compensated. Commonly, the control output is simply divided by the rated V_{dc}, as it is shown in the detailed control diagram in Figure 17.11. The decoupling control stage may also include compensation for the impact of remote voltage V_s.

17.7.3 Current Feedback Control

Replacing Eq. (17.21) in Eq. (17.19) and Eq. (17.22) in Eq. (17.20) the following is obtained:

$$V_{Sd} = M_d^c + R_s I_{gd} + (L_s + L_T)\frac{dI_{gd}}{dt} \tag{17.23}$$

$$V_{Sq} = M_q^c + R_s I_{gq} + (L_s + L_T)\frac{dI_{gq}}{dt} \tag{17.24}$$

In the above equations, there are two control signal M^C_d and M^C_q controlling two independent variables I_{gd} and I_{gq} under external disturbances V_{Sd} and V_{Sq}. Therefore the controller design is simplified to two independent first order systems.

In high-power systems, the line resistance R_s is typically very small and $X_s/R_s > 10$. Assuming as the first approximation $R_s = 0$, the system in Eqs (17.23) and (17.24) has dominant integral behaviour. An integrator-system is normally controlled with a proportional differential (PD) type controller to achieve good performance and zero tracking error, as a first-order overall system. The dynamic equations of the PD controller in Laplace domain is therefore in the following form:

$$M_d^c = \left(k_{p1} + \frac{sk_{d1}}{T_d s + 1}\right)\left(I_{gdref} - I_{gd}\right) \tag{17.25}$$

$$M_q^c = \left(k_{p1} + \frac{sk_{d1}}{T_d s + 1}\right)\left(I_{gqref} - I_{gq}\right) \tag{17.26}$$

where s is LaPlace operator, k_{p1} and k_{d1} are the controller proportional and differential gains, respectively and T_d is the filter constant associated with differential term (very small). Figure 17.6 shows the schematic of the decoupled system with the inner control loops.

Replacing Eqs (17.26) and (17.25) in Eqs (17.23) and (17.24), and neglecting disturbances, the closed loop system is obtained:

$$I_{gd} = \frac{(k_{p1}T_d + k_{d1})s + k_{p1}}{(L_s + L_T)T_d s^2 + (L_s + L_T + k_{p1}T_d + k_{d1})s + k_{p1}} I_{gdref} \tag{17.27}$$

$$I_{gq} = \frac{(k_{p1}T_d + k_{d1})s + k_{p1}}{(L_s + L_T)T_d s^2 + (L_s + L_T + k_{p1}T_d + k_{d1})s + k_{p1}} I_{gqref} \tag{17.28}$$

As the differential time constant is small compared with the dominant dynamics, it is possible to further assume $T_d = 0$, and Eqs. (17.27) and (17.28) become:

$$I_{gd} = \frac{k_{d1}s + k_{p1}}{(L_s + L_T + k_{d1})s + k_{p1}} I_{gdref} \tag{17.29}$$

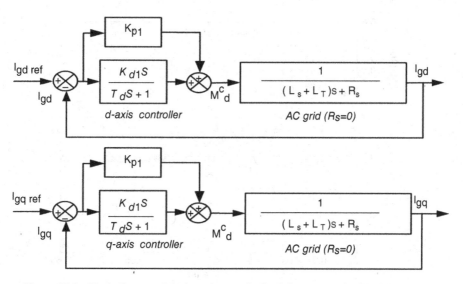

Figure 17.6 Block diagram of the inner current feedback loops assuming ideal decoupling.

$$I_{gq} = \frac{k_{d1}s + k_{p1}}{(L_s + L_T + k_{d1})s + k_{p1}} I_{gqref} \qquad (17.30)$$

17.7.4 Controller Gains

In the above system it is feasible to determine the controller gains. The initial value of the step response is given by k_{d1}, and the time constant is $T_I = (L_s + L_T + k_{d1})/k_{p1}$. Therefore, for a desired speed of response, (T_I) k_{p1} and k_{d1} can be determined. Clearly, k_{p1} increases speed of transient response whereas k_{d1} slows the speed of responses. However k_{d1} increases the initial value of the response. The speed of response is typically limited by the dynamics of feedback filters for current measurement in the decoupling loops in Eqs (17.21) and (17.22). As an example, if $L_s + L_t = 0.11$ H, $T_d = 0.0004$ s, and desired time constant is $T_I = 4$ ms, then good choice for control gains is $k_{d1} = 0.6$, $k_p = 200$.

If there is a finite system resistance R_s, the above controller will give a steady-state error equal to $R_s/(R_s + k_{p1})$. To eliminate this error, an integral term, k_{I1}, is typically added in the inner control loop. The nonzero R_s will also increase the speed of response and the new time constant will be $T_{I1} = (L_s + L_T + k_{d1})/(k_{p1} + R_s)$.

The inner control can be based on estimation of current signals (rather than measurements), which reduces the impact of noise in current measurements, but such control has shown to be less robust to parameter variations.

Figure 17.7 shows the testing of inner control loops on a detailed simulation model. A step up and then step down inputs are applied on each control channel (at 0.3 s $I_{gdref} = 0 \rightarrow 0.5$, at 0.4 s $I_{gqref} = 0 \rightarrow 0.5$, at 0.5 s $I_{gqref} = 0.5 \rightarrow 0$, at 0.6 s $I_{gdref} = 0.5 \rightarrow -1$) and both reference current and actual currents are shown. It is seen that the response is fast and that there is only a small coupling between the two channels. The same test system as in previous sections is employed: $P = 1000$ MW, $L_s = 0.01$ H, $R_s = 0.398\ \Omega$, $L_t = 0.1$ H $(X_t = 0.2$ pu$)$, $k_d = 0.6$, $k_p = 200$, $k_i = 16\,000$, current feedback filter frequency is 1200 Hz.

Figure 17.6 shows that the feedback gain and time constant depend on the system parameters (R_s) and (L_s). Some HVDCs (like the Caprivi link in Namibia) operate with a low and variable short circuit ratio (SCR), which may cause problems in the inner fast control loops. They will use some feedback control gain scheduling to compensate for the variation in the system parameters.

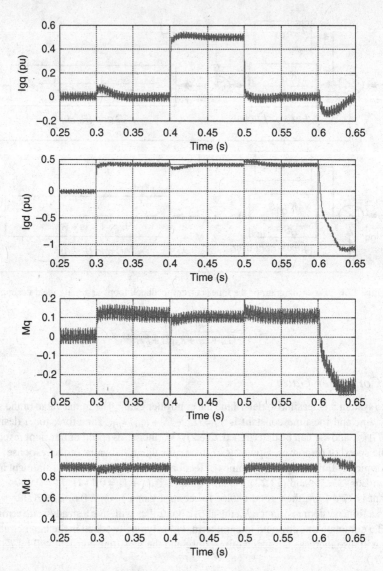

Figure 17.7 Step responses of the inner control loops on a detailed model: i_{gqref} step at 0.4 and 0.5 s, i_{gdref} step at 0.3 and 0.6 s.

17.8 Outer Controller Design

The inner controller loops can be viewed as a fast, first-order system, considering dominant dynamics in Eqs (17.29) and (17.30). The outer control loops manipulate current references (i_{gdref} and i_{gqref}) to achieve higher control goals. From Eq. (17.18) the d current component affects the active power and the q current component affects the reactive power. The d current is therefore used for controlling either power transfer or DC voltage. The q current is typically manipulated to regulate the reactive power exchange or AC voltage level.

The outer controllers are typically of the proportional integral (PI) type in order to ensure satisfactory speed of response and to eliminate tracking error.

17.8.1 AC Voltage Control

The AC voltage control using VSC converters may be attractive with very weak AC systems – that is, systems with high impedance. Figure 17.8 shows the outer AC voltage control system assuming that only the dominant dynamics from the inner q-current control shown in Figure 17.8 are present.

From Figure 17.8, the transfer function is given by:

$$V_g = \frac{\left[k_{I2}k_{p1} + s\left(k_{p2}k_{p1} + k_{d1}k_{I2}\right) + s^2 k_{p2}k_{d1}\right]\left(L_s + L_T\right)}{\left(k_{p1}k_{I2}\right)\left(L_s + L_T\right) + s\left[k_{p1} + \left(k_{p2}k_{p1} + k_{d1}k_{I2}\right)\left(L_s + L_T\right)\right] + s^2\left[k_{p2}k_{d1}\left(L_s + L_T\right) + k_{d1} + L_s + L_T\right]} V_{gref}$$

(17.31)

The above system is a conventional second-order filter. Therefore for a desired damping ratio ζ_2 and frequency ω_2 of the AC voltage control system, the two controller parameters (gains k_{p2} and k_{I2}) can be determined:

$$\omega_2^2 = \frac{\left(k_{p1}k_{I2}\right)\left(L_s + L_T\right)}{k_{p2}k_{d1}\left(L_{AC} + L_T\right) + k_{d1} + L_s + L_T}$$

$$2\zeta_2\omega_2 = \frac{k_{p1} + \left(k_{p2}k_{p1} + k_{d1}k_{I2}\right)\left(L_s + L_T\right)}{k_{p2}k_{d1}\left(L_s + L_T\right) + k_{d1} + L_s + L_T}$$

(17.32)

17.8.2 Power Control

Figure 17.9 shows the outer power control system, assuming that only the dominant dynamics from the inner d-current control shown in Figure 17.9 are present.

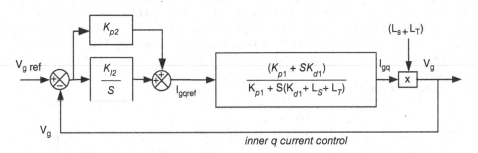

Figure 17.8 Designing outer AC voltage controller.

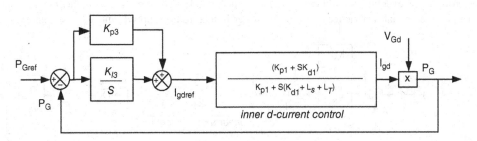

Figure 17.9 Designing outer power controller.

The transfer function is derived from Figure 17.9:

$$P_g = \frac{\left[k_{I3}k_{p1} + s\left(k_{p3}k_{p1} + k_{d1}k_{I3}\right) + s^2 k_{p3}k_{d1}\right]V_{gd}}{\left(k_{p1}k_{I3}\right)V_{gd} + s\left[k_{p1} + \left(k_{p3}k_{p1} + k_{d1}k_{I3}\right)V_{ACd}\right] + s^2\left[k_{p3}k_{d1}V_{gd} + k_{d1} + L_s + L_T\right]}P_{gref} \quad (17.33)$$

Therefore, for a desired damping ratio ζ_3 and frequency ω_3, the gains k_{p3} and k_{I3} can be determined:

$$\omega_3^2 = \frac{\left(k_{p1}k_{I3}\right)V_{gd}}{k_{p3}k_{d1}V_{gd} + k_{d1} + L_s + L_T}$$

$$2\zeta_3\omega_3 = \frac{k_{p1} + \left(k_{p3}k_{I1} + k_{d1}k_{I3}\right)V_{sd}}{k_{p3}k_{d1}V_{gd} + k_{d1} + L_s + L_T}$$

$$(17.34)$$

17.8.3   DC Voltage Control

Considering schematic in Figure 17.5, the DC voltage equation is given as:

$$C_{dc}\frac{dV_{dc}}{dt} = I_{dcc} - I_{dc} \quad (17.35)$$

As power equations are simpler than current equations, Eq. (17.35) is converted to a power equation by multiplying by V_{DC}:

$$\frac{1}{2}C_{dc}\frac{d(V_{dc})^2}{dt} = P_{dcc} - P_{dc} \quad (17.36)$$

The power flow is therefore directly proportional to the differential of square of the DC voltage. This equation also shows the capacitor energy dynamics, since From Eq. (12.4), the DC capacitor energy is directly proportional to the square of the DC voltage. Considering the significance of the capacitor energy and better control response, the DC voltage controller is frequently designed to control square of the DC voltage (rather than directly DC voltage). A simple DC voltage controller is shown in Figure 17.10.

The transfer function for the system in Figure 17.10 is derived as:

$$V_{dc}^2 = \frac{\left[k_{I4}k_{p1} + s\left(k_{p4}k_{p1} + k_{d1}k_{I4}\right) + s^2 k_{p4}k_{d1}\right]2V_{sd}}{2\left(k_{p1}k_{I4}\right)V_{gd} + s\left(k_{p4}k_{p1} + k_{d1}k_{I4}\right)2V_{ACd} + s^2\left(2k_{p4}k_{d1}V_{gd} + k_{p1}\right) + s^3\left(k_{d1} + L_s + L_T\right)}V_{dcref}^2 \quad (17.37)$$

The above system has dominant integral dynamics, which can be explained considering the magnitude of various terms in the denominator in Eq. (17.37). Consequently, it can be controlled with a PD-type controller. However because of the importance of accurate DC voltage regulation, the integral controller gain is essential in practice. The PI controller is therefore proposed with gains k_{I4} and k_{p4}. The

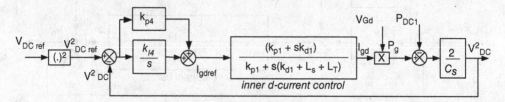

Figure 17.10   Designing outer DC voltage controller.

controller gains are typically tuned using a root-locus technique. Firstly, the controller zero $z_c = k_{I4}/k_{p4}$ is determined from the controller transfer function:

$$I_{gdref} = \frac{k_{p4}s + k_{I4}}{s}\left(V_{dcref}^2 - V_{dc}^2\right) = k_{p4}\frac{s + z_c}{s}\left(V_{dcref}^2 - V_{dc}^2\right) \tag{17.38}$$

The zero is placed close to the dominant system pole, or close to origin for integral systems. In the next design steps, the proportional gain is varied and the location of poles is observed in the root locus.

17.8.4 AC Grid Support

The power reference in Figure 17.9 is typically constant and maintained at a scheduled level for a particular system/station. Alternatively, the power reference can be made dependent on frequency deviations in AC grids. There are two reasons for this additional control function:

- Large VSC converters in the inverter mode should be treated as any large conventional power plant. It is normal that large generators use frequency droop feedback as an additional signal with power regulation. The rectifiers can be treated like any AC system load.
- In the case of weak, low-inertia AC systems it might be beneficial to modulate VSC power exchange in response to AC system frequency deviation. This control function applies equally to rectifiers and inverters. Such control has been used with conventional and VSC HVDC. In particular, with VSC HVDC the frequency support can be extended to regulate frequency on dead AC networks.

The droop gain for AC grid support can be calculated as:

$$k_{gdroop} = \frac{2P_{dc\_max}}{f_{g\,max} - f_{g\,min}} \tag{17.39}$$

where f_{max} and f_{min} are the allowed limits for the AC system frequency deviation and P_{dc_max} is the maximum converter power deviation allowed for system support. Where dynamic support and AC grid stabilization are needed, a dynamic feedback loop can be developed instead of static gain k_{gdroop}. Such a feedback loop can be designed as a frequency-dependent transfer function to enhance stability (phase or gain margin) – in particular frequency bandwidth.

With conventional point-to-point HVDC there is usually one stronger AC system that can dynamically support the other system. Such stabilization is achieved through HVDC controls but power is essentially drawn from the remote (stronger) AC system.

There are other possible high-level power modulation methods which are discussed in Chapter 21.

17.9 Complete VSC Converter Controller

The complete VSC converter controller is shown in Figure 17.11. The schematic includes the inner current and decoupling control loops, the outer loops with droop feedback, the PLL synchronization and also the switch firings (SPWM at switching frequency f_s). The outer control loops each have a selection switch to provide top-level control choice. The controller outputs are the firing signals for six IGBT switches (S_1–S_6).

Additional modulation signals on power reference is shown in the control diagram which stabilizes the AC grid. If the converter is part of a DC grid, then the DC voltage droop feedback (k_{gdroop}) can be included in order to enhance the stability of the DC grid, as discussed in Part III, section 25.5.

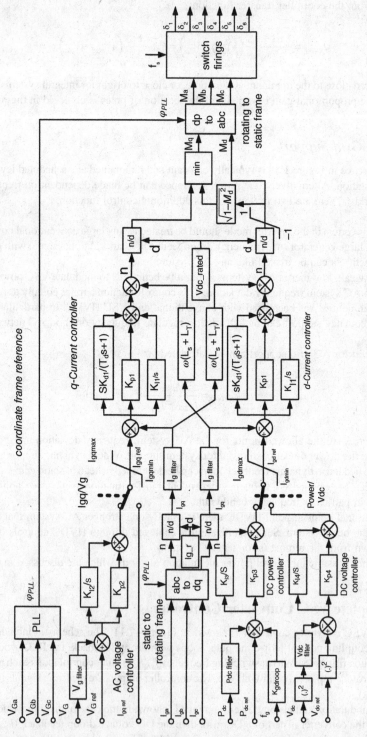

Figure 17.11 VSC converter controller with inner and outer control loops.

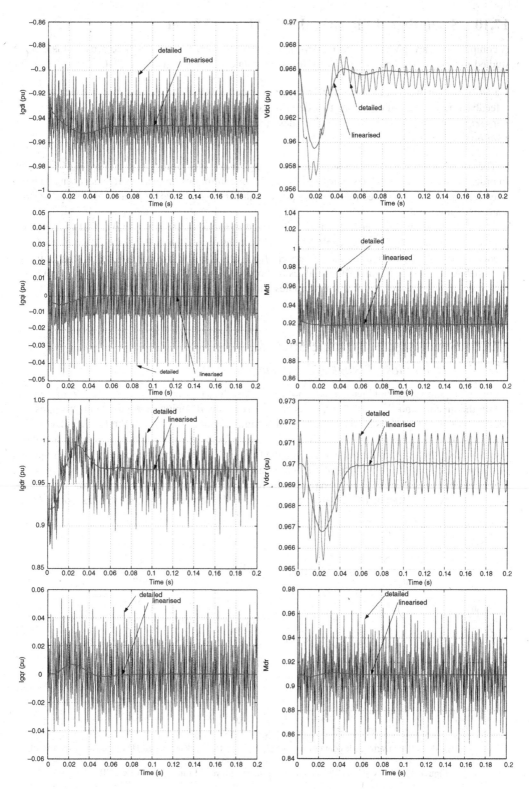

Figure 17.12   Verification of small-signal linearized VSC HVDC model.

17.10   Small-Signal Linearized VSC HVDC Model

A complete two-level VSC HVDC model can be derived by connecting the three subsystems that are studied in this chapter and in Part I with line-commutated converter (LCC) HVDC:

- a *DC system model* that includes a DC line, rectifier controller and PLL, inverter controller and PLL, feedback filters and all equations that describe coupling with AC systems;
- a *rectifier AC system* that is modelled as described with LCC HVDC;
- an *inverter AC system*.

The above modelling principles support developing an average-value nonlinear VSC HVDC model. In order to perform eigenvalue studies and to use major control design theories it is necessary to use a linearized model. A linearized model is obtained by linearizing all nonlinear terms around the steady-state operating point.

Figure 17.12 shows the verification of a linearized dynamic model for a 1 GW, 640 kV two-level HVDC system with 100 km DC cable. Both AC systems have $SCR = 40$, $X/R = 10$. The linearized model shows excellent accuracy for a 5% step input on I_{dref}. The model shows good matching for all variables, including AC and DC variables on both sides (rectifier and inverter). It is important that the model represents well all physical variables and parameters in order to enable accurate dynamic, parametric studies and control design.

17.11   Small-Signal Dynamic Studies

17.11.1   Dynamics of Weak AC Systems

Figure 17.13 shows the root locus of the dominant eigenvalues for the reduction in AC system strength, (a) on the inverter and (b) on the rectifier side. It is seen that damping of dominant eigenvalues 1, 2 reduces as the system strength reduces, which is a similar conclusion as with the LCC HVDC. However, VSC HVDC is less susceptible to SCR reduction and, even at $SCR = 2$, the eigenvalue damping is still around 0.5, which is much better than with the LCC HVDC. The dynamic instabilities are expected at frequencies around 11–15 Hz for the inverter and 6–10 Hz for the rectifier.

The rectifier and inverter terminals have similar robustness to SCR reduction, despite the fact that the outer controls are different (I_{gd} control at rectifier and V_{dc} control at inverter). This points to the

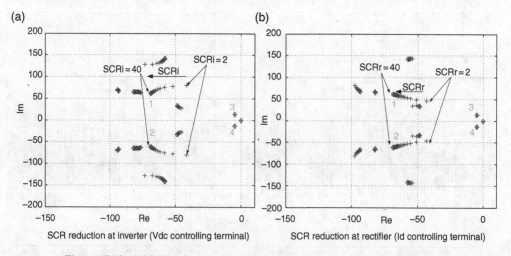

Figure 17.13   VSC HVDC system dynamics with reduction in SCR ($2 < SCR < 40$).

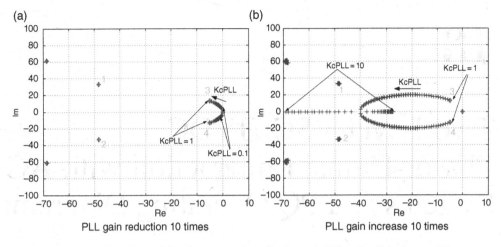

Figure 17.14   VSC HVDC system dynamics with change in PLL gains ($0.1 < k_{cPLL} < 10$).

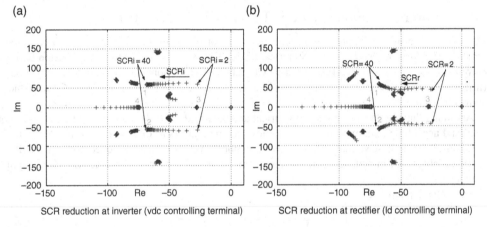

Figure 17.15   VSC HVDC system dynamics with reduced SCR ($2 < SCR < 40$), using increased PLL gains, $k_{cPLL} = 10$.

conclusion that inner control loops (which are identical for the rectifier and inverter) have a significant impact on stability.

17.11.2   Impact of PLL Gains on Robustness

The original PLL gains were set as $k_{pPLL} = 10$, $k_{iPLL} = 200$, $k_{cPLL} = 1$. Figure 17.14 indicates that dynamics are generally improving as PLL gains are increased further. However, beyond a certain point dynamics deteriorate with large PLL gains. Phase-locked loop gains have most impact at low frequencies, in the region of a few Hertz.

Higher PLL gains do not improve much dynamic of HVDC with weak AC systems. With very high PLL gains dynamics with weak system may actually deteriorate. This is illustrated in Figure 17.15, which shows the same root locus conditions as in Figure 17.13, but higher PLL gains are used (k_{cPLL} = 10), as it is seen by the improved position of eigenvalues 3, 4. Root locus branches are longer, indicating more sensitivity to changes in the AC system strength.

18

Two-Level VSC HVDC Phasor-Domain Interaction with AC Systems and PQ Operating Diagrams

18.1 Power Exchange between Two AC Voltage Sources

The first section of this chapter will review the theory of power exchange between two AC systems, which is important to establish the theoretical limits. The study is similar to that in section 9.3, with current-source line-commutated converters (LCCs), but in this section there are two AC voltage sources (instead of current sources), which mimic voltage source converters (VSCs). The system under study is shown in Figure 18.1.

Assuming that the AC system is symmetrical and balanced, considering that AC frequency is constant and neglecting dynamics, all AC variables become phasors as discussed in Appendix C. In this notation, all three-phase variables become two-component vectors (the zero sequence is neglected) with RMS variables. All impedances will become static complex numbers with reactances as imaginary components.

The voltage phasors are:

$$\overline{V_{sdq}} = V_s \angle \varphi_{Vs} = V_{sd} + jV_{sq} \tag{18.1}$$

$$\overline{V_{gdq}} = V_g \angle \varphi_{Vg} = V_{gd} + jV_{gq} \tag{18.2}$$

where, $\overline{V_{gdq}}, \overline{V_{sdq}}$ are the phasors, V_g and V_s are the magnitudes (line-neutral RMS) and $\varphi_{Vg}, \varphi_{Vs}$ are phase angles of respective voltages. The subscripts d and q denote corresponding phasor components. The grid current phasor is:

$$\overline{I_g} = \frac{\overline{V_s} - \overline{V_g}}{Z_s}$$

$$I_g \angle \varphi_I = \frac{V_s \angle \varphi_{Vs} - V_g \angle \varphi_{Vg}}{Z_s \angle \varphi_Z} \tag{18.3}$$

High-Voltage Direct-Current Transmission: Converters, Systems and DC Grids, First Edition.
Dragan Jovcic and Khaled Ahmed.
© 2015 John Wiley & Sons, Ltd. Published 2015 by John Wiley & Sons, Ltd.

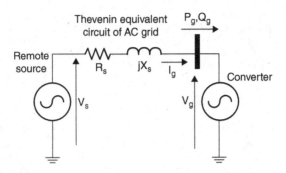

Figure 18.1   Test system for studying power exchange between two AC voltage sources.

Assuming that the coordinate frame is linked with the remote source voltage V_s:

$$I_{gd} + jI_{gq} = \frac{V_s - V_{gd} - jV_{gq}}{R_s + jX_s}$$ (18.4)

It is also assumed that the magnitudes of the two voltages are equal and constant: $V_s = V_g = const$ as this is the case of most interest in transmission engineering:

$$I_{gd} + jI_{gq} = V_s \frac{1 - \cos\varphi_{Vg} - j\sin\varphi_{Vg}}{R_s + jX_s}$$ (18.5)

Rearranging Eq. (18.5), the current expression is obtained:

$$I_{gd} + jI_{gq} = \frac{V_s}{R_s^2 + X_s^2} \left[R_s \left(1 - \cos\varphi_{Vg}\right) - X_s \sin\varphi_{Vg} - j\left(R_s \sin\varphi_{Vg} + X_s \left(1 - \cos\varphi_{Vg}\right)\right)\right]$$ (18.6)

The condition for maximal real current is calculated by equating the first derivative of real part with zero:

$$\left(tg\varphi_{Vg}\right)_{\max Pg} = \frac{X_s}{R_s}$$ (18.7)

The condition for maximal reactive current is:

$$\left(tg\varphi_{Vg}\right)_{\max Qg} = \frac{R_s}{X_s}$$ (18.8)

In the case that $R_s = 0$, the maximum current is obtained for $\varphi_{Vg} = 90°$, which is a known case from transmission system studies, and which equals:

$$I_{gd} + jI_{gq} = \frac{V_s}{X_s} + j\frac{V_s}{X_s} = 1\,pu + j1\,pu$$ (18.9)

where the pu unit system is based on the short-circuit level as discussed in section 9.3.

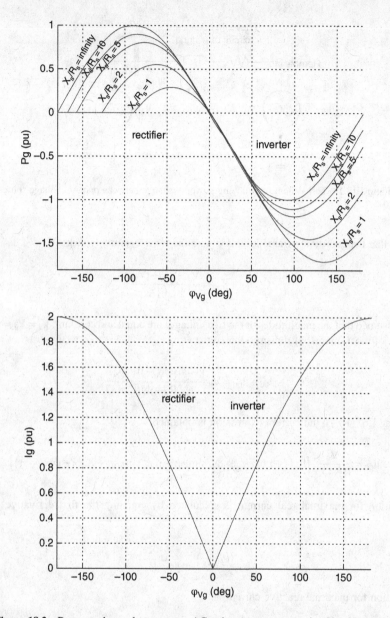

Figure 18.2   Power exchange between two AC voltage sources assuming $V_s = V_g = \text{const.}$

Figure 18.2 shows the power exchange between the two AC systems as the phase angle of V_g is changing. Considering firstly the case $X_s/R_s = infinity$, it is seen that the maximum power transfer is 1 pu, for both rectifier and for inverter. This implies that the minimal theoretical short circuit ratio (SCR), according to definition Eq. (9.3) is 1. If there is some resistive part in the grid impedance, then maximum transfer limit reduces for rectifier but increases for inverter.

Figure 18.3 shows the power transfer limits assuming that the current is in phase with voltage, V_g, as this is a common operating mode for many converters. In this case, the magnitude of V_g is allowed to vary in order to compensate for reactive current. When $X_s/R_s = infinity$, the maximum power transfer is

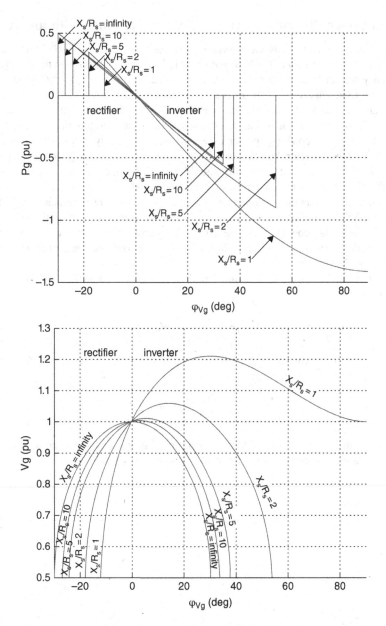

Figure 18.3   Power exchange between two AC voltage sources assuming that current is in phase with voltage V_g.

0.5 pu for both rectifier and for inverter. This implies that the minimal theoretical SCR is 2. However, if a restriction is placed on V_g variation to 0.9 pu then the maximum rectifier or inverter power is 0.3 pu (corresponding to minimal SCR or 3.3).

If $X_s/R_s = 5$ then the maximum power that a rectifier can receive is 0.4 pu (giving a minimal SCR of 2.5) and the maximum power that an inverter can deliver to the grid is 0.65 pu (giving a minimal SCR of 1.54). Further, if V_g variation is limited to 0.9 pu, then the maximum rectifier power is only 0.2 pu while the inverter power is limited to 0.5 pu.

18.2   Converter Phasor Model and Power Exchange with an AC System

The study of power exchange between two voltage sources now expands to include realistic limitations associated with a VSC converter. The VSC converter is exchanging power with the AC grid through an interconnecting reactor (transformer reactance), by controlling the converter AC voltage V_c, as illustrated in Figure 18.4. This is a slightly different case from that in the previous section because an additional reactance is included. The VSC converter is also synchronized with the point of common coupling (PCC) voltage V_g using a phase-locked loop (PLL) and therefore the converter control angle is referenced to V_g. The converter power exchange is consequently controlled using two voltages across the interfacing reactor but the delivered power must ultimately be exchanged with the remote source V_s. The interfacing reactance should be added to the grid reactance (total reactance is $X_s + X_t$) if the study method from the previous section is used.

It is important to note that the grid operators are interested in the power exchange at the PCC. On the other hand, the VSC manufacturers will adequately rate the converter to meet the PCC power specifications, considering the converter voltage and current capabilities. As the converter rating must be observed at all times, PQ operating limits at the PCC are of particular interest as the power transfer and operating modes change.

The PCC voltage V_g and remote source voltage V_s are defined in Eqs (18.1) and (18.2). The converter voltage V_c is:

$$\overline{V_{cdq}} = V_c \angle \varphi_{Vc} = V_{cd} + jV_{cq} \tag{18.10}$$

As the PLL is used to synchronize the controller with the grid PCC voltage, it is more convenient to assume that V_g is located on the d-axis ($\varphi_{Vg} = 0$):

$$\overline{V_{gdq}} = V_{gd} = V_g, \quad V_{gq} = 0 \tag{18.11}$$

The converter voltage V_c is controllable, and assuming a simple sinusoidal pulse width modulation (SPWM) modulation, the converter phasor voltage components are:

$$V_{cd} = M_d \times \frac{V_{dc}}{2\sqrt{2}} \tag{18.12}$$

$$V_{cq} = M_q \times \frac{V_{dc}}{2\sqrt{2}} \tag{18.13}$$

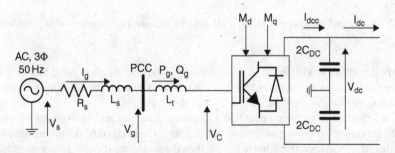

Figure 18.4   Two-level VSC converter connected to an AC grid for phasor interaction study.

and the modulation index magnitude is also expressed as:

$$M = \sqrt{M_d^2 + M_q^2} < 1 \tag{18.14}$$

where M_d and M_q are DQ components of sinusoidal PWM (pulse width modulation) control signal and V_{dc} is the DC voltage. The converter voltage rating is defined by the DC voltage. Assuming a maximum modulation ratio $M = 1$, the converter rated AC voltage can be determined from Eqs (18.12)–(18.14).

The VSC converter current rating is defined by the magnitude of the current through the semiconductors. Either AC or DC current can be specified, but grid operators typically specify AC current requirements. The basic current equation for the circuit in Figure 18.4 is:

$$\overline{I_{gdq}} = \frac{\overline{V_g} - \overline{V_c}}{jX_t} = \frac{V_g - V_{cd} - jV_{cq}}{jX_t} \tag{18.15}$$

where X_t is transformer reactance $(L_t = X_t/(2\pi f)$ and grid frequency is $f = 50\,\text{Hz})$. Expressing Eq. (18.15) using current components:

$$I_{gd} + jI_{gq} = \frac{-V_{cq}}{X_t} + j\frac{V_{cd} - V_g}{X_t} \tag{18.16}$$

This is further represented using two scalar equations as:

$$I_{gd} = \frac{-V_{cq}}{X_t} \tag{18.17}$$

$$I_{gq} = \frac{V_{cd} - V_g}{X_t} \tag{18.18}$$

The PCC complex power can be calculated using Eq. (18.16) as:

$$\overline{S_g} = 3\overline{V_g} \times \overline{I_g}^* = 3V_g \left[-\frac{V_{cq}}{X_t} + j\frac{V_g - V_{cd}}{X_t} \right] \tag{18.19}$$

where $(\cdot)^*$ stands for conjugate complex, which is required to ensure consistent sign for the reactive power. Separating Eq. (18.19) into real and imaginary parts:

$$P_g = -3V_g\frac{V_{cq}}{X_t} \tag{18.20}$$

$$Q_g = 3V_g\frac{V_g - V_{cd}}{X_t} \tag{18.21}$$

or in terms of the magnitude and angle of control signals:

$$P_g = -3V_g\frac{MV_{dc}}{X_t 2\sqrt{2}}\sin(\varphi_{Vc}) \tag{18.22}$$

$$Q_g = 3\frac{V_g^2 - V_g\dfrac{MV_{dc}}{2\sqrt{2}}\cos(\varphi_{Vc})}{X_t} \tag{18.23}$$

18.3   Phasor Study of VSC Converter Interaction with AC System

18.3.1   Test System

Table 18.1 shows the test system parameters in this chapter. The converter is designed considering the requirement for PCC maximum active and reactive power export. For the given grid voltage the rated converter current can be determined. A transformer will normally be used to adjust the ratio between grid voltage and DC voltage but it is omitted here for simplicity.

18.3.2   Assumptions and Converter Limits

In the study in the sections below, various AC system variables and also the VSC converter internal variables are examined as the converter active power is changing, for the test case in Table 18.1. The most important converter operating modes will be separately investigated.

The following assumptions are employed:

- Coordinate frame is linked with PCC voltage: $\overline{V}_g = V_{gd} = V_g$.
- Magnitude of remote source is constant: $V_s = const$ (infinite bus): considering test system data with full VSC power in inverter mode, this voltage is calculated as $V_{s\_ll} = 369$ kV.
- DC voltage is constant $V_{dc} = 636$ kV: a remote VSC converter controls DC voltage.
- Modulation index cannot exceed 1: converter controller will prevent over modulation:

$$M = \sqrt{M_d^2 + M_q^2} \le 1 \tag{18.24}$$

- The converter current cannot exceed the rated value: the converter inner current controller will prevent current exceeding rated value:

$$I_g = \sqrt{I_{gd}^2 + I_{gq}^2} \le I_{grated} = 1540 \text{ A} \tag{18.25}$$

- The converter q-voltage, V_{cq}, is an independent variable in the study. This variable is determined directly from the desired active power flow. Then, knowing V_{cq}, from Eq. (18.16) the active current can be determined as in Eq. (18.17).

Table 18.1   Test system parameters.

Rated S_g	1000 MVA
Design point P_g, Q_g	$Pg = -995$ MW, $Qg = -100$ MVAr
Rated converter current I_{grated}	1540 A
Rated PCC AC voltage V_{g_ll}	375 kV
Rated converter AC voltage V_{c_ll}	390 kV
Rated DC voltage V_{dc}	636 kV
Remote source voltage V_{s_ll}	369 kV
Transformer	$X_{tpu} = 0.2$ pu, $St = 1000$ MVA, $V_{t\_ll} = 375$ kV, $n = 1$ ($Xt = 28$ Ω, $Lt = 0.0895$ H)
AC grid	$SCR = 10$, $X/R = 10$, ($R_s = 1.4$ Ω, $X_s = 14.1$ Ω, $L_s = 0.0448$ H)

18.3.3   Case 1: Converter Voltages are Known

The case where converter voltage, V_{cd}, is also known is examined first, as this case will give system variables when the converter is on the maximum voltage limit ($M = 1$). The current equation for the circuit in Figure 18.4 is:

$$I_{gd} + jI_{gq} = \frac{V_{sd} + jV_{sq} - V_g}{Z_s} \tag{18.26}$$

This equation can be converted into two scalar equations, which are squared, then added and replacing $V_s^2 = V_{sd}^2 + V_{sq}^2$:

$$\left(R_s I_{gd} - X_s I_{gq} + V_g\right)^2 + \left(X_s I_{gd} + R_s I_{gq}\right)^2 = V_s^2 \tag{18.27}$$

Replacing currents from Eqs (18.17), (18.18) into Eq. (18.27) a quadratic equation is obtained:

$$
\begin{aligned}
&V_g^2 a_1 + V_g b_1 + c_1 = 0 \\
&a_1 = R_s^2 + \left(X_s + X_t\right)^2 \\
&b_1 = 2R_s\left(-V_{cd}R_s + V_{cq}X_s\right) + 2\left(X_s + X_t\right)\left(-V_{cd}X_s - V_{cq}R_s\right) \\
&c_1 = \left(-V_{cd}R_s + V_{cq}X_s\right)^2 + \left(-V_{cd}X_s - V_{cq}R_s\right)^2 - V_s^2 X_t^2
\end{aligned}
\tag{18.28}
$$

Equation (18.28) enables direct calculation of V_g for given converter voltages V_{cd} and V_{cq}. Once V_g is known I_{gq} is determined using Eq. (18.18).

18.3.4   Case 2: Converter Currents are Known

The next case assumes that the converter current magnitude I_g is given, as it would happen if the converter current limit is reached. The reactive current I_{gq} can be determined from Eqs (18.18) and (18.25). Therefore the remote source voltage and PCC voltage can be determined from Eq. (18.26):

$$
\begin{aligned}
&V_{sq} = X_s I_{gd} + R_s I_{gq} \\
&V_{sd} = \sqrt{V_s^2 - V_{sq}^2} \\
&V_g = V_{sd} - R_s I_{gd} + X_s I_{gq}
\end{aligned}
\tag{18.29}
$$

At the final stage the converter voltages can be calculated from Eq. (18.16).

18.3.5   Case 3: PCC Voltage is Known

If the PCC voltage is known, $V_g = const$, as is the case if the converter is operated to control the PCC voltage, then from Eq. (18.26) the following quadratic equation is obtained:

$$
\begin{aligned}
&V_{cd}^2 a_2 + V_{cd} b_2 + c_2 = 0 \\
&a_2 = R_s^2 + X_s^2 \\
&b_2 = -2R_s\left(V_g R_s + V_{cq}X_s\right) - 2X_s\left(-V_g(X_s + X_t) - V_{cq}R_s\right) \\
&c_2 = \left(V_g R_s + V_{cq}X_s\right)^2 + \left(-V_g(X_s + X_t) - V_{cq}R_s\right)^2 - V_s^2 X_t^2
\end{aligned}
\tag{18.30}
$$

Equation (18.30) enables calculation of the converter voltage V_{cd}, when the PCC voltage is known.

18.4　Operating Limits

Figure 18.5 shows the four quadrant operating diagrams for the VSC converter. The sign convention applies equally to active and reactive power: positive power means that converter (DC system) draws power from AC grid. The curves show operating limits considering converter voltage rating ($M = 1$ limit at rated DC voltage) and current rating (I_g limit).

The top left diagram shows the control variables M_d and M_q, and the operating point can be anywhere inside the curve. The upper curve is the limit on the maximum modulation index, which will represent maximum converter voltage. In practice this is the VSC reactive power export limit. Eq. (18.21) shows that, if V_{cd} is larger than V_g, the converter will export reactive power to the grid, and vice versa. The lower curve represents low modulation index and in such case the operating limit is derived considering the converter maximum current. The negative M_q corresponds to positive active power, which represents power delivered to the DC system.

The available converter current envelope is also of interest. In the reactive power export mode (positive I_{gq}) the maximum current curve is not symmetrical as the converter can export more reactive current when importing active power. This is understandable because V_g is lower when power is transferred from AC system to DC.

The diagram with the PCC voltage V_g as the converter operates on the limit curves is particularly significant for grid operators. As the test AC grid is quite strong ($SCR = 10$) the AC voltage changes

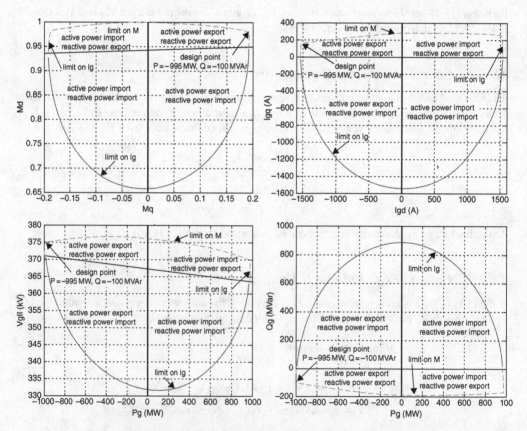

Figure 18.5　VSC converter operating limits: SCR = 10, X/R = 20, $X_t = 0.2$ pu.

by only 5 kV as the operating point moves from full active power export to full import, along the maximal reactive power export.

The PQ diagram in this figure is the most commonly used indicator of HVDC power exchange by the manufacturers. It shows the maximum PQ envelope at PCC for a converter. The converter is designed for $P_g = -995$ MW and $Q_g = -100$ MVAr ($S_g = 1000$ MVA). It is seen that when $Q_g = 0$, the converter can export around -990 MW. However, the maximum import power is slightly lower (around 970 MW) as V_g is lower when power is transferred to the DC system. If $P_g = 0$, the converter exchanges only reactive power and operates as a STATCOM. As a STATCOM, this VSC can export around -180 MVAr and theoretically import up to 880 MVAr. This maximum reactive power import may not be achievable in practice since PCC voltage would be unacceptably low.

18.5   Design Point Selection

For a given MVA power at the PCC, the converter nominal operating point can be selected according to grid operator requirements. This selection allows some tradeoff between active and reactive power within the converter limits. Some VSC HVDC operate with weak AC grids (like the Caprivi HVDC in Namibia) and these AC grids require strong VSC reactive power capability at all active power levels.

Figure 18.6 shows the diagrams of operating limits for three different design points. Selecting larger reactive power at the design point brings the advantage of a wider reactive power export region. In

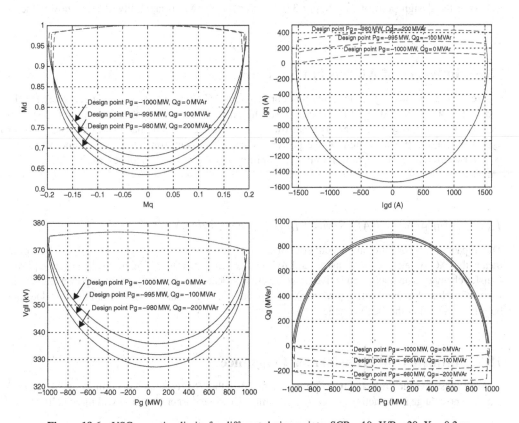

Figure 18.6   VSC operating limits for different design points: SCR = 10, X/R = 20, $X_t = 0.2$ pu.

Table 18.2 System variables for different design points (Sg = 1000 MVA, Vg = 375 kV).

Design Pg (MW)	Design Qg (MVAr)	Ig (A)	Vg (kV)	Vdc (kV)	Pg (MW) at Qg = 0
−1000	0	1540	375	625	−1000
−995	−100	1540	375	636	−990
−980	−200	1540	375	648	−980
−954	−300	1540	375	660	−965

practice this larger reactive power capability is achieved by changing the transformer ratio (increasing DC voltage if no transformer is used). On the downside, selecting a design point with larger reactive power implies the following:

• The converter current or DC voltage increases and therefore the converter rating will increase. Losses will also increase.
• The active power capability at zero reactive power is reduced.
• Operation at zero reactive power will require a lower modulation index, bringing more losses and harmonics.

Table 18.2 summarizes the difference between four operating points for the same apparent power S_g = 1000 MVA and the same grid voltage V_g = 375 kV. As an example, selecting −300 MVAr at the design point would imply 3.6% lower active power capability at unity power factor operation.

18.6 Influence of AC System Strength

Figure 18.7 shows the operating limits for four different AC strengths: SCR = 2, SCR = 5, SCR = 10 and SCR = 30. As the AC grid becomes weaker the power exchange capability with the DC system reduces. With weak systems there is a narrow range of voltage which enables full converter power export or import. The maximum active power exchange can occur only for the maximum reactive power export. The reactive power import capability also significantly reduces as the AC strength reduces.

18.6.1 Influence of AC System Impedance Angle (X_s/R_s)

The impact of the X/R ratio on the converter operating curves is shown in Figure 18.8. A low X/R ratio means that AC system is more resistive and AC voltage will generally swing more as the active power transfer changes. The maximum active power exchange occurs with a small reactive power transfer in such case. Note that the reactive power export capability increases at a low X/R ratio but this does not help in maintaining the PCC voltage.

18.7 Influence of Transformer Reactance

Figure 18.9 shows the impact of transformer reactance on the VSC converter operating curves. There is virtually no change in the active power transfer limits. However larger reactance will marginally reduce

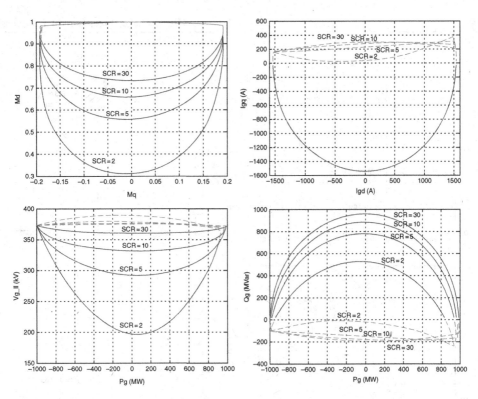

Figure 18.7 VSC operating limits for variable SCR; X/R = 10, Xt = 0.2 pu.

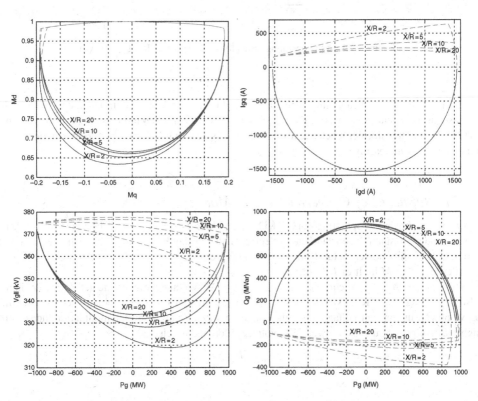

Figure 18.8 VSC operating limits for variable X/R; SCR = 10, Xt = 0.2 pu.

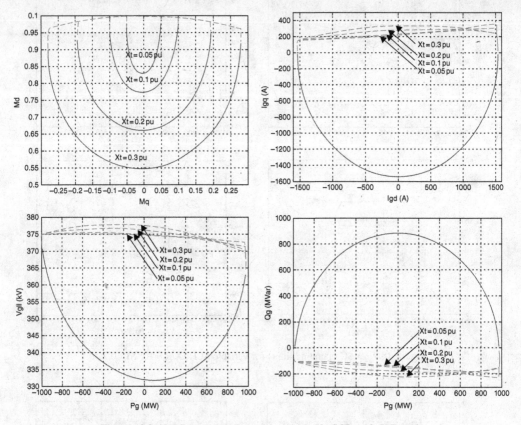

Figure 18.9 VSC operating limits for variable X_t; SCR = 10, X/R = 10.

the reactive power export capability. The main impact is on the range of control signal variation, which increases as the reactance increases. This implies that a larger converter rating is required for a given DC voltage. Higher transformer reactance also affects the system dynamics as open-loop gain decreases as the reactance is increasing.

18.7.1 Influence of Converter Control Modes

Figure 18.10 shows the operating curves with closed loop HVDC control. The two commonly used control modes are studied: constant reactive current and constant PCC voltage. Note that operation with $I_g = 0$, gives a unity power factor. If the converter controller cannot achieve required PCC voltage then it settles on either M limit or I_g limit.

When a constant reactive current mode is used, the AC voltage varies as the active power changes. Also depending on the reactive current reference, the converter may not be able to achieve maximum active power. In constant voltage mode the converter will commonly hit the modulation index limit at higher active power, unless a lower PCC voltage reference is selected.

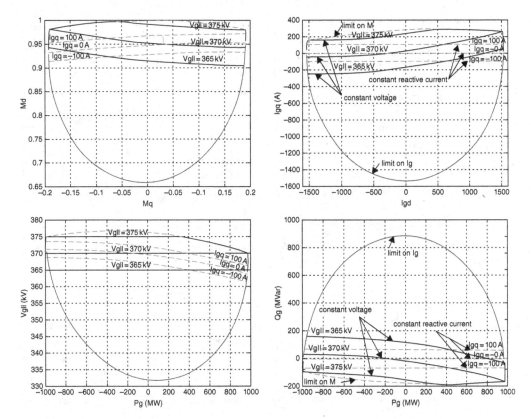

Figure 18.10   VSC operating diagrams in different control modes; SCR = 10, X/R = 10, Xt = 0.2 pu.

18.8   Operation with Very Weak AC Systems

Figure 18.11 shows the operating curves for different control modes when the AC system is very weak ($SCR = 1.8$). It is seen that many operating restrictions are present. The main concerns are the variations in PCC voltage and limitations on the power transfer. The PCC voltage varies significantly unless voltage control is used. The maximum active power import is limited to 940 MW and this is achieved for a very narrow reactive power range. If the reactive current is zero, the maximum import power is only 830 MW, but the AC voltage would in such case drop to 310 kV (0.83 pu), which is unacceptable. Therefore it is not possible to operate with $SCR = 1.8$ and $I_{gq} = 0$. Lower SCR can be achieved if current is not in phase with voltage, which is analysed in more depth in section 18.1.

In practice the converter will need to use a large portion of the rated current to supply reactive power in order to stabilize the voltage on a weak AC system.

This study assumed that the remote source voltage is constant ($V_s = const.$). In practice this voltage may vary, in particular with weak AC networks. Lower grid voltage will further reduce power exchange capability.

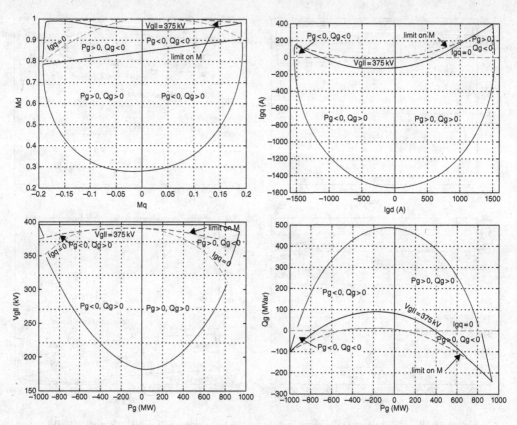

Figure 18.11 VSC converter operating diagrams in different control modes with very weak AC system; SCR = 1.8, X/R = 10, Xt = 0.2 pu.

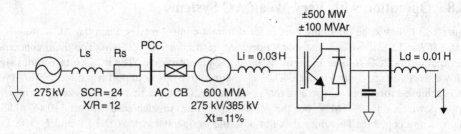

Figure 18.12 VSC HVDC terminal in Example 18.1.

Example 18.1

Consider the VSC HVDC terminal as shown in Figure 18.12. Assume that the PCC bus voltage is controlled to 275 kV.

a. *Determine the converter AC voltage for the following rectifier operating point: Pg = 500 MW, Qg = 100 MVAr. Refer the converter voltage to the grid side and determine the percentage voltage swing.*

b. *Determine the converter AC voltage for the maximum power export: Pg = −500 MW, Qg =*
   100 MVAr. Refer the converter voltage to the grid side and determine the percentage
   voltage swing.
c. *Assume that the phase reactance is increased four times because of DC fault problems. Determine*
   the maximum AC voltage swing.

Solution

a. *Converter AC voltage for rectifier*

$$V_{cd} = -\frac{Q_g(X_t + X_i)}{V_g} + V_g = \frac{100e6\left(0.11\dfrac{275\ 000^2}{600e6} + 2\pi \times 50 \times 0.03 \times (275/385)^2\right)}{275\ 000} + 275\ 000 = 263\ \text{kV}$$

$$V_{cq} = -\frac{P_g(X_t + X_i)}{V_g} = -\frac{500e6\left(0.11\dfrac{275\ 000^2}{600e6} + 2\pi \times 50 \times 0.03 \times (275/385)^2\right)}{275\ 000} = -60\ \text{kV}$$

$$V_c = \sqrt{V_{cd}^2 + V_{cq}^2} = \sqrt{60\ 000^2 + 26\ 3000^2} = 269.76\ \text{kV}$$

$$\Delta V\% = \frac{V_g - V_c}{V_g}100 = \frac{275\ \text{kV} - 269.76\ \text{kV}}{275\ \text{kV}}100 = 1.7\%$$

b. *Converter AC voltage for maximum power export*

$$V_{cd} = -\frac{Q_g(X_t + X_i)}{V_g} + V_g = \frac{-100e6\left(0.11\dfrac{275\ 000^2}{600e6} + 2\pi \times 50 \times 0.03 \times (275/385)^2\right)}{275000} + 275\ 000 = 287\ \text{kV}$$

$$V_{cq} = -\frac{P_g(X_t + X_i)}{V_0} = -\frac{-500e6\left(0.11\dfrac{275\ 000^2}{600e6} + 2\pi \times 50 \times 0.03 \times (275/385)^2\right)}{275\ 000} = 60\ \text{kV}$$

$$V_{c\max} = \sqrt{V_{cd}^2 + V_{cq}^2} = \sqrt{60\ 000^2 + 287\ 000^2} = 293.3\ \text{kV}$$

$$\Delta V_{\max}\% = \frac{V_g - V_{c\max}}{V_g}100 = \frac{275\ \text{kV} - 293.3\ \text{kV}}{275\text{kV}}100 = -3.2\%$$

The above voltage swing is quite small.
c. *The above voltage calculation is repeated with $L_i = 0.12\ H$:*

$$\Delta V\% = 1.9\%$$

$$\Delta V_{\max}\% = -6.6\%$$

The total voltage swing is 8.6%, which is acceptable, but should be taken into account when dimen-
sioning equipment.

Example 18.2

A 200 kV, 1000 A, HVDC is being considered as an interconnection between two AC systems. Power flow is unidirectional. The HVDC converter ratings are 200 MVA. The 320 kV, 50 Hz inverter grid draws a total of 150 MVA at a 0.75 power factor, while the rectifier VSC draws the power requirement from the 230 kV, 60 Hz network at the unit power factor. Sinusoidal pulse width modulation (SPWM) switching is used with a 1350 Hz triangular carrier wave. The 1000 mm^2 copper DC cable has a resistance per km of $R_{ocukm1000} = 1.39 \times 10^{-2}$ Ω/km. The total distance between terminals is 50 km.
Determine:

a. *Transformer ratio for rectifier and inverter transformers, assuming that transformer leakage reactances are: $X_{t50Hzpu} = 0.1$ pu, $X_{t60Hzpu} = 0.16$ pu.*
b. *DC cable losses.*
c. *Active and reactive power delivered at inverter side.*
d. *Active and reactive power drawn at rectifier side.*

Solution

1. *It is assumed that inverter station controls DC voltage:*

$$V_{dc50Hz} = 200 \ kV$$

$$\mathsf{V}_{c50Hz} = \frac{V_{dc50Hz}}{2\sqrt{2}} = 70.71 \ kV, \ \mathsf{V}_{c50Hz\_LL} = \sqrt{3}\mathsf{V}_{c50Hz} = 122.47 \ kV$$

$$X_{t50Hz} = X_{t50Hzpu}\frac{\mathsf{V}^2_{50HzLL}}{S_t} = 0.10\frac{320\ 000^2}{200\ 000\ 000} = 51.2 \ \Omega$$

$$I_{L50Hz} = \frac{S_{50Hz}}{\sqrt{3} \times \mathsf{V}_{50HzLL}} = \frac{200 \ MVA}{\sqrt{3} \times 320 \ kV} = 360.84 \ A$$

$$\mathsf{V}_{g50Hz\_est} = \mathsf{V}_{c50Hz} + jX_{tr}\mathsf{I}_{L50Hz}$$

$$= 320/\sqrt{3} + j51.2 \times 360.84 = 184.75 \ kV + j18.475 \ kV = 185.671 \ kV$$

$$\mathsf{V}_{g50Hz\_LL\_est} = \sqrt{3}\mathsf{V}_{g50Hz\_est} = 321.59 \ kV$$

Transformer turns ratio: 322/122 = 2.63.

Rectifier station:

$$R_{dc} = 2R_{ocu}l_c = 2 \times 1.39 \times 10^{-2} \times 50 = 1.39 \ \Omega$$

$$V_{dc60Hz} = V_{dc50Hz} + R_{dc}I_{dc} = 200\ 000 + 1.39 \times 1000 = 201.39 \ kV$$

$$\mathsf{V}_{c60Hz} = \frac{V_{dc60Hz}}{2\sqrt{2}} = 71.2 \ kV, \ \mathsf{V}_{cLL\_60Hz} = \sqrt{3}V_{60Hz} = 123.32 \ kV$$

$$X_{t60Hz} = X_{t60Hzpu}\frac{\mathsf{V}^2_{g60Hz\_LL}}{S_t} = 0.16\frac{230\ 000^2}{200\ 000\ 000} = 42.32 \ \Omega$$

$$I_{L60\,Hz} = \frac{S_{60\,Hz}}{\sqrt{3} \times V_{g60\,HzLL}} = \frac{200\ MVA}{\sqrt{3} \times 230\ kV} = 502.04\ A$$

$$V_{g60\,Hz\_est} = V_{g60\,Hz} - jX_{t60\,Hz}I_L$$
$$= 230/\sqrt{3} - j42.32 \times 502.04 = 132.79\ kV - j21.246\ kV = 134\ kV$$

$$V_{g60\,HzLL\_est} = \sqrt{3}V_{g60\,Hz\_est} = 233\ kV$$

Transformer ratio: $233/123 = 1.89$.

2. *Losses:*

$$P_{dcloss} = I_{dc}^2 R_{dc} = 1000^2 \times 1.39 = 1.39\ MW = 0.7\%$$

3. *The total active power at the inverter is:*

$$P_{50Hz} = S_{50Hz} \times pf_{50Hz} = 150\ MVA \times 0.75 = 112.5\ MW$$

The inverter reactive power delivered is:

$$Q_{50Hz} = \sqrt{S_{50Hz}^2 - P_{50Hz}^2} = \sqrt{150^2 - 112.5^2} = 99.21\ MVAr$$

4. *At the rectifier side:* $P_{60Hz} = 112.5\ MW + 2.78\ MW = 115.28\ MW$, $Q_{60Hz} = 0$.

Example 18.3
A ±150 kV VSC HVDC cable transmission system connects a 230 kV, 60 Hz three-phase AC system to a 220 kV, 50 Hz three-phase AC system, some 100 km apart. The VSC converters are interfaced by a Δ-Y transformer with 25 mH of leakage reactance. The DC cable is rated at 1000 A with a total resistance of 2 Ω. Consider the operating point where the 60 Hz inverter grid receives a total of 300 MVA at a 0.85 power factor, while the rectifier VSC draws the power requirement from the 50 Hz network at unit power factor.

Determine the phasor diagram for:

a. *The 60 Hz receiving end*
b. *The 50 Hz transmitting end.*

Solution

1. *The total active power at the inverter is:*

$$P_{60Hz} = S_{60Hz} \times pf_{60Hz} = 300\ MVA \times 0.85 = 255\ MW$$

The 60 Hz reactive power is:

$$Q_{60Hz} = \sqrt{S_{60Hz}^2 - P_{60Hz}^2} = \sqrt{300^2 - 255^2} = 158\ MVAr$$

The 60 Hz AC line current is:

$$I_{L60Hz} = \frac{S_{60Hz}}{\sqrt{3} \times V_{60HzLL}} = \frac{300 \ MVA}{\sqrt{3} \times 230 \ kV} = 753.1 \ A$$

The reference 60 Hz phase voltage is:

$$V_{g60Hz} = \frac{V_{g60HzLL}}{\sqrt{3}} = \frac{230 \ kV}{\sqrt{3}} = 132.8 \ kV$$

The angle between the line current and the phase voltage is:

$$\varphi = \cos^{-1} 0.85 = 31.8°$$

The transformer reactance at the 60 Hz system is:

$$X_{L60Hz} = 2\pi f L = 2\pi \times 60 \times 0.025 = 9.42 \ \Omega$$

The 60 Hz converter line to neutral voltage is:

$V_{c60Hz} = V_{g60Hz} + jX_{L60Hz}I_{L60Hz}$
$\quad = 132.8 + j9.42 \times 753.1 \ \angle -31.8° = 132.8 \ kV + 7.1 \ kV \ \angle \ 58.2° = 136.54 + j6.03 = 136.67 \ kV \ \angle \ 2.5°$

2. *The ±150 kV DC-link has a pole-to-pole voltage of 300 kV. The DC current is:*

$$I_{dc} = \frac{255MW}{300kV} = 850 \ A$$

The DC cable voltage drop is:

$$V_{\Delta dc} = I_{dc} \times R_{dc} = 850 \times 2 = 1.7 \ kV$$

The VSC DC voltage at the 50 Hz transmitting end is:

$$V_{dc50Hz} = 300 + 2 \times 1.7 = 303.4 \ kV$$

The pole DC voltage is 151.7 kV.
The 50 Hz VSC power delivered to the DC link is:

$$P_{dc} = I_{dc}V_{dc50Hz} = 850 \times 303.4 = 257.9 \ MW$$

With unity power factor at the 50 Hz sending end, P = S, and the line current is:

$$I_{L50Hz} = \frac{S_{50Hz}}{\sqrt{3} \times V_{50HzLL}} = \frac{257.9 \ MVA}{\sqrt{3} \times 220 \ kV} = 676.8 \ A$$

$$V_{g50Hz} = \frac{V_{g50HzLL}}{\sqrt{3}} = \frac{220 \ kV}{\sqrt{3}} = 127 \ kV$$

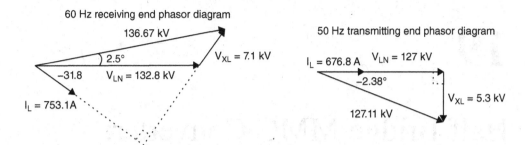

Figure 18.13   Phasor diagram in Example 18.3.

The transformer reactance at the 50 Hz end is:

$$X_{L50Hz} = 2\pi fL = 2\pi \times 50 \times 0.025 = 7.85 \ \Omega$$

The VSC line to neutral voltage at the 50 Hz end is:

$$V_{c50Hz} = V_{g50Hz} - jX_{t50Hz}I_{L50Hz} = 127 \ kV - j7.85 \times 676.8 = 127 \ kV - j5.3 \ kV = 127.11 \ kV$$

Figure 18.13 *shows the phasor diagram for both AC systems.*

19

Half Bridge MMC Converter: Modelling, Control and Operating PQ Diagrams

19.1 Half Bridge MMC Converter Average Model in ABC Frame

The modular multilevel converter (MMC) voltage source converter (VSC) high-voltage direct current (HVDC) will be modelled using principles similar to those discussed in section 7.1, with the aim of developing a phasor model for power flow study and also accurate analytical models in ABC and DQ frames. The average value modelling in an ABC frame is most often used for system studies on electromagnetic transient simulators and it will be addressed first.

The average value MMC modelling replaces the string of cells on each arm with an equivalent voltage source. It is assumed that the AC voltage on each arm (v_p and v_n) in Figure 16.7 is of an ideal *sine* waveform, which is achieved in practice if the number of MMC levels is large (or switching frequency is sufficiently high) and all modules have an equal voltage. Higher harmonics caused by cell switching and cell voltage variations are neglected.

Figure 19.1 shows the structure of the ABC frame average value model for one phase of an MMC converter. L_{arm} is the arm inductance while R_{arm} represents the internal resistance of the inductor and the switches. The labels for individual phases are omitted for simplicity, and whole study is done per phase. The 'IGBT trip' switch is normally connected to v_p and v_n signals and therefore diodes are disconnected in normal operation. When IGBTs are tripped, for example in case of large currents (for DC faults), then v_p and v_n are disabled and the diodes are connected, which converts the half-bridge MMC into a three-phase diode bridge configuration. Compared with two-level VSC ABC model in Figure 17.1, the MMC model has no DC current source as two AC voltage sources (v_P and v_N) provide a direct link between the AC and DC sides.

The mathematical derivation for the equivalent arm voltages (v_P and v_N) is the main modelling challenge, as analysed below. By inspection from Figure 19.1 the AC grid current i_g is the sum of positive and negative arm currents:

$$i_g = i_P + i_N \tag{19.1}$$

High-Voltage Direct-Current Transmission: Converters, Systems and DC Grids, First Edition.
Dragan Jovcic and Khaled Ahmed.
© 2015 John Wiley & Sons, Ltd. Published 2015 by John Wiley & Sons, Ltd.

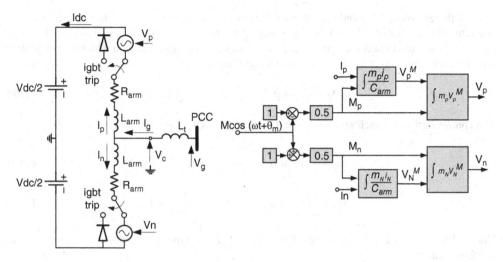

Figure 19.1 One phase of the average value nonlinear dynamic model for the MMC converter.

The voltage equations for the positive and negative poles of the DC circuit are:

$$v_c = \frac{V_{DC}}{2} + \left(R_{arm} + L_{arm}\frac{d}{dt}\right)i_P - v_P \tag{19.2}$$

$$v_c = -\frac{V_{DC}}{2} + \left(R_{arm} + \frac{d}{dt}L_{arm}\right)i_N + v_N \tag{19.3}$$

Also, considering the current loop driven by the full DC voltage:

$$v_P + v_N - 2\left(R_{arm} + \frac{d}{dt}L_{arm}\right)i_{diff} - V_{DC} = 0 \tag{19.4}$$

In the above equation, the differential current i_{diff} is the equivalent DC current in one phase, which will sum with the other two phases to build the DC current:

$$I_{dc} = i_{diff\_a} + i_{diff\_b} + i_{diff\_c} \tag{19.5}$$

The differential current also contains harmonics, which will be discussed later. Considering Eqs (19.2)–(19.4), it is concluded that the differential (also called circulating) current is:

$$i_{diff} = (i_P - i_N)/2 \tag{19.6}$$

Therefore, from Eqs (19.1) to (19.6), it is possible to obtain a link between arm currents, phase current and differential current:

$$i_P = i_g/2 + i_{diff} \tag{19.7}$$

$$i_N = i_g/2 - i_{diff} \tag{19.8}$$

Each of the AC voltages v_P and v_N, represent instantaneous voltage across n series connected cells in one arm. The number of cells in ON state (inserted capacitors) at any instant depends on the control signal and can vary between 0 and n. The control signal m is a *sine* function which is synchronized with reference coordinate frame using phase-locked loop (PLL):

$$m = M\cos(\omega t + \theta_m) \quad 0 < M < 1, \quad -1 < m < 1 \tag{19.9}$$

The control signals for each arm are derived to satisfy the DC voltage balance in Eq. (19.4):

$$m_P = \frac{1}{2}(1 - M\cos(\omega t + \theta_m)), \quad 0 < m_P < 1 \tag{19.10}$$

$$m_N = \frac{1}{2}(1 + M\cos(\omega t + \theta_m)), \quad 0 < m_N < 1 \tag{19.11}$$

Considering sinusoidally inserted capacitors, the equivalent dynamic equation for the positive and negative arms are:

$$\frac{dv_P^M}{dt} = \frac{i_P}{C_{arm}/m_P} \tag{19.12}$$

$$\frac{dv_N^M}{dt} = -\frac{i_N}{C_{arm}/m_N} \tag{19.13}$$

where C_{arm} is the total capacitance in the arm when all capacitors are connected (when all cells are ON):

$$C_{arm} = C_{cell}/n \tag{19.14}$$

and assuming that C_{cell} is a single cell capacitance, while n is the total number of cells in each arm. The expressions C_{arm}/m_P and C_{arm}/m_N are the instantaneous capacitance values in the positive and negative arms respectively. The voltages v_P^M, v_N^M are the maximum available arm voltages (when all cells are ON). Therefore, using Eqs (19.7), (19.8), (19.12) and (19.13), the differential equations for maximal arm voltages can be obtained:

$$\frac{dv_P^M}{dt} = \frac{m_P}{C_{arm}}\left(i_{diff} + \frac{i_g}{2}\right) \tag{19.15}$$

$$\frac{dv_N^M}{dt} = \frac{m_N}{C_{arm}}\left(i_{diff} - \frac{i_g}{2}\right) \tag{19.16}$$

In each of these two equations there is a multiplication of two *sine* signals (i_g and m), which will produce DC plus a second harmonic. The second harmonic voltage generates a circulating differential current and therefore this model will accurately present characteristic even-order harmonics in a MMC converter.

The actual instantaneous AC arm voltages, v_P, v_N, depend on the number of inserted cells and can be represented as:

$$v_P = m_P v_P^M \tag{19.17}$$

$$v_N = m_N v_N^M \tag{19.18}$$

The above voltages are AC waveforms $0 < v_p < V_{dc}$, $0 < v_N < V_{dc}$, as illustrated in Figure 16.8. The schematic diagram of the above dynamic expressions for the equivalent arm voltages is shown in Figure 19.1. The arm voltage expression uses only arm current (i_p or i_n) and phase control voltage $m = M\cos(\omega t + \theta_m)$ as the inputs.

The equivalent single MMC source voltage for one leg (v_{ce}) is now derived as:

$$v_{ce} = \frac{v_P - v_N}{2} \tag{19.19}$$

This is some imaginary voltage that replaces two AC voltage sources per leg with a single one and which enables simple MMC converter model connection with the AC grid model, as analysed later in section 19.2. Therefore, by paralleling the passive parts of the positive and negative arms:

$$v_c = i_g \frac{1}{2} R_{arm} + \frac{di_g}{dt} \frac{1}{2} L_{arm} + v_{ce} \tag{19.20}$$

The above model is very accurate for all power flow, dynamic and fault studies. The second harmonic circulation is also properly represented. The accuracy is better than that with two-level VSC tested in Figure 17.2 since MMC has less harmonics. The limitations of this MMC model are:

- Very high-order harmonics caused by switching individual cells are not represented.
- Cell-level events (faults) cannot be represented.
- Capacitor balancing is not represented.
- As with all ABC frame models, it uses oscillating AC variables and supports only trial-and-error simulations. Studies with multiple parameters/gains are difficult.
- Phasor modelling is not possible.
- Linearization is not possible with oscillating signals in ABC frame.

19.2 Half-Bridge MMC Converter-Static DQ Frame and Phasor Model

19.2.1 Assumptions

DQ frame modelling is essential for power flow, control design and dynamic studies. Firstly, considering modelling complexity, the static model will be derived in the DQ frame, which is the basis for the MMC phasor model required for power-flow studies. The full dynamic DQ model will be derived at a later stage. As with previous DQ frame modelling, it is first assumed that the system is symmetrical and balanced.

The control signal is further assumed to contain only the fundamental component:

$$m = M\cos(\omega t + \varphi_m) = M_d \cos(\omega t) - M_q \sin(\omega t) \tag{19.21}$$

where M_d and M_q are corresponding components in DQ frame rotating at fundamental frequency (ω). Considering the definition for arm-control signals in Eqs (19.10) and (19.11), they will also contain a zero sequence M_{p0} and M_{n0}, and can be represented as

$$m_P = \frac{1}{2}(1 - M\cos(\omega t + \varphi_m))$$

$$= M_{P0} + M_{Pd}\cos(\omega t) - M_{Pq}\sin(\omega t) = \left(\frac{1}{2}\right)_0 + \left(-\frac{M_d}{2}\right)_d \cos(\omega t) - \left(-\frac{M_q}{2}\right)_q \sin(\omega t) \tag{19.22}$$

$$m_N = \frac{1}{2}\left(1 + M\cos(\omega t + \varphi_m)\right)$$

$$= M_{N0} + M_{Nd}\cos(\omega t) - M_{Nq}\sin(\omega t) = \left(\frac{1}{2}\right)_0 + \left(\frac{M_d}{2}\right)_d \cos(\omega t) - \left(\frac{M_q}{2}\right)_q \sin(\omega t) \tag{19.23}$$

where subscripts $(\cdot)_0$, $(\cdot)_d$ and $(\cdot)_q$ denote 0, d and q components in the coordinate frame DQ rotating at fundamental frequency, and all notations are elaborated in Appendices A and B.

The AC grid current is assumed to be a clean *sine* signal at fundamental frequency:

$$i_g = I_g\cos(\omega t + \varphi_i) = I_{gd}\cos(\omega t) - I_{gq}\sin(\omega t) \tag{19.24}$$

whereas differential current (in each phase) consists of zero sequence and the second harmonic:

$$i_{diff} = I_{diff0} + I_{diff2}\cos\left(2\omega t + \varphi_{idiff2}\right) = I_{diff0} + I_{diffd2}\cos(2\omega t) - I_{diffq2}\sin(2\omega t) \tag{19.25}$$

where subscripts $(\cdot)_{d2}$ and $(\cdot)_{q2}$ denote d and q components in the coordinate $D2Q2$ frame rotating at twice the fundamental frequency (2ω). The MMC converter AC circuit modelling requires equations in three coordinate frames: zero sequence, the DQ frame at fundamental frequency and the $D2Q2$ frame at the second harmonic, since the coordinate frames are closely interconnected and each coordinate frame plays an important role in the power flow solution and the system dynamics.

The above signals will be replaced in the basic MMC equations in section 19.1 and each equation will result in several equations in the three different coordinate frames which should be separately analysed.

19.2.2   Zero Sequence Model

This section will analyse zero sequence MMC equations after all variables from section 19.2.1 are replaced in equations in section 19.1. The zero sequence DC-side equation can be derived from Eq. (19.4):

$$0 = 2R_{arm}I_{diff0} - V_{P0} - V_{N0} + V_{DC} \tag{19.26}$$

As the system is assumed to be symmetrical and balanced, the above equation becomes:

$$V_{P0} = V_{N0} = R_{arm}I_{diff0} + \frac{1}{2}V_{DC} \tag{19.27}$$

Using the definition for zero-sequence terms from Appendix B, the two zero-sequence AC voltage equations from Eqs (19.15) to (19.16) are derived as:

$$0 = \left(m_P i_{diff}\right)_0 + \left(\frac{m_P}{2}i_g\right)_0 \tag{19.28}$$

$$0 = \left(m_N i_{diff}\right)_0 - \left(\frac{m_N}{2}i_g\right)_0 \tag{19.29}$$

Similarly, zero-sequence expressions for Eqs (19.22) and (19.23) are derived:

$$\left(\frac{m_P i_g}{2}\right)_0 = -\frac{1}{8}M_d I_{gd} - \frac{1}{8}M_q I_{gq} \tag{19.30}$$

$$-\left(\frac{m_N i_g}{2}\right)_0 = -\frac{1}{8}M_d I_{gd} - \frac{1}{8}M_q I_{gq} \tag{19.31}$$

Using Eqs (19.28) and (19.30), it is possible to derive the equation for zero-sequence differential current:

$$I_{diff0} = \frac{1}{4}M_d I_{gd} + \frac{1}{4}M_q I_{gq} \tag{19.32}$$

The above model is valid for any phase, and since the DC current can be derived as the sum of three differential currents in Eq. (19.5), under the assumption of a symmetrical and balanced system it gives:

$$I_{dc} = 3I_{diff0} \tag{19.33}$$

It is evident that the above model in Eqs (19.32) and (19.33) is consistent with the two-level VSC converter model for DC current, previously derived in Eq. (17.7).

19.2.3 Fundamental Frequency Model in DQ Frame

This section will develop fundamental frequency MMC equations after all variables from section 19.2.1 are replaced in equations in Section 19.1. The main equation for the maximal arm voltage Eq. (19.15), can be represented in a DQ frame using the methods given in Appendix B. The d-component of this equation is:

$$\left(C_{arm}\frac{dv_P^M}{dt}\right)_d = \left(m_P i_{diff}\right)_d + \left(m_P \frac{i_g}{2}\right)_d \tag{19.34}$$

Using further assumptions for the signals m_p and i_g, and the rules for product of DQ signals, Eq. (19.34) gives:

$$C_{arm}\omega V_{Pq}^M = -\frac{M_d}{2}I_{diff0} - \frac{M_d}{4}I_{diffd2} - \frac{M_q}{4}I_{diffq2} + \frac{I_{gd}}{4} \tag{19.35}$$

Therefore, from the above equation, the expression for maximal positive pole arm q-component voltage is:

$$V_{Pq}^M = \frac{1}{C_{arm}\omega}\left(-\frac{M_d}{2}I_{diff0} - \frac{M_d}{4}I_{diffd2} - \frac{M_q}{4}I_{diffq2} + \frac{I_{acd}}{4}\right) \tag{19.36}$$

Similarly, the q component of Eq. (19.15) is:

$$\left(C_{arm}\frac{dv_P^M}{dt}\right)_q = \left(m_P i_{diff}\right)_q + \left(m_P \frac{i_g}{2}\right)_q \tag{19.37}$$

Using the same approach, Eq. (19.37) is expanded as:

$$-C_{arm}\omega V_{Pd}^M = -\frac{M_q}{2}I_{diff0} + \frac{M_q}{4}I_{diffd2} - \frac{M_d}{4}I_{diffq2} + \frac{I_{gq}}{4} \tag{19.38}$$

Therefore the expression for maximal positive pole-arm d-component voltage:

$$V_{Pd}^M = \frac{1}{C_{arm}\omega}\left(\frac{M_q}{2}I_{diff0} - \frac{M_q}{4}I_{diffd2} + \frac{M_d}{4}I_{diffq2} - \frac{I_{gq}}{4}\right)$$

(19.39)

The basic expression for zero sequence positive arm voltage is derived using the definition for zero sequence of a product from Appendix B:

$$V_{P0} = \left(m_P v_P^M\right)_0 = \left(M_{P0}V_{CP0}^M + \frac{M_{Pd}V_{Pd}^M}{2} + \frac{M_{Pq}V_{Pq}^M}{2}\right)_0$$

(19.40)

Finally, replacing Eqs (19.36), (19.39) and (19.22), (19.23) in Eq. (19.40), the expression for the zero sequence of the positive-arm voltage becomes:

$$V_{P0} = \frac{V_{P0}^M}{2} + \frac{1}{\omega C_{arm}}\left(-\frac{M_d}{4}\left[\frac{M_q I_{diff0}}{2} - \frac{M_q I_{diffd2}}{4} + \frac{M_d I_{diffq2}}{4} - \frac{I_{gq}}{4}\right] - \frac{M_q}{4}\left[-\frac{M_d I_{diff0}}{2} - \frac{M_d I_{diffd2}}{4} - \frac{M_q I_{diffq2}}{4} + \frac{I_{gd}}{4}\right]\right)$$

(19.41)

However, the goal is to get the zero sequence of the maximal arm voltage $V_{CP0}{}^M$ in the above equation, which is required for the further derivations. Therefore equating Eq. (19.41) with Eq. (19.27), the expression for the zero sequence of the maximal arm voltage is derived:

$$V_{P0}^M = 2R_{arm}I_{diff0} + V_{DC} - \frac{1}{\omega C_{arm}}\left(\frac{M_d I_{gq}}{8} - \frac{M_q I_{gd}}{8} + \frac{M_d M_q}{4}I_{diffd2} + \frac{-M_d^2 + M_q^2}{8}I_{diffq2}\right)$$

(19.42)

A similar approach is used when considering negative-pole fundamental frequency modelling and the following equations are derived:

$$V_{Nq}^M = \frac{1}{C_{arm}\omega}\left(\frac{M_d}{2}I_{diff0} + \frac{M_d}{4}I_{diffd2} + \frac{M_q}{4}I_{diffq2} - \frac{I_{gd}}{4}\right)$$

(19.43)

$$V_{Nd}^M = \frac{1}{C_{arm}\omega}\left(-\frac{M_q}{2}I_{diff0} + \frac{M_q}{4}I_{diffd2} - \frac{M_d}{4}I_{diffq2} + \frac{I_{gq}}{4}\right)$$

(19.44)

$$V_{N0}^M = 2R_{arm}I_{diff0} + V_{DC} - \frac{1}{\omega C_{arm}}\left(\frac{M_d I_{gq}}{8} - \frac{M_q I_{gd}}{8} + \frac{M_d M_q}{4}I_{diffd2} + \frac{-M_d^2 + M_q^2}{8}I_{diffq2}\right)$$

(19.45)

From the above derivations it is evident that Eqs (19.42) and (19.45) are identical. At the final stage, the equations for the actual positive pole arm voltages are obtained using Eq. (19.17):

$$V_{Pd} = \left(m_P v_P^M\right)_d$$

(19.46)

Replacing Eqs (19.22), (19.23) in Eq. (19.46), the d component of the positive pole-arm voltage is found:

$$V_{Pd} = -\frac{1}{2}M_d V_{P0}^M + \frac{1}{2}V_{Pd}^M - \frac{1}{4}M_d V_{Pd2}^M - \frac{1}{4}M_q V_{Pq2}^M$$

(19.47)

Similarly, it is possible to obtain the q component of the positive pole voltage:

$$V_{Pq} = -\frac{1}{2}M_q V_{P0}^M + \frac{1}{2}V_{Pq}^M + \frac{1}{4}M_q V_{Pd2}^M - \frac{1}{4}M_d V_{Pq2}^M \tag{19.48}$$

The negative pole voltages can also be derived in the same way:

$$V_{Nd} = \frac{1}{2}M_d V_{N0}^M + \frac{1}{2}V_{Nd}^M + \frac{1}{4}M_d V_{Nd2}^M + \frac{1}{4}M_q V_{Nq2}^M \tag{19.49}$$

$$V_{Nq} = \frac{1}{2}M_q V_{N0}^M + \frac{1}{2}V_{Nq}^M - \frac{1}{4}M_q V_{Nd2}^M + \frac{1}{4}M_d V_{Nq2}^M \tag{19.50}$$

Therefore the fundamental frequency phasor MMC model is given by Eqs (19.42), (19.45), (19.47)–(19.50). The model also depends on the zero-sequence variables and second harmonic variables and clearly there is cross coupling between the three coordinate frames.

19.2.4   Second Harmonic Model in D2Q2 Coordinate Frame

This section will analyse the D2Q2 variables in the MMC equations after all variables from section 19.2.1 are replaced in equations in section 19.1. Eq. (19.15) is firstly represented in the DQ frame rotating at 2ω, using the methods given in Appendix B. The d2-component of this equation is:

$$\left(C_{arm}\frac{dv_P^M}{dt}\right)_{d2} = \left(m_P i_{diff}\right)_{d2} + \left(m_P \frac{i_g}{2}\right)_{d2} \tag{19.51}$$

Using further assumptions for the signals m_p and i_g and the rules for product of dq signals:

$$2C_{arm}\omega V_{Pq2}^M = \frac{1}{2}I_{diffd2} - \frac{M_d I_{gd}}{8} + \frac{M_q I_{gq}}{8} \tag{19.52}$$

Therefore, from Eq. (19.52), the $q2$ component of the positive pole second harmonic maximal arm voltage is:

$$V_{Pq2}^M = \frac{1}{2C_{arm}\omega}\left(\frac{1}{2}I_{diffd2} - \frac{M_d I_{gd}}{8} + \frac{M_q I_{gq}}{8}\right) \tag{19.53}$$

Similarly, $q2$ component of Eq. (19.15) is:

$$\left(C_{arm}\frac{dv_P^M}{dt}\right)_{q2} = \left(m_P i_{diff}\right)_{q2} + \left(m_P \frac{i_g}{2}\right)_{q2} \tag{19.54}$$

Using the same approach, Eq. (19.54) is expanded as:

$$-2C_{arm}\omega V_{Pd2}^M = \frac{1}{2}I_{diffq2} - \frac{M_q I_{gd}}{8} - \frac{M_d I_{gq}}{8} \tag{19.55}$$

Therefore from Eq. (19.55) the $d2$ component of the positive pole second harmonic maximal arm voltage is:

$$V_{Pd2}^M = \frac{1}{2C_{arm}\omega}\left(-\frac{1}{2}I_{diffd2} + \frac{M_q I_{gd}}{8} + \frac{M_d I_{gq}}{8}\right) \tag{19.56}$$

The basic expression for the $d2$ component of the second harmonic arm voltage is obtained from Eq. (19.17):

$$(v_P)_{d2} = \left(m_P v_P^M\right)_{d2} \tag{19.57}$$

Replacing Eqs (19.36), (19.39), (19.56) and (19.22), (19.23) in Eq. (19.57):

$$V_{Pd2} = \frac{1}{\omega C_{arm}}\left(-\frac{M_d M_q}{4}I_{diff0} + \frac{3M_d I_{gq}}{32} + \frac{3M_q I_{gd}}{32} - \frac{1}{8}I_{diffq2}\right) \tag{19.58}$$

Similarly, the expression for $q2$ component of positive pole second harmonic is obtained from Eq. (19.18):

$$V_{Pq2} = \frac{1}{\omega C_{arm}}\left(+\frac{M_d^2 - M_q^2}{8}I_{diff0} + \frac{3M_q I_{gq}}{32} - \frac{3M_d I_{gd}}{32} + \frac{1}{8}I_{diffd2}\right) \tag{19.59}$$

Considering the negative pole, using the same approach, the following expressions are obtained:

$$V_{Nq2}^M = \frac{1}{2\omega C_{arm}}\left[\frac{1}{2}I_{diffd2} - \frac{M_d I_{gd}}{8} + \frac{M_q I_{gq}}{8}\right] \tag{19.60}$$

$$V_{Nd2}^M = \frac{1}{2\omega C_{arm}}\left[-\frac{1}{2}I_{diffq2} + \frac{M_q I_{gd}}{8} + \frac{M_d I_{gq}}{8}\right] \tag{19.61}$$

$$V_{Nd2} = \frac{1}{\omega C_{arm}}\left(-\frac{M_d M_q}{4}I_{diff0} + \frac{3M_d I_{gq}}{32} + \frac{3M_q I_{gd}}{32} - \frac{1}{8}I_{diffq2}\right) \tag{19.62}$$

$$V_{Nq2} = \frac{1}{\omega C_{arm}}\left(+\frac{M_d^2 - M_q^2}{8}I_{diff0} + \frac{3M_q I_{gq}}{32} - \frac{3M_d I_{gd}}{32} + \frac{1}{8}I_{diffd2}\right) \tag{19.63}$$

19.3 Differential Current at Second Harmonic

This section examines the second harmonic differential current I_{diffd2} and I_{diffq2} as this variable appears in a number of equations in the fundamental and second harmonic model. The $d2$ component of the dynamic equation for the differential current Eq. (19.4) is represented in the $D2Q2$ frame as:

$$\left(2\omega L_{arm}I_{diffq2}\right)_{d2} = \left(-R_{arm}I_{diffd2}\right)_d - \frac{1}{2}(V_{CP} + V_{CN})_{d2} \tag{19.64}$$

The above equation directly gives the $q2$ component of the second harmonic differential current:

$$I_{diffq2}2\omega L_{arm} = -R_{arm}I_{diffd2} - \frac{1}{2}V_{CPd2} - \frac{1}{2}V_{CNd2} \qquad (19.65)$$

Similarly, the $d2$ component is obtained from Eq. (19.4) as:

$$\left(-2\omega L_{arm}I_{diffd2}\right)_{q2} = \left(-R_{arm}I_{diffq2}\right)_q - \frac{1}{2}\left(V_{CP}+V_{CN}\right)_{q2} \qquad (19.66)$$

$$I_{diffd2}2\omega L_{arm} = R_{arm}I_{diffq2} + \frac{1}{2}V_{CPq2} + \frac{1}{2}V_{CNq2} \qquad (19.67)$$

Therefore the system model at the second harmonic is given by: Eqs (19.58), (19.59), (19.62), (19.63), (19.65) and (19.67). This model is interlinked with the fundamental frequency model as many fundamental frequency variables are involved in Eqs (19.58), (19.59), (19.62) and (19.63).

19.4   Complete MMC Converter DQ Model in Matrix Form

Considering the above modelling, the following observations can be made:

$$V_{Pd}^M = -V_{Nd}^M, \quad V_{Pq}^M = -V_{Nq}^M$$
$$V_{Pd2}^M = V_{Nd2}^M, \quad V_{Pq2}^M = V_{Nq2}^M \qquad (19.68)$$

Therefore several equations can be eliminated to avoid duplication. The complete MMC DQ static model is expressed in matrix form as:

$$Ax = Bu$$
$$y = Cx \qquad (19.69)$$

where the model states are:

$$x^T = \begin{bmatrix} V_{Pd2}^M & V_{Pq2}^M & V_{Pd}^M & V_{Pq}^M & V_{P0}^M & | & I_{diff0} & I_{diffd2} & I_{diffq2} \end{bmatrix} \qquad (19.70)$$

The model inputs are the AC current components from the AC side and DC voltage from the DC side:

$$u^T = \begin{bmatrix} I_{gd} & I_{gq} & V_{dc} \end{bmatrix} \qquad (19.71)$$

The outputs for the basic power-flow study are DC current and the converter AC voltages:

$$y^T = \begin{bmatrix} I_{dc} & V_{ced} & V_{ceq} \end{bmatrix} \qquad (19.72)$$

The matrix A is derived as:

$$\mathbf{A} = \begin{bmatrix} A_{11} & A_{12} \\ A_{21} & A_{22} \end{bmatrix}, \mathbf{A}_{11} = \mathbf{I}_{5\times 5}, \mathbf{A}_{22} = \frac{1}{2\omega L_{arm}} \begin{bmatrix} 0 & 0 & 0 \\ 0 & 0 & R_{arm} \\ 0 & -R_{arm} & 0 \end{bmatrix}$$

$$\mathbf{A}_{12} = \frac{1}{16\omega C_{arm}} \begin{bmatrix} 0 & 0 & -4 \\ 0 & 4 & 0 \\ 8M_q & -4M_q & 4M_d \\ -8M_d & -4M_d & -4M_q \\ -32R_{arm}\omega C_{arm} & 4M_d M_q & -M_d^2 + M_q^2 \end{bmatrix} \tag{19.73}$$

$$\mathbf{A}_{21} = \frac{1}{8\omega L_{arm}} \begin{bmatrix} 0 & 0 & 0 & 0 & 0 \\ 0 & 2 & -M_q & -M_d & 0 \\ -2 & 0 & M_d & -M_q & 0 \end{bmatrix}$$

while matrices B and C are:

$$\mathbf{B}^T = \frac{1}{16\omega C_{arm}} \begin{bmatrix} M_q & -M_d & 0 & 4 & 2M_q & 4M_d & 0 & 0 \\ M_d & M_q & -4 & 0 & -2M_d & 4M_q & 0 & 0 \\ 0 & 0 & 0 & 0 & 0 & 16\omega C_{arm} & 0 & 0 \end{bmatrix} \tag{19.74}$$

$$\mathbf{C} = \frac{1}{4} \begin{bmatrix} M_d & M_q & -2 & 0 & 2M_d & 0 & 0 & 0 \\ M_q & M_d & 0 & -2 & 2M_q & 0 & 0 & 0 \\ 0 & 0 & 0 & 0 & 0 & 3 & 0 & 0 \end{bmatrix} \tag{19.75}$$

The above model enables the accurate calculation of fundamental component MMC voltages and second harmonics, for particular control signals (M_d and M_q) and for given inputs: grid current and DC voltage.

19.5　Second Harmonic Circulating Current Suppression Controller

The second harmonic circulating current in each arm of MMC is quite large and can cause problems. It has been shown in Figure 16.8 that the second harmonic in the three phases will cancel when summed both on the AC and the DC side, under balanced conditions. However a large second-harmonic current will still be circulating in the converter arm components, which will cause higher peak voltages and currents and lead to increased losses and heating.

Figure 19.2 shows the peak amplitude of the circulating second harmonic versus arm inductance value for a typical 1 GW (640 kV, 1500 A) MMC converter. It is seen that the arm inductance affects the circulating current directly but, even with very large arm inductors, the circulating current is still significant. Increasing arm inductance is not an effective method of reducing second harmonics, while large arm inductance has other negative consequences for converter rating and control.

The differential current second harmonic can be eliminated using a control signal injection at the second harmonic. This can be confirmed by considering Eq. (19.21) and assuming that the

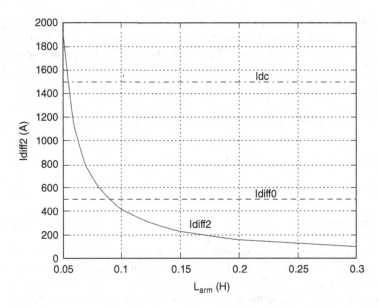

Figure 19.2 Second-harmonic circulating current versus arm inductance.

control signal includes a second harmonic of magnitude M_2, with d and q components given by M_{d2} and M_{q2}:

$$m = M_1\cos(\omega t + \varphi_m) + M_2\cos(2\omega t + \varphi_{m2}) = M_{d1}\cos(\omega t) - M_{q1}\sin(\omega t) + M_{d2}\cos(2\omega t) + M_{q2}\sin(2\omega t)$$

$$m_P = \frac{1}{2}(1-m) = \left(\frac{1}{2}\right)_0 + \left(-\frac{M_{d1}}{2}\right)_d\cos(\omega t) - \left(-\frac{M_{q1}}{2}\right)_q\sin(\omega t) + \left(-\frac{M_{d2}}{2}\right)_{d2}\cos(2\omega t)$$

$$- \left(-\frac{M_{q2}}{2}\right)_{q2}\sin(2\omega t)$$

$$m_N = \frac{1}{2}(1+m) = \left(\frac{1}{2}\right)_0 + \left(\frac{M_{d1}}{2}\right)_d\cos(\omega t) - \left(\frac{M_{q1}}{2}\right)_q\sin(\omega t) + \left(\frac{M_{d2}}{2}\right)_{d2}\cos(2\omega t) - \left(\frac{M_{q2}}{2}\right)_{q2}\sin(2\omega t)$$

$$(19.76)$$

Replacing Eq. (19.76) into equations for maximal arm voltages Eqs. (19.53) and (19.56):

$$V_{Pq2}^M = \frac{1}{2\omega C_{arm}}\left[\frac{1}{2}I_{diffd2} - \frac{M_{d2}I_{diff0}}{2} - \frac{M_{d1}I_{gd}}{8} + \frac{M_{q1}I_{gq}}{8}\right]$$

$$V_{Pd2}^M = \frac{1}{2\omega C_{arm}}\left[-\frac{1}{2}I_{diffq2} + \frac{M_{q2}I_{diff0}}{2} + \frac{M_{q1}I_{gd}}{8} + \frac{M_{d1}I_{gq}}{8}\right]$$

$$(19.77)$$

The above equations show that the second harmonic on the control signal (M_{d2} and M_{q2}) will generate a second harmonic on the arm voltages, which confirms steady-state controllability; this method can therefore be used for reducing the second-harmonic circulating current. The control gain with this method is a zero-sequence differential current I_{diff0} (the term that multiplies M_{d2} and M_{q2} in Eq. (19.77)), which has significant values. It is good that this control approach has a large gain, as small control inputs can be very effective in reducing circulating current harmonics.

Following the above derivation for the arm voltages on both poles and the circulating current, and equating the circulating current second harmonic with zero ($I_{diffd2} = 0$, $I_{diffq2} = 0$), it is possible to obtain the required magnitude of control signals M_{d2} and M_{q2} to eliminate the second harmonic current:

$$\begin{bmatrix} M_{d2} \\ M_{q2} \end{bmatrix} = \frac{1}{Z} \begin{bmatrix} (3-2M_d^2)M_q & (3-2M_q^2)M_d \\ (M_d^2-M_q^2-3)M_d & (M_d^2-M_q^2+3)M_q \end{bmatrix} \begin{bmatrix} I_{gd} \\ I_{gq} \end{bmatrix} \quad I_{diffd2}=I_{diffq2}=0 \qquad (19.78)$$

where:

$$Z = 8\omega C_{arm} \left(2V_{DC} - R_{arm}(M_d I_{gd} + M_q I_{gq})\right) - 3\left(M_d I_{gq} - M_q I_{gd}\right) \qquad (19.79)$$

The above expressions for the second harmonic control signals can be used to develop a feedforward (open-loop) second-harmonic controller.

An alternative control method to eliminate second-harmonic circulating current – and this is more popular in practice – is to use a feedback controller. The circulating-current suppression controller (CCSC) is shown in Figure 19.3. The signal φ_{PLL} is the same 50 Hz coordinate frame reference used for the main control loops. This signal is multiplied by −2 to get a reference in the negative sequence coordinate frame rotating at 2ω. The differential currents are measured on each phase and converted to $d2$ and $q2$ components, which are processed separately through a proportional integral (PI) controller at each channel. The outputs are converted to static frame (m_{a2}, m_{b2}, m_{c2}) and added to the control signals from the main control loops at the fundamental frequency (m_{a1}, m_{b1}, m_{c1}).

The impact of activation of differential current suppression control is shown in Figure 19.4. The system is operating rated at 1 GW power in an open loop with $M_{d1} = 0.95$ and $M_{q1} = -0.03$, and at 2.0 s the CCSC is activated. Before activation of CCSC the second harmonic is almost half the DC current, which is very large. The CCSC can completely eliminate the second harmonic current in a short time, and the control effort (magnitude of M_{d2} and M_{q2}) is small and typically below 5% relative to the main control signals (M_{d1} and M_{q1}).

However, it should be noticed that the power flow changes significantly after the activation of the CCSC (as seen by reduced I_{gd} and I_{gq}), despite the fact that M_{d1} and M_{q1} are unchanged. This is a result

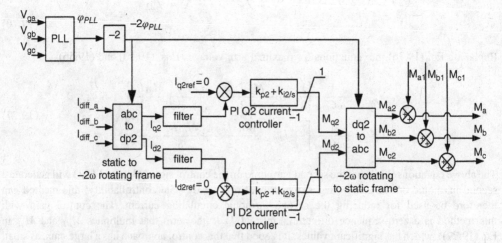

Figure 19.3 Second-harmonic circulating current suppression controller (CCSC).

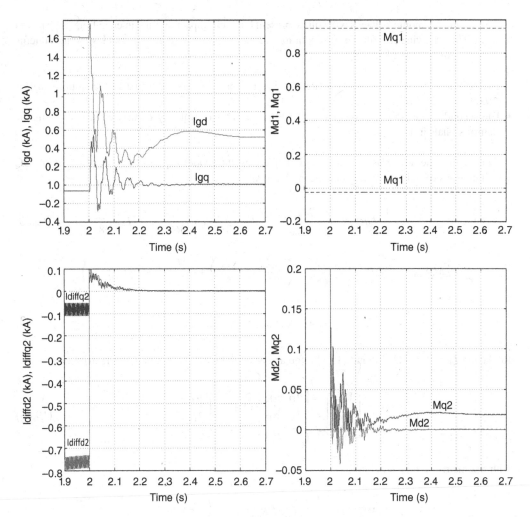

Figure 19.4   Activation of second-harmonic circulating current suppression controller at 2 s.

of the impact of second harmonic variables on the fundamental frequency power flow, which is seen in the Eqs (19.42), (19.47) and others. Second harmonic currents actually increase power flow at fundamental frequency. As the result of this interaction, M_{d1} and M_{q1} should be increased in order to return to the required power level after CCSC activation.

A dynamic interaction between control loops at the first and second harmonics is also present. This is a consequence of feedback control dynamics for eliminating the second harmonic circulating current.

19.6   DQ Frame Model of MMC with Circulating Current Controller

As the CCSC can completely eliminate second-harmonic currents, it is possible to delete the last two equations in the model Eq. (19.69). Consequently, the second harmonic currents are deleted from zero-sequence voltages in Eqs (19.42) and (19.45) and this will have an effect on the reduction in the power flow.

The second-harmonic voltages will still be present but their impact on fundamental frequency power is small. The following two further assumptions will significantly simplify the model while introducing a modest error (below 5%):

- The second-harmonic arm voltages are neglected. This eliminates the first two equations in Eqs (19.69) and (19.70).
- The second-harmonic control signals are neglected. This completely eliminates second-harmonic control signals in Eq. (19.77).

Considering the above assumptions and simplifying the model Eq. (19.69), the MMC static DQ which consists of only three equations, can be derived:

$$V_{ced} = \frac{1}{2}M_d V_{DC} + \frac{1}{3}M_d R_{arm} I_{dc} + \frac{M_d}{\omega C_{arm}}\left(\frac{M_d I_{gq}}{16} - \frac{M_q I_{gd}}{16}\right) + \frac{M_d}{\omega C_{arm}}\left(\frac{M_q}{12}I_{dc} - \frac{I_{gd}}{8}\right) \tag{19.80}$$

$$V_{ceq} = \frac{1}{2}M_q V_{DC} + \frac{1}{3}M_q R_{arm} I_{dc} + \frac{M_q}{\omega C_{arm}}\left(\frac{M_d I_{gq}}{16} - \frac{M_q I_{gd}}{16}\right) + \frac{M_q}{\omega C_{arm}}\left(-\frac{M_d}{12}I_{dc} + \frac{I_{gd}}{8}\right) \tag{19.81}$$

$$I_{dc} = \frac{3}{2}M_d I_{gd} + \frac{3}{2}M_q I_{gq} \tag{19.82}$$

Replacing I_{dc} in Eqs (19.80) and (19.81), the above model can be represented in matrix form as:

$$\begin{bmatrix} V_{ced} \\ V_{ceq} \\ I_{dc} \end{bmatrix} = \begin{bmatrix} a_{11} & a_{12} \\ a_{21} & a_{22} \\ a_{31} & a_{32} \end{bmatrix}\begin{bmatrix} I_{gd} \\ I_{gq} \end{bmatrix} + \begin{bmatrix} b_d \\ b_q \\ b_{dc} \end{bmatrix}V_{dc} \tag{19.83}$$

with the matrix elements defined as:

$$a_{11} = \frac{M_d^2 R_{arm}}{4}, \quad a_{12} = \frac{M_d M_q R_{arm}}{4} - \frac{8 - 3\left(M_d^2 + M_q^2\right)}{64\omega C_{arm}},$$

$$a_{21} = \frac{M_d M_q R_{arm}}{4} + \frac{8 - 3\left(M_d^2 + M_q^2\right)}{64\omega C_{arm}}, \quad a_{22} = \frac{M_q^2 R_{arm}}{4}, \tag{19.84}$$

$$a_{31} = \frac{3}{2}M_d, \quad a_{32} = \frac{3}{2}M_q, \quad b_d = \frac{1}{2}M_d, \quad b_q = \frac{1}{2}M_q, \quad b_{dc} = 0,$$

The following further assumptions can be introduced in order to simplify the model:

$$\frac{M_d M_q R_{arm}}{4} << \frac{8 - 3\left(M_d^2 + M_q^2\right)}{64\omega C_{arm}}$$

$$\frac{M_d^2 R_{arm}}{4} = 0 \tag{19.85}$$

$$\frac{M_q^2 R_{arm}}{4} = 0$$

In the final form, the DQ frame static model of MMC with circulating current control is:

$$V_{ced} = -\frac{1}{\omega C_{MMC}}I_{gq} + \frac{1}{2}M_d V_{dc}$$

$$V_{ceq} = \frac{1}{\omega C_{MMC}}I_{gd} + \frac{1}{2}M_q V_{dc} \tag{19.86}$$

$$I_{dc} = \frac{3}{2}M_d I_{gd} + \frac{3}{2}M_q I_{gq}$$

where the equivalent MMC capacitance C_{MMC} is:

$$C_{MMC} = \frac{64 C_{arm}}{8 - 3\left(M_d^2 + M_q^2\right)} \tag{19.87}$$

In the above model, it is observed that the MMC AC voltage has two components:

- A two-level VSC model voltage represented by $1/2 MV_{dc}$.
- A component that shows dependence on the converter currents, given by gain $1/(\omega C_{MMC})$. By inspection, this component resembles an equivalent capacitor and hence C_{MMC} label.

19.7 Phasor Model of MMC with Circulating Current Suppression Controller

The DQ frame MMC model in Eq. (19.86) can be converted into complex number vector equation to obtain the phasor MMC model:

$$\overline{V_{ce}} = -jX_{MMC}\overline{I_g} + \frac{1}{2}\overline{M}V_{dc} \tag{19.88}$$

where the MMC reactance X_{MMC} is:

$$X_{MMC} = \frac{1}{\omega C_{MMC}}, \tag{19.89}$$

The MMC AC-side model therefore includes internal regulation, which will cause AC voltage variation depending on the loading and the equivalent capacitance, C_{MMC}. The equivalent capacitance predominantly depends on the arm capacitance but it is also a function of control signals. Considering the practical range for the control signals $0 < (M_d^2 + M_q^2)^{0.5} < 1$, the equivalent capacitance range is:

$$\frac{64 C_{arm}}{5} < C_{MMC} < \frac{64 C_{arm}}{8} \tag{19.90}$$

In most operating conditions the control signal magnitude will be just below one, and therefore it is possible to use a constant $C_{MMC} = 64 C_{arm}/5$ in power-flow calculations without significant loss in accuracy.

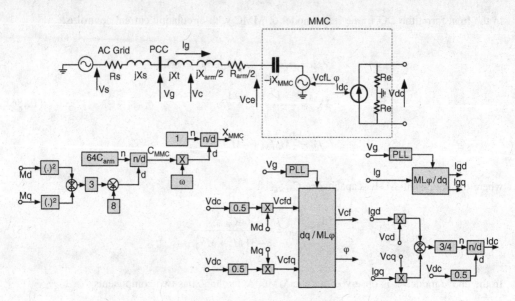

Figure 19.5   MMC phasor model using equivalent series capacitor C_{MMC}.

Figure 19.5 shows the MMC phasor model schematic. It is in a simple form, which is suitable for fundamental frequency studies with symmetrical and balanced systems. The large resistors R_e are placed only to provide middle DC ground-point reference.

19.8   Dynamic MMC Model Using Equivalent Series Capacitor C_{MMC}

The above MMC model in Eq. (19.86) can be converted to the dynamic model by writing the dynamic equation for the MMC equivalent series capacitor. The equivalent model is shown in Figure 19.6, and can be represented in state-space nonlinear format in a static ABC frame as:

$$\frac{d}{dt}v_{CMMC} = \frac{1}{C_{MMC}}i_g$$

$$v_{ce} = v_{CMMC} + \frac{1}{2}mV_{dc} \tag{19.91}$$

$$I_{dc} = \left(v_{cea}i_{ga} + v_{ceb}i_{gb} + v_{ceca}i_{gc}\right)/V_{dc}$$

The diodes are not relevant for steady-state operation. They are included to provide a direct current path for DC faults, in which case it is assumed that V_{ce} is disabled (IGBTs tripped). This model is not as accurate as the average non-linear model given in Figure 19.1, and a more accurate dynamic model is derived below in section 19.9.

However, this model has the following advantages:

- It is considerably simpler, and assuming that C_{MMC} is constant it will be in a linear form.
- The model can be represented in DQ rotating frame, which is very convenient for linearization.

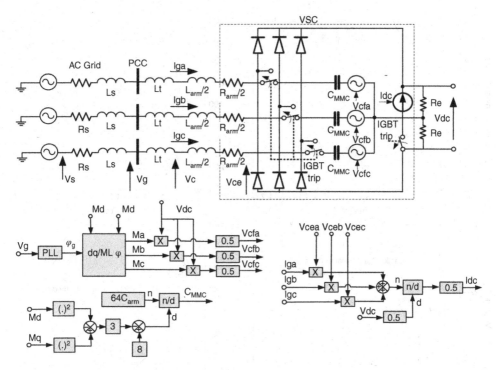

Figure 19.6   MMC average dynamic model using equivalent series capacitor C_{MMC}.

Table 19.1   MMC converter test-system data and comparison with two-level VSC.

	Two-level VSC	MMC VSC
Rated S_g	1000 MVA	1000 MVA
Rated converter current I_g *(RMS)*	1520 A	1520 A
Rated PCC AC voltage V_{g_ll}	380 kV	380 kV
Rated converter AC voltage V_c	395 kV	395 kV
Rated DC voltage V_{dc}	644 kV	633 kV
Arm inductor	–	$X_{arm} = 0.14$ pu
Energy to power ratio E_s	–	$E_s = 30$ kJ/MW
Transformer	$X_{tpu} = 0.2$ pu	$X_{tpu} = 0.13$ pu
AC grid	SCR = 2, X/R = 10, $V_s = 390$ kV	

- The model represents the MMC with a second harmonic suppression control. If the model in Figure 19.1 is used, then the second harmonic suppression control loops must be included.

Figure 19.7 shows the verification of the above two models (phasor and dynamic) against the nonlinear ABC frame MMC model from Figure 19.1 for a typical 1 GW MMC system with the test system data given in Table 19.1. Open-loop operation is assumed (steps are applied on M_d and M_q).

At 1 s, a step is applied on M_q, which raises the active power close to the rated value. The second step is applied on M_d at 1.4 s, which changes the reactive power. The label 'detailed' refers to the

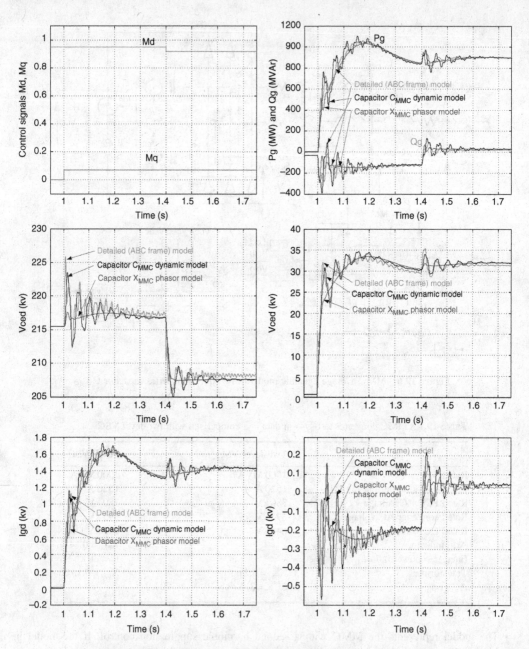

Figure 19.7 Verification of MMC model with series capacitor C_{MMC}.

model in Figure 19.1; the label 'capacitor C_{MMC} dynamic' is the model from Figure 19.6, while the label 'capacitor X_{MMC} phasor' is same as in Figure 19.6 but capacitor is represented using a reactance X_{MMC} as in Eq. (19.88) (no capacitor dynamics). It is seen that both models represent very well the MMC steady-state responses and therefore they are suitable for power flow study. The phasor model ignores system dynamics and dynamic accuracy is poor, as expected. The model with dynamic series capacitance gives a reasonable indication of the dominant oscillatory mode, which is around 25 Hz in this test system. This model can therefore be employed for the initial screening of MMC dynamic instabilities.

19.9 Full Dynamic Analytical MMC Model

A complete dynamic model for MMC with CCSC can be derived using initial dynamic Eqs (19.4), (19.12) and (19.13) and following the rules for signal multiplications in DQ frames. The second harmonic controller dynamics must also be included, as in Figure 19.3, because M_{d2} and M_{q2} affect the dynamics at fundamental frequency. The complete MMC dynamic nonlinear model is of the 10th order, and includes three coordinate frames as given by the following final expression in state space:

$$
\frac{d}{dt}
\begin{bmatrix}
V_{Pd2}^M \\
V_{Pq2}^M \\
V_{Pd}^M \\
V_{Pq}^M \\
V_{P0}^M \\
I_{diff0} \\
I_{diffd2} \\
I_{diffq2} \\
M_{d2} \\
M_{q2}
\end{bmatrix}
=
\begin{bmatrix}
\begin{matrix}
0 & 2\omega & 0 & 0 & 0 & 0 \\
-2\omega & 0 & 0 & 0 & 0 & 0 \\
0 & 0 & 0 & \omega & 0 & 0 \\
0 & 0 & -\omega & 0 & 0 & 0 \\
0 & 0 & 0 & 0 & 0 & 0 \\
0 & 0 & 0 & 0 & 0 & -\dfrac{R_{arm}}{L_{arm}}
\end{matrix} & 0_{6\times4} \\[2em]
0_{4\times6} &
\begin{matrix}
-\dfrac{R_{arm}}{L_{arm}} & 2\omega & 0 & 0 \\
-2\omega & -\dfrac{R_{arm}}{L_{arm}} & 0 & 0 \\
k_{I2}-k_{P2}\dfrac{R_{arm}}{L_{arm}} & 2k_{P2}\omega & 0 & 0 \\
-2k_{P2}\omega & k_{I2}-k_{P2}\dfrac{R_{arm}}{L_{arm}} & 0 & 0
\end{matrix}
\end{bmatrix}
\begin{bmatrix}
V_{Pd2}^M \\
V_{Pq2}^M \\
V_{Pd}^M \\
V_{Pq}^M \\
V_{P0}^M \\
I_{diff0} \\
I_{diffd2} \\
I_{diffd2} \\
M_{d2} \\
M_{q2}
\end{bmatrix}
$$

$$
+
\begin{bmatrix}
\left(M_{d1}I_{gd}-M_{q1}I_{vq}-4I_{diffd2}+4M_{d2}I_{diff0}\right)/8C_{arm} \\[0.8em]
\left(M_{q1}I_{gd}+M_{d1}I_{gq}-4I_{diffq2}+4M_{q2}I_{diff0}\right)/8C_{arm} \\[0.8em]
\left(2M_{d1}I_{diff0}+M_{d1}I_{diffd2}+M_{q1}I_{diffQ2}-I_{gd}+\dfrac{1}{2}M_{d2}I_{gd}+\dfrac{1}{2}M_{q2}I_{gq}\right)/4C_{arm} \\[0.8em]
\left(2M_{q1}I_{diff0}-M_{q1}I_{diffd2}+M_{d1}I_{diffQ2}-I_{gq}-\dfrac{1}{2}M_{d2}I_{gq}+\dfrac{1}{2}M_{q2}I_{gd}\right)/4C_{arm} \\[0.8em]
\left(M_{d1}I_{gd}+M_{q1}I_{gq}-4I_{diff0}-2M_{d2}I_{diffd2}-2M_{q2}I_{diffq2}\right)/8C_{arm} \\[0.8em]
\left(-2V_{DC}-M_{d1}V_{Pd}^M-M_{q1}V_{Pq}^M+2V_{P0}^M-M_{d2}V_{Pd2}^M-M_{q2}V_{Pq2}^M\right)/4L_{arm} \\[0.8em]
\left(-M_{d1}V_{Pd}^M+M_{q1}V_{Pq}^M+2V_{Pd2}^M-2M_{d2}V_{P0}^M\right)/4L_{arm} \\[0.8em]
\left(-M_{d1}V_{Pq}^M-M_{q1}V_{Pq}^M+2V_{Pq2}^M-2M_{q2}V_{P0}^M\right)/4L_{arm} \\[0.8em]
k_{P2}\left(-M_{d1}V_{Pd}^M+M_{q1}V_{Pq}^M+2V_{Pd2}^M-2M_{d2}V_{P0}^M\right)/4L_{arm} \\[0.8em]
k_{P2}\left(-M_{d1}V_{Pq}^M-M_{q1}V_{Pq}^M+2V_{Pq2}^M-2M_{q2}V_{P0}^M\right)/4L_{arm}
\end{bmatrix}
$$

$$(19.92)$$

while the output nonlinear equations are:

$$\begin{bmatrix} V_{ced} \\ V_{ceq} \end{bmatrix} = \frac{1}{4} \begin{bmatrix} 2M_{d1}V_{P0}^M + M_{d1}V_{Pd2}^M + M_{q1}V_{Pq2}^M - 2V_{Pd}^M + \frac{1}{2}M_{d2}V_{Pd}^M + \frac{1}{2}M_{Q2}V_{Pq}^M \\ 2M_{q1}V_{P0}^M + M_{d1}V_{Pq2}^M - M_{q1}V_{Pd2}^M - 2V_{Pq}^M + \frac{1}{2}M_{q2}V_{Pd}^M - \frac{1}{2}M_{d2}V_{Pq}^M \end{bmatrix} \tag{19.93}$$

$$I_{DC} = -3I_{diff0}$$

At the final step the nonlinear multiplication terms are linearized around the operating point of interest in the usual manner. The final model is in linear form represented in state-space in the DQ frame, which is convenient for eigenvalue studies and controller design.

Figure 19.8 shows verification of the 10th order dynamic linearized MMC model, for the same 1 GW, 640 kV test system as in previous tests but at a different operating point. The model shows excellent accuracy for all variables. Comparing with 2nd order model using C_{MMC} in Figure 19.7, it is concluded that the 10th-order model is more accurate and should be used for designing inner fast-control loops.

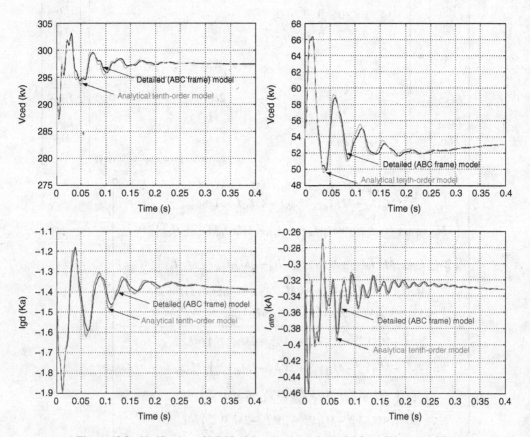

Figure 19.8 Verification of MMC 10th order dynamic model for a 5% step on M_q.

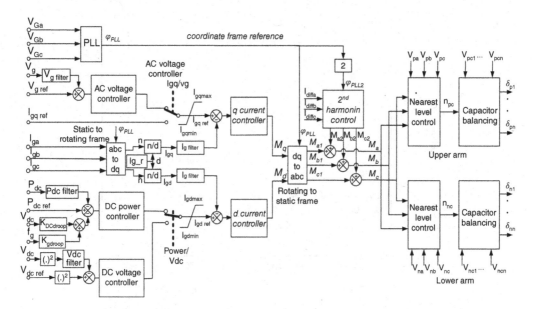

Figure 19.9   MMC converter controller structure.

19.10   MMC Converter Controller

Figure 19.9 shows the main MMC VSC converter controller structure. The upper level controls are same as with two-level VSC converter and they have been discussed in Figure 17.11. The second-harmonic circulating current suppression is studied in Figure 19.3. The nearest level controller is explained in section 16.6.6, while capacitor balancing is discussed in section 16.6.2.

Only the control loops that dominate MMC dynamics are shown in the above figure, while a practical MMC controller will involve many more control loops, like arm, phase and pole balancing, energy-charge control and AC-waveform tracking. The total number of control loops in a single MMC converter may approach 20.

19.11   MMC Total Series Reactance in the Phasor Model

This section employs the MMC phasor model from Figure 19.5 to study the interfacing reactance and make comparison with a two-level VSC converter. The total series reactance on the converter side of point of common coupling (PCC) for a MMC converter is:

$$X_{MMCt} = X_t + X_{arm}/2 - X_{MMC} \tag{19.94}$$

Under DC fault conditions, X_{MMC} will be bypassed by the diodes and the total reactance on converter-side of PCC will be $X_t + X_{arm}/2$, as shown in Figure 19.6. In the converter design stage the DC fault reactance is very important because it determines the rating of the converter diodes.

It is of interest to compare the MMC converter responses with equivalent two-level VSC. A two-level VSC converter has X_t (which includes the series reactor) on the converter side of the PCC. It will be

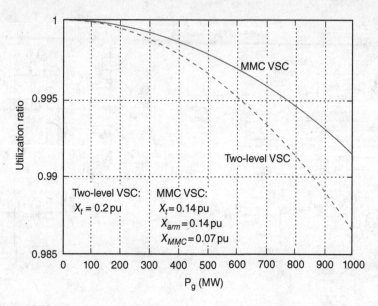

Figure 19.10 Comparison of MMC VSC and two-level VSC converter utilization ratio.

assumed in the study below that MMC $X_t + X_{arm}/2$ will have a similar value as the transformer reactance X_t in a traditional two-level VSC. For the same fault reactance, therefore, the total MMC reactance in normal operation, as given in Eq. (19.94), will be lower than with the equivalent two-level VSC. This brings advantages to the MMC converter, since lower interfacing reactance implies that converter rating will be lower and there will be less voltage swing at PCC. Figure 19.10 compares the converter utilization ratio (power at PCC divided by converter power) for two-level VSC and MMC VSC, assuming that DC fault reactance is same. Because of the larger interface reactance, the two-level VSC will have somewhat less favourable power factor at the converter terminals, which implies marginally larger converter current and possibly also larger DC voltage.

In the next step, the expressions for per unit values for MMC reactance X_{MMCpu} and the arm reactance X_{armpu} are studied. Considering Eq. (19.89), the pu MMC reactance is:

$$X_{MMCpu} = \frac{X_{MMC}}{X_{base}} = \frac{S_{MMC}}{\omega C_{MMC} V_{g\_ll}^2} \tag{19.95}$$

where S_{MMC} is the converter rating and V_{gll} is the line-line PCC AC voltage referred to the converter side. Replacing Eqs (19.87) and (16.7) in Eq. (19.95):

$$X_{MMCpu} = \frac{15 V_{cell}^2 n^2}{\omega 64 E_s V_{g\_ll}^2} \tag{19.96}$$

where energy to power ratio E_s is defined in section 16.6.4. Considering that the number of cell is approximately $n = V_{dc}/V_{cell}$, and replacing the equation for converter AC voltage in Eq. (19.96):

$$X_{MMCpu} = \frac{5}{8\omega E_s} \tag{19.97}$$

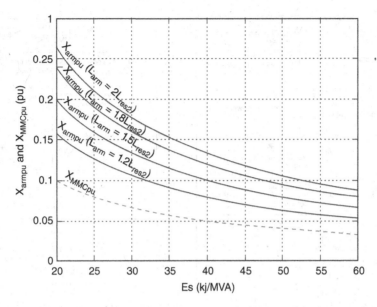

Figure 19.11   Arm reactance X_{armpu} and MMC capacitive reactance X_{MMCpu} as the function of energy to power converter ratio E_s.

The pu MMC capacitive reactance therefore depends only on the energy-to-power ratio, and it should be similar for most converters regardless of DC voltage or power rating.

The arm inductance for harmonic resonance is given in Eq. (16.9), which can be simplified assuming the most important second harmonic, $h = 2$, and maximum modulation index $m_a = 1$, as:

$$L_{arm\_res2} = \frac{1}{C_{arm}\omega^2}\frac{10}{96} \tag{19.98}$$

To avoid resonance, the arm inductance is selected to be higher than the above resonant value. Assuming that a 50% higher inductance is chosen, replacing Eqs (19.98) and (19.87) into Eq. (19.98), the arm reactance expressed in pu is:

$$X_{armpu} = \frac{1}{X_{base}}\omega\frac{3}{2}L_{arm\_res2} = 2X_{MMCpu} \tag{19.99}$$

This result shows that, under the above assumptions, MMC equivalent capacitance will exactly cancel half the arm inductance in Eq. (19.94), and the total MMC reactance (on the converter side) will consist of solely transformer reactance $X_{MMCt} \approx X_t$.

Figure 19.11 shows the pu values for arm reactance and for MMC series capacitive reactance. Under the above assumptions, these values should equally apply to MMC converters of any rating.

19.12   MMC VSC Interaction with AC System and PQ Operating Diagrams

The interaction with the AC system and operating diagrams can be obtained in similar manner as with the two-level VSC studied in section 18.3. Essentially, MMC VSC will behave similarly to a two-level

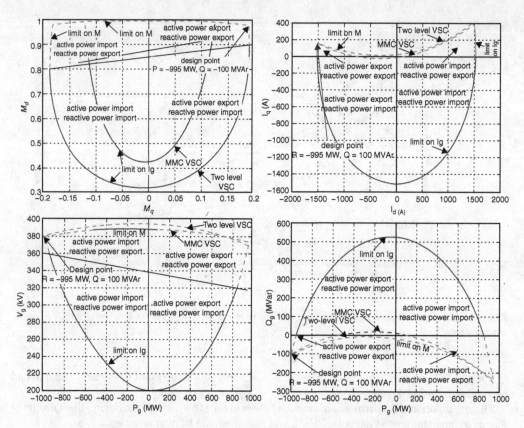

Figure 19.12   MMC VSC PQ diagram and comparison with two-level VSC (SCR = 2, X/R = 10).

VSC with reduced transformer reactance. However, the MMC equivalent series reactance also depends on the control signals and loading, which should be considered for accurate studies.

Table 19.1 shows the data for the grid interaction study, also considering a two-level VSC for comparison. As no transformer is used (a 1 : 1 transformer with leakage reactance X_t is considered for simplicity), it is seen that for the same PCC power, a two-level VSC will require slightly larger DC voltage because of the larger interface reactance. A weak AC system (*short circuit ratio*, *SCR* = 2) is assumed to better illustrate the nature of interactions.

Figure 19.12 shows the PQ diagrams for a MMC and compared with a similar two-level VSC. The MMC PQ diagrams resemble two-level VSC diagrams, except that the MMC requires less control effort and the regulation of converter voltage is lower.

20

VSC HVDC under AC and DC Fault Conditions

20.1 Introduction

This chapter will examine high-voltage direct-current (HVDC) system responses to AC and DC faults from a theoretical perspective and using simulation models. The test system is 1000 MW, 640 kV HVDC, as in previous chapters, but the AC systems are strengthened (short circuit ratio (SCR) = 30, X/R = 10) in order to illustrate better the issues with DC faults. The analysis will consider all three HVDC topologies: two-level voltage source converter (VSC), half-bridge modular multilevel converter (MMC) and full-bridge MMC.

The test system is shown in Figure 20.1, where the location of faults is also illustrated. The investigation is concerned with system responses for faults located at the converter DC or AC bus which is the worst-case fault location, and it is representative also for faults further away from the converter. It is presumed that faults will not happen closer to the converter, because this area is located in the valve hall and faults are unlikely. If faults happen closer to the converter then the fault current will be too fast and strong for the normal protection methods and insulated gate bipolar transistors (IGBTs) will be tripped by the internal switch-level protection.

20.2 Faults on the AC System

The converter current under an extreme AC system fault can be obtained from Eq. (18.15) by replacing $V_g = 0$:

$$I_{gd} = \frac{-V_{cq}}{X_t} = \frac{-M_q V_{dc}}{2X_t} \tag{20.1}$$

$$I_{gq} = \frac{V_{cd}}{X_t} = \frac{M_d V_{dc}}{2X_t} \tag{20.2}$$

High-Voltage Direct-Current Transmission: Converters, Systems and DC Grids, First Edition.
Dragan Jovcic and Khaled Ahmed.
© 2015 John Wiley & Sons, Ltd. Published 2015 by John Wiley & Sons, Ltd.

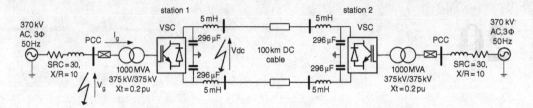

Figure 20.1   A VSC-HVDC transmission system with a location for AC and DC fault.

It is seen that, in general, the converter can control an AC fault current by reducing voltage V_c through current control loops (using M_d and M_q). This current reduction is achieved by reducing significantly the modulation index M_d, which is typically much larger than M_q. The DC voltage may remain at rated value during the AC fault. However, the converter control loop has limited speed as it is based on feedback measurements and involves signal processing. In the first instant after the fault, assuming that converter DC voltage is at rated value, only the reactor/transformer (X_t) limits the current.

Figure 20.2 shows the HVDC inverter response to a 60 ms (0.7–0.76 s) three-phase fault at point of common coupling (PCC) (V_g). The converter is exporting full power ($I_d = -1$ pu) before the fault. There is an initial increase in the current but the controller reduces the current to the rated value within 40 ms. The initial current peak is around 1.8 pu, which is satisfactory. It is important to keep the AC current peaks below 2–3 pu in all operating conditions, in order to avoid IGBT tripping by overcurrent protection. IGBT tripping should be avoided because recovery after converter blocking would cause significant delays. Starting a blocked converter requires undergoing full startup sequence involving synchronization, capacitor charging and numerous cross checks. A VSC HVDC is normally designed to ride through all AC faults without tripping. If an AC fault persists for long time, the converter is tripped.

The current peak depends primarily on the speed of inner control loops but it is also significantly influenced by the transformer reactance and this is one of the main requirements in selecting Xt.

20.3   DC Faults with Two-Level VSC

Figure 20.3 shows a two-level VSC converter under a DC fault. The fault current path in one switch pair is shown corresponding to the interval of 1/6 of the cycle. During a DC fault, the VSC converter behaves like three step-up converters under high-voltage faults. In the positive half cycle the diode connected to the positive rail (D1, D3 and D5) will experience high fault level currents and current returns through the diode connected to largest negative voltage on negative rail. The IGBTs on the opposite arms also take large phase currents.

At the DC fault instant the currents will increase quickly though IGBTs and diodes. The IGBTs will be tripped by the feedback overcurrent protection in the main control system, which is set to 2–3 pu. This protection operates with normal current feedback sensors and microcontrollers.

High-power IGBTs also have hard-wired local overcurrent protection, which will serve as backup protection. This is a last line of protection, which is implemented at driver level (mounted on the switch) and it is capable of tripping the IGBT within a few microseconds. This type of protection is based only on local current measurement (collector-emitter voltage measurement in the ON state), operates on individual switches and it is highly inaccurate and indiscriminative. Activating this protection level would result in disabling the module.

Once the IGBTs are blocked, the VSC converter becomes a VSC diode bridge as shown in Figure 20.4. The diodes will experience the full fault current and they cannot be controlled.

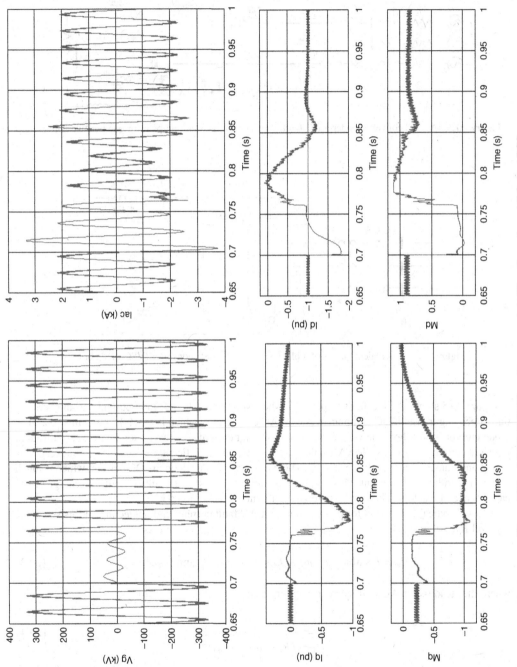

Figure 20.2 A VSC HVDC inverter response to a 60 ms three-phase AC fault at PCC.

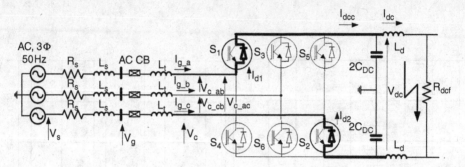

Figure 20.3   A VSC converter under a DC fault, showing the fault current path for one positive half cycle.

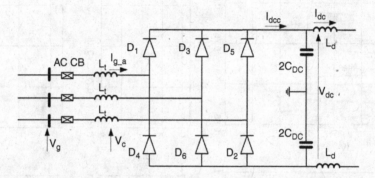

Figure 20.4   Equivalent circuit of a blocked two-level VSC converter (VSC diode bridge).

Figure 20.5 shows the DC fault response for the test system. The AC fault current has an exponentially decaying transient DC component and an AC component, which is around 5 pu in this case. The diode currents will follow the waveform of the AC fault currents, but each diode will have a different transient current wave shape. The AC voltage is not reducing significantly in this case as the AC grid is strong and transformer impedance is large. There is a large first peak on the DC current, which is much larger than with the AC current. This first peak is short (2–10 ms) and results from the discharge of DC capacitors and reactors. The steady-state DC fault current is around 9 kA.

A full analytical expression for phase a transient AC fault current, and neglecting, at this stage, L_d, C_{dc} and R_{dcf} is:

$$i_{ga} = I_g \sin(\omega t + \alpha - \varphi) + \left[I_{g0}\sin(\alpha - \varphi_0) - I_g \sin(\alpha - \varphi)\right]e^{-t/\tau} \tag{20.3}$$

where the steady-state AC fault current magnitude is:

$$I_g = \frac{V_s}{\sqrt{R_s^2 + (\omega(L_s + L_t))^2}} \tag{20.4}$$

while $\varphi = \arctan(\omega(L_s + L_t)/R_s)$, time constant is $\tau = (L_s + L_t)/R_s$, and I_{g0}, φ_0 are current magnitude and phase at the fault instant. This expression is important for dimensioning diodes and it shows that the AC fault current consists of the steady-state oscillating component (first term) and transient

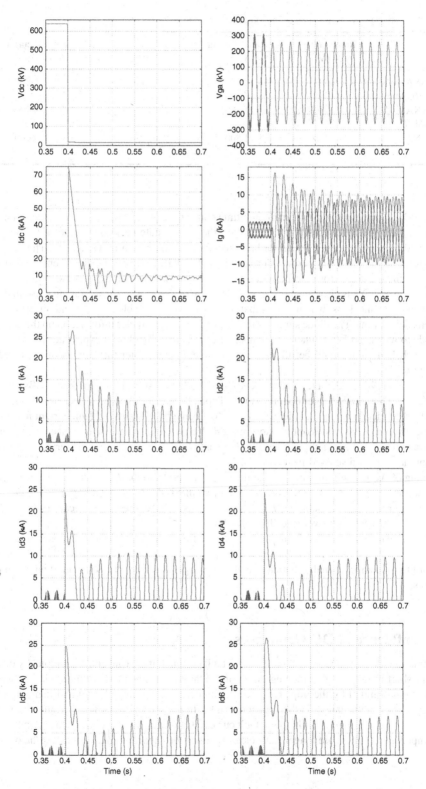

Figure 20.5 VSC converter variables for a permanent DC fault.

Table 20.1 High-power IGBT data of relevance for protection.

IGBT	Rated voltage V_{CES} (V)	Rated current I_C (A)	Peak current I_{CM} (A)	Surge diode current I_{FSM} (A) (10 ms, half sine)
ABB 5SNA 0750G650300	6 500	750	1 500	8 000
ABB 5SNA 1200G450300	4 500	1 200	2 400	9 000
ABB 5SNA 1500E330300	3 300	1 500	3 000	14 000

decaying DC component (second term) with time constant τ. In the worst-case fault, the initial peak of the fault current can reach $2\sqrt{2}I_g$, which is twice the steady-state peak magnitude $\sqrt{2}I_g$.

Table 20.1 shows the data relevant for fault tolerance for some high-power IGBTs. The peak current, I_{CM}, is relevant for the IGBT turnoff during faults. The IGBT protection should ensure that the gate signals are interrupted before the fault current exceeds I_{CM}. Typically, I_{CM} is close to twice the DC current rating and this is also relevant for the protection of AC faults. The averaging constant for IGBT parameters is around 1 ms, which means that the device might be able to withstand higher currents for durations below 1 ms. The diode surge current capability I_{FSM} provides information on the capability of the diode to withstand DC faults for a 10 ms half cycle. Note that diodes experience high temperature during such fault and should be offline for a sufficient period after the fault in order to reduce the temperature.

With VSC HVDC, the DC faults are cleared with an AC circuit breakers (CBs) (shown in Figure 20.4). These have longer opening times. Typically the high-voltage CB clearing time is around 50–100 ms. Evidently a single diode from Table 20.1 cannot withstand DC fault current for 50–100 ms and typically additional diodes or thyristors are added in parallel, to increase current carrying capability. There are two module design topologies, involving additional diodes or parallel thyristors depending on the manufacturer, as discussed in section 12.7.

Opening of AC CB will interrupt DC fault current infeed from the AC system. However diodes still represent a short circuit for any DC circulating current that is discharging energy from the DC line (line capacitance or inductance including L_d). This can potentially cause issues because of the long time required to clear an arc on overhead lines.

The DC side reactor (L_d) contributes to reducing the first peak of the diode and capacitor fault current, as may be seen from Figure 20.6. This reactor has no influence on the steady-state DC fault current. The value of this reactance is small – typically below 5 mH per pole, in order to avoid dynamic interaction problems.

20.4   Influence of DC Capacitors

In general, the shunt capacitors will increase the fault current but the capacitors are typically discharged in a very short time interval at the beginning of the fault. The magnitude of the initial capacitor fault current is only limited by the impedance in the fault path (cable impedance) and damping resistors/ inductors if they are installed on the capacitor banks. In the short period after the fault, the shunt capacitors can contribute significantly to the fault current. Capacitors may also excite oscillations with DC cable impedance. If there are longer fault-clearing times the capacitors may not affect the overall fault energy.

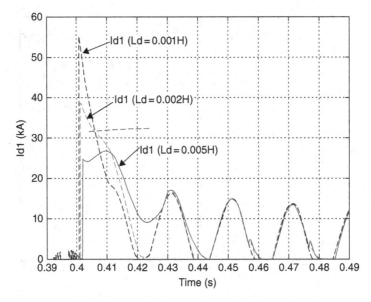

Figure 20.6 Influence of DC inductor L_d on the diode fault current.

The full analytical expression for the DC capacitor C_{dc} transient discharge through L_d into a fault of resistance R_{dcf} (neglecting the contribution from the AC grid) is:

$$v_{dc} = \frac{V_{dc0}\omega_{LC0}}{\omega_{LC}}e^{-t/\tau}\sin(\omega_{LC}t+\beta) - \frac{I_{dc0}}{\omega_{LC}C_{dc}}e^{-t/\tau}\sin(\omega_{LC}t)$$

$$i_{dc} = -\frac{I_{dc0}\omega_{LC0}}{\omega_{LC}}e^{-t/\tau}\sin(\omega_{LC}t-\beta) + \frac{V_{dc0}}{2\omega_{LC}L_d}e^{-t/\tau}\sin(\omega_{LC}t)$$

(20.5)

where $\omega_{LC} = 1/\sqrt{2L_dC}$, $\beta = \arctan(\omega_{LC}2L_d/R_{dcf})$, $\tau = 2L_d/R_{dcf}$, $\omega_{LC0} = \sqrt{\omega_{LC}^2 + 1/\tau^2}$, I_{dc0} and V_{dc0} are the initial values at the fault instant.

The primary issue with DC capacitors is the initial peak current, which theoretically can be infinity if $L_d = 0$. A widely adopted solution is to use DC-side inductors, L_d, which will limit the initial fault current as shown in Eq. (20.5) and in Figure 20.6.

The complete DC-side fault current will consist of the sum of positive half cycles of the AC fault current from Eq. (20.3) for all three phases, and the capacitor discharge current from Eq. (20.5). This current is important for dimensioning DC CBs and it can have a substantial first-peak magnitude as seen in Figure 20.5.

20.5 VSC Converter Modelling under DC Faults and VSC Diode Bridge

20.5.1 VSC Diode Bridge Average Model

This section examines the VSC converter analytical modelling, assuming significantly reduced DC voltage – that is, a DC fault as shown in Figure 20.3. Under significantly depressed DC voltage the current will increase and IGBTs will be blocked, which implies that a VSC converter becomes a VSC diode

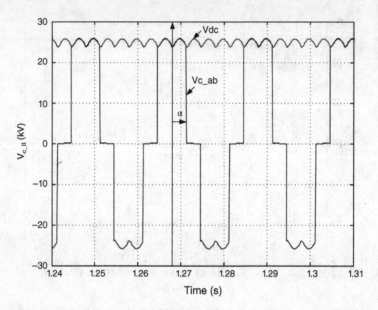

Figure 20.7   VSC diode bridge AC voltage (line-line) and DC voltage under a DC fault.

bridge as shown in Figure 20.4. Such a diode bridge behaves differently form the common current source diode bridge, which is studied in section 3.1. A VSC diode bridge responds as a voltage source on the AC side (because of DC capacitors) and a corresponding average value model will be derived in this section.

During normal operating conditions, IGBT switches conduct according to the firing pulses and the antiparallel diodes conduct complimentary to the switches. When IGBT firings are blocked, diodes do not conduct either. As the DC voltage reduces (or, alternatively, AC voltage increases) to a certain level, diodes start conducting regardless of the converter control signals. This DC voltage is labelled critical DC voltage, V_{dc_cr}. The VSC diode bridge will start conducting if the DC voltage reduces below the grid open-circuit line-line peak voltage, V_{g_ll}. The VSC converter line-line peak AC voltage can be determined using Eq. (17.5) and assuming maximum modulation index, $M = 1$. Therefore the critical DC voltage is:

$$V_{dc\_cr} = V_{c\_ll} = \frac{\sqrt{3}}{2} V_{dc} = 0.865\, V_{dc} \tag{20.6}$$

Figure 20.7 shows the VSC diode bridge AC voltage waveform for one phase under a severe DC fault. The AC waveform consists of 120° conducting intervals for each diode (the conducting interval for a diode pair is $\alpha = 60°$). During the conducting interval the AC voltage is clipped to the value of the DC voltage. This is valid for all voltage source converters. The peak value of the line-line AC voltage fundamental component is obtained using Fourier series expansion:

$$V_{c\_ll} = \frac{2}{\pi} \int_{-\alpha}^{+\alpha} V_{dc} \cos(\omega t)\, d(\omega t) \tag{20.7}$$

$$V_{c\_ll} = \frac{4}{\pi} V_{dc} \sin\alpha \tag{20.8}$$

The diodes start conducting uncontrollably when the DC voltage reduces below a critical value and, as the DC voltage reduces further, the conduction angle increases according to the following formula:

$$\cos\alpha = \frac{V_{dc}}{V_{dc\_cr}} \qquad (20.9)$$

For a three-phase system, the maximum conduction angle for each diode is $360/3 = 120°$ ($\alpha = 60°$). Therefore, from Eq. (20.9), if the DC voltage drops below $\tfrac{1}{2}V_{dc\_cr}$, the conduction angle remains at $120°$.

The VSC diode bridge voltage in Eq. (20.8) can be analysed for two cases: partial conduction and full conduction. In the full-conduction interval:

$$V_{c\_ll} = \frac{2}{\pi}V_{dc} \quad for \quad V_{dc} < \frac{1}{2}V_{dc\_cr} \qquad (20.10)$$

In the partial conduction interval, $0 < \alpha < \pi/3$, diodes experience conducting and nonconducting intervals. While diodes are conducting, using Eqs (20.8) and (20.9), the AC voltage is:

$$V_c = \frac{4}{\pi\sqrt{3}}V_{dc}\sqrt{1 - \left(\frac{V_{dc}}{V_{dc\_cr}}\right)^2} \quad while \ conducting \ for \quad \frac{1}{2}V_{dc\_cr} < V_{dc} < V_{dc\_cr} \qquad (20.11)$$

While diodes are not conducting, the converter AC voltage equals the open circuit remote AC voltage:

$$V_c = V_s \quad while \ not \ conducting \ for \quad \frac{1}{2}V_{dc\_cr} < V_{dc} < V_{dc\_cr} \qquad (20.12)$$

The converter voltage in partial conduction is the average of the two voltages in Eqs (20.11) and (20.12):

$$V_c = \left(\alpha\frac{4}{\pi\sqrt{3}}V_{dc}\sqrt{1 - \left(\frac{V_{dc}}{V_{dc\_cr}}\right)^2} + (\pi/3 - \alpha)V_s\right)/(\pi/3) \quad for \quad \frac{1}{2}V_{dc\_cr} < V_{dc} < V_{dc\_cr} \qquad (20.13)$$

In any diode bridge, diodes start conducting as soon as they become forward biased, without any delay angle, and therefore the AC current will be in phase with the AC voltage. This is true during full conduction only. During partial conduction the current may be lagging voltage because of the extinction angle and commutation overlap but these factors will not be discussed here considering that full conduction is most important in practice. The study also neglects a small current delay, which will occur because of the presence of harmonics. Assuming that the current angle is the same as voltage angle:

$$\frac{V_{cd}}{V_c} = \frac{I_{gd}}{I_g}, \quad \frac{V_{cq}}{V_c} = \frac{I_{gq}}{I_g} \qquad (20.14)$$

Expanding Eq. (20.14) and considering Eq. (20.10), for full conduction the diode bridge AC voltages are:

$$V_{cd} = \frac{2}{\pi}V_{dc}\frac{I_{gd}}{\sqrt{I_{gd}^2 + I_{gq}^2}}, \quad V_{cq} = \frac{2}{\pi}V_{dc}\frac{I_{gq}}{\sqrt{I_{gd}^2 + I_{gq}^2}} \qquad (20.15)$$

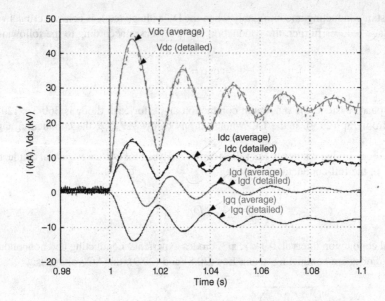

Figure 20.8 VSC diode bridge average model verification for a DC fault.

The power-balance equation for the VSC diode bridge is:

$$\frac{3}{2}\left(V_{cd}I_{gd} + V_{cq}I_{gq}\right) = V_{dc}I_{dc} \tag{20.16}$$

Replacing Eq. (20.15) into Eq. (20.16) the link is obtained between the AC current magnitude and the DC current:

$$I_{dc} = \frac{3}{\pi}I_g = \frac{3\sqrt{2}}{\pi}\mathsf{I_g} \tag{20.17}$$

where I_g is the peak amplitude of the AC fault current and $\mathsf{I_g}$ is the RMS value. Where there is partial conduction, expression (20.13) can be used to obtain the DC current formula.

Figure 20.8 shows the verification of the above VSC diode bridge average model for a severe DC fault on the 640 kV HVDC system. Accuracy is excellent, even during transients.

20.5.2 Phasor Model of VSC Converter under DC Faults

The above VSC diode-bridge model is now replaced in the AC grid phasor model to obtain power flow under DC faults. This study will also enable calculation of the steady-state DC fault current. Assuming that the AC grid is represented as an impedance z_s behind an equivalent AC source, V_{s}, as shown in Figure 20.3, the AC current phasor equation is:

$$\overline{\mathsf{I_g}} = \frac{\overline{\mathsf{V_s}} - \overline{\mathsf{V_c}}}{\overline{z_s} + \mathsf{jX_t}} \tag{20.18}$$

It is most convenient to locate the coordinate frame at the converter voltage V_c in this case:

$$\overline{V}_c = V_c, \quad \overline{I}_g = I_g \qquad (20.19)$$

Considering Eq. (20.19), when expanded Eq. (20.18) gives:

$$\overline{I}_g = \frac{V_{sd} + jV_{sq} - V_c}{R_s + jX_s + jX_t} \qquad (20.20)$$

Eq. (20.20) is converted into two scalar equations:

$$\begin{aligned} I_g R_s + V_c &= V_{sd} \\ I_g (X_s + X_t) &= V_{sq} \end{aligned} \qquad (20.21)$$

Taking the square of each of the equations in Eq. (20.21) and summing:

$$I_g^2 \left(R_s^2 + (X_s + X_t)^2 \right) + 2I_g R_s V_c + V_c^2 = V_s^2 \qquad (20.22)$$

Assuming that the total DC-side resistance is R_{dcf} as shown in Figure 20.3, the DC voltage is:

$$V_{dc} = I_{dc} R_{dcf} \qquad (20.23)$$

Replacing the AC voltage with DC voltage from Eq. (20.10), DC voltage with DC current using Eq. (20.23), DC current with AC current using Eq. (20.17), and simplifying in Eq. (20.22), the expressions are derived for AC and DC fault currents (considering only the most important case where $V_{dc} < \frac{1}{2}V_{dc\_cr}$):

$$I_g = \frac{V_s}{\sqrt{R_s^2 + (X_s + X_t)^2 + \dfrac{12 R_{dcf} R_s}{\pi^2} + \dfrac{36 R_{dcf}^2}{\pi^4}}} \qquad (20.24)$$

$$I_{dc} = \frac{3 V_s}{\pi \sqrt{R_s^2 + (X_s + X_t)^2 + \dfrac{12 R_{dcf} R_s}{\pi^2} + \dfrac{36 R_{dcf}^2}{\pi^4}}} \qquad (20.25)$$

Figure 20.9 shows the VSC diode bridge DC current, considering different values of transformer impedance and AC system SCR. The basic case ($Xt = 0.2$ pu, $SCR = 30$, $X/R = 10$) gives a steady-state DC fault current of 9 kA for $V_{dc} = 0$, which is in excellent agreement with the results on the detailed model in Figure 20.5. In conclusion, the transformer impedance makes a significant impact on the fault current magnitude. The DC fault current will reach significant values, and the fault current does not change much with the DC voltage magnitude. The fault currents at low DC voltage are similar to the values at $V_{dc} = 0$ even for DC voltage of 0.2 pu. This implies that the third and fourth terms in the denominator of Eq. (20.24) are much smaller than the first two terms. For practical purposes the converter AC current under DC faults can be approximated with a simpler expression:

$$I_g = \frac{V_s}{\sqrt{R_s^2 + (X_s + X_t)^2}} = \frac{V_s}{Z_{s+t}} \qquad (20.26)$$

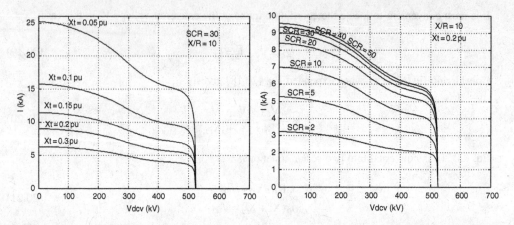

Figure 20.9   VSC diode bridge DC currents as DC voltage is reducing, with Xt and SCR as parameters.

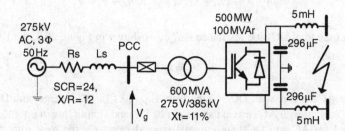

Figure 20.10   *VSC HVDC terminal in Example 20.1.*

Example 20.1

Figure 20.10 shows one terminal of a 500 MW VSC HVDC. The converter is rated for P_c = 500 MW and Q_c = 100 MVAr.

a. *Determine the RMS current in the switches.*
b. *Determine the steady-state RMS current in the diodes for a DC fault near the DC bus (as indicated). Comment on the result considering the converter switches.*
c. *Determine the current in the DC cable for the same DC fault.*

Solution

a. *Total converter power:*

$$S_c = \sqrt{P_c^2 + Q_c^2} = \sqrt{(500e6)^2 + (100e6)^2} = 510 \ MVA$$

Converter current is:

$$I_c = \frac{S_c}{\sqrt{3}V_c} = \frac{510 \ MVA}{\sqrt{3} \times 385e3} = 765 \ A$$

b. *The AC grid impedance:*

$$SCR = \frac{V_s^2}{z_s P_{dc}} \Rightarrow z_s = \frac{V_s^2}{SCR \times P_{dc}} = \frac{275\,000^2}{24 \times 500e6} = 6.3\,\Omega$$

$$R_s = \frac{z_s}{\sqrt{1+(X/R)^2}} = \frac{6.3}{\sqrt{1+(12)^2}} = 0.5234\,\Omega,\;\; X_s = R_s \times X/R = 0.5234\,\Omega \times 12 = 6.28\,\Omega$$

Total impedance referred to grid side

$$\overline{z_{s+t}} = R_s + j(X_s + X_t) + jX_i$$

$$\overline{z_{s+t}} = 0.5234 + j\left(6.28 + 0.11\frac{275\,000^2}{600e6}\right) + j2\pi \times 50 \times 0.03 \times (275/385)^2 = 0.5234 + j24.95\,\Omega$$

$$z_{s+t} = 24.959\,\Omega$$

Note that DC-side inductor has no impact on the fault current.
The converter steady-state RMS current for DC fault is:

$$I_{cf} = \frac{V_{s\_\parallel}/\sqrt{3}}{z_s} \times (275/385) = \frac{275\,000/\sqrt{3}}{24.959} \times (275/385) = 4,494\,A$$

This fault current is over six times the rated current. Typical IGBT reverse diodes would take this current for perhaps 10 ms but with a substantial temperature rise. Additional parallel diodes/ thyristors are therefore required in order to sustain the DC fault current long enough for the AC CB to open the circuit.

c. *The DC fault current is:*

$$I_{dcf} = \frac{3\sqrt{2}}{\pi} I_{cf} = \frac{3\sqrt{2}}{\pi} 4494 = 6069\,A$$

20.5.3 Simple Expression for VSC Diode Bridge Steady-State Fault Current Magnitude

The above calculation of the DC fault current is based on the accurate VSC diode bridge model. However it has been shown that the fault current does not vary significantly for a range of low DC voltage values. This facilitates derivation of a very simple expression for the diode bridge fault current, which uses only the AC system SCR and the transformer leakage reactance.

The DC current can be expressed using the SCR and the short-circuit level, which are defined in Chapter 9, and which are commonly given for an AC system:

$$I_{dc} = \frac{SCL}{SCR\,V_{dc}} = \frac{SCL\sqrt{3}\,n_t}{SCR\,2\sqrt{2}\,V_{s\_\parallel}} \tag{20.27}$$

where n_t is the transformer ratio. Under an extreme DC fault at the converter terminals, $(R_{dcf} = 0)$, the AC current is approximately expressed as:

$$\overline{I_{gf}} = \frac{V_s}{Z_s + jX_t} \tag{20.28}$$

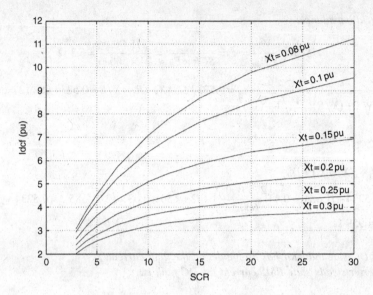

Figure 20.11 VSC converter DC fault current (in pu relative to nominal DC current) as a function of AC grid strength.

Assuming further that transformer rating is same as nominal DC power ($S_t = P_{dc}$):

$$I_{gf} = \frac{V_s}{z_s + z_s SCR\, X_{tpu}} \qquad (20.29)$$

Replacing Eq. (20.17) into Eq. (20.29), the first approximation for DC fault current is obtained:

$$I_{dcf} = \frac{n_t \sqrt{6}\, SCL}{\pi\, V_{s\_ll}\left(1 + SCR X_{tpu}\right)} \qquad (20.30)$$

Dividing Eq. (20.30) by Eq. (20.27), the DC fault current in pu relative to the nominal DC current is calculated:

$$I_{dcf\_pu} = \frac{4\, SCR}{\pi\left(1 + SCR X_{tpu}\right)} \qquad (20.31)$$

Eq. (20.31) gives a simple relationship between the AC grid SCR, transformer impedance and the steady-state DC fault current magnitude for a VSC converter (transformer ratio has no influence). This relationship is illustrated graphically in Figure 20.11. The value $SCR = 30$, $Xt = 0.2$ gives $I_{dcfpu} = 5.8$, which is in good agreement with more detailed modelling in Figures 20.5 and 20.9. It is also evident that the transformer impedance (which includes any series reactor) has a large influence and can be used to limit the fault current magnitude.

20.6   Converter-Mode Transitions as DC Voltage Reduces

Figure 20.12 shows the VSC converter-mode transitions as the DC voltage is reducing. The VSC converter model, as presented in Chapter 17, can be used, while the diode-bridge model is given in section 20.5. Note that the active power control has priority in this model whereas the reactive power control signal is limited to prevent modulation index magnitude exceeding 1.

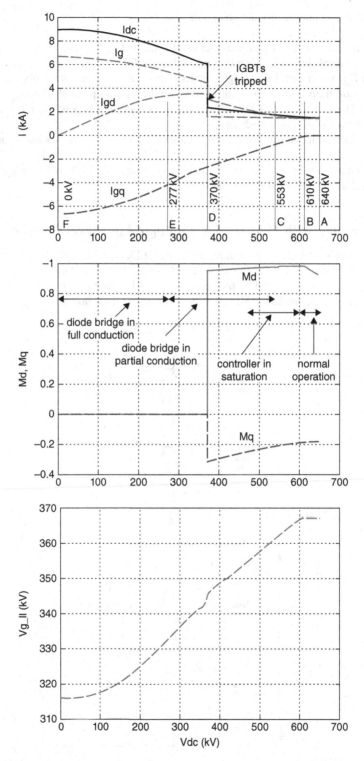

Figure 20.12 VSC converter mode transitions as the DC voltage is reducing (currents in RMS values).

The following operating modes are observed:

- *A–B:* Normal operation. I_{gd} is maintained at rated value and I_{gq} is zero.
- *B–C:* Control signals hit limits. Currents increase as the DC voltage drops.
- *C–D:* Point C is the critical DC voltage. Diodes start conducting uncontrollably.
- *D–E:* IGBTs are tripped on overcurrent (assumed to be 2 pu) at point D. Diodes remain in partial conduction. The exact value of the trip current magnitude will depend on the particular design and how fast the DC voltage is reducing.
- *E–F:* VSC diode bridge in full conduction. At point F, grid current reaches maximum value.

Depending on the speed of protection operation, at some stage, AC, CBs will be tripped and the converter will be disconnected from the AC system.

20.7 DC Faults with Half-Bridge Modular Multilevel Converter

During an MMC converter DC fault, as shown in Figure 20.13, the current will flow from the AC grid to the DC fault, in a similar way as with two-level VSC converters. The IGBTs must be blocked since otherwise they would be destroyed by the large current and capacitors would experience destructive currents and voltages. A blocked MMC converter becomes a diode bridge.

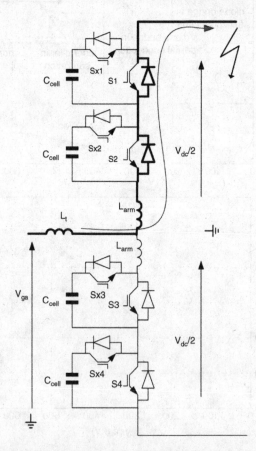

Figure 20.13   One phase of a half-bridge MMC during DC fault.

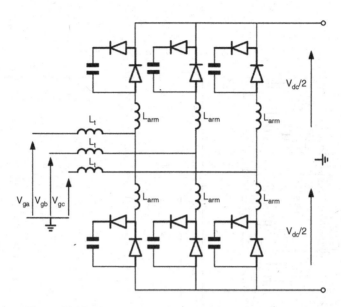

Figure 20.14 Equivalent circuit of a blocked half-bridge MMC.

Figure 20.14 shows the equivalent circuit for a blocked half-bridge MMC. The module capacitors will not discharge to the fault, which is an advantage compared with two-level topology. The module capacitors will maintain their charge during the fault, and this will result in less current stress (inrush current) and faster responses during the system recovery in transient DC faults. The arm inductors will be in series with the transformer reactance and they will limit the current raise and magnitude during fault conditions in the same way as phase reactors do with two-level converters.

Figure 20.15 shows the simulation of a severe DC fault with a half-bridge MMC converter, assuming similar parameters as the two-level VSC test system. It is seen that the response is similar as for two-level VSC in Figure 20.5. The MMC converter will not discharge the cell capacitors into the fault and, as there are no DC capacitors, the initial transient of DC current will be lower than with two-level VSC.

20.8   DC Faults with Full-Bridge Modular Multilevel Converter

The full-bridge MMC has the ability to operate with reduced DC voltage and also to operate with either positive or negative DC voltage. This implies that full-bridge MMC need not be blocked during DC faults. There are a number of operational advantages at grid level if full-bridge MMC maintains operation during DC faults:

- Very fast postfault recovery is possible (no need for synchronization or capacitor charging): which is particularly important for transient DC faults.
- DC voltage polarity can be reversed during the DC fault: in order to extinguish arcing in the same way as with a line-commutated converter (LCC) HVDC as discussed in section 10.2. This is very beneficial, with overhead DC lines exposed to frequent lightning.
- The fault will not be transferred to the AC side (AC current will be low).
- Reactive power can be controlled during DC fault conditions, which is important with weak AC systems.

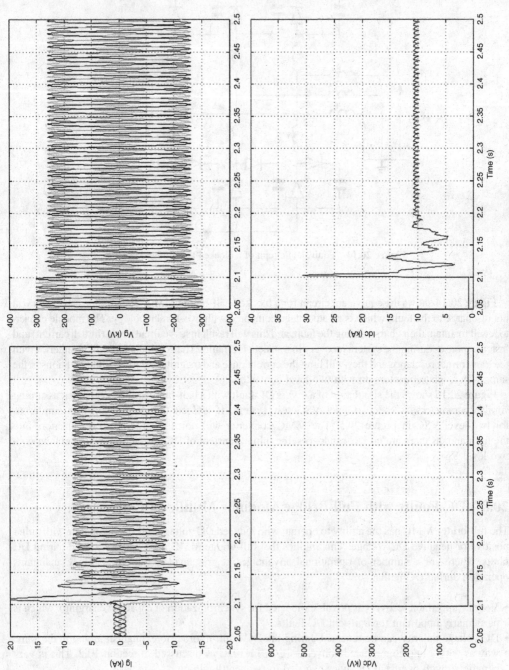

Figure 20.15 Simulation of a DC fault with a half-bridge MMC.

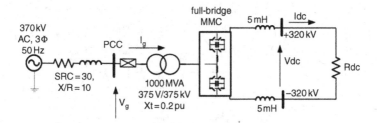

Figure 20.16   Full-bridge MMC test system.

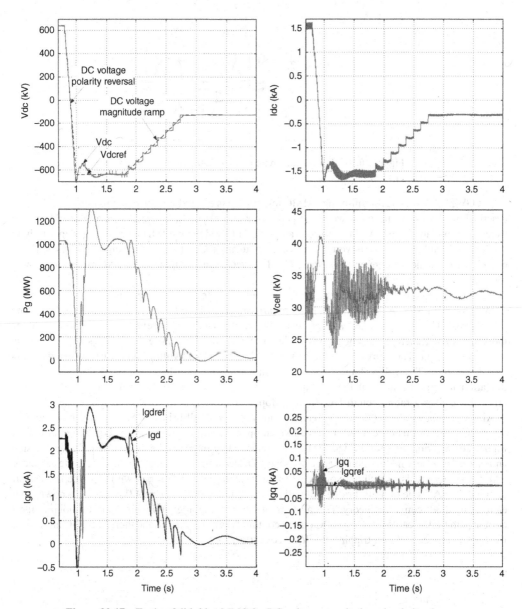

Figure 20.17   Testing full-bridge MMC for DC voltage magnitude and polarity change.

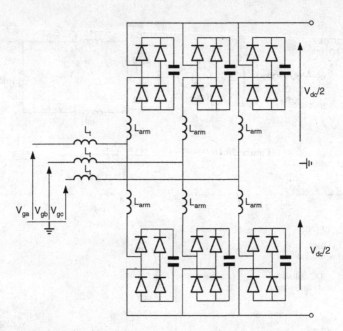

Figure 20.18   Equivalent circuit of a blocked full-bridge MMC.

A 1 GW, 640 kV 21-level full-bridge MMC test system is created, as illustrated in Figure 20.16. In a practical 640 kV converter there would be perhaps over 400 levels, but such a system would require excessive simulation time. All the arm cells are of the full-bridge type, which according to the study in section 16.7 implies that the converter can operate with any DC voltage in the range $-640\,\text{kV} < V_{dc} <$ 640 kV. A resistance is connected on the DC side to facilitate voltage/current reversal but not power reversal.

Figure 20.17 shows the test-system responses. Initially the system is operated as rectifier with a 1 GW power transfer. At 0.8 s, DC voltage polarity reversal is ordered while maintaining the same power level. This results in DC voltage reversal and also DC current reversal, while power retains the same sign. Reactive current I_{gq} is regulated at zero through all tests. At 1.8 s DC voltage is reduced to 0.2 pu. The converter operates well with reduced DC voltage and it keeps reactive current I_{gq} at zero. The cell voltage V_{cell} is maintained at the rated level although transient voltage deviation is large because of the low number of cells. The low cell number (only 20 cells per arm) also causes a rather rough DC voltage slope.

Although this would not be the preferred control action, a full-bridge MMC may be blocked during a DC fault. In this case it will behave like an open circuit, as shown in the equivalent circuit in Figure 20.18. Tripping a full-bridge MMC will immediately interrupt the fault current on the AC side and on the DC side. The full diode bridge cells have DC terminals, which are connected to the capacitor and the AC terminals, which are connected to the external circuit in series with other cells. A short circuit on the AC side of a full-bridge cell will not cause uncontrolled current flow, as it is known from elementary diode converter theory. This also implies that the topology in Figure 20.18 cannot provide a current path for DC cable discharge (DC circuit continuity is not available and therefore a DC arc will be interrupted immediately on tripping the VSC converter).

21

VSC HVDC Application for AC Grid Support and Operation with Passive AC Systems

21.1 VSC HVDC High-Level Controls and AC Grid Support

The main application drivers for high-voltage direct current (HVDC) in transmission systems are discussed in some depth in Chapter 1 and they also apply to voltage-source converter high-voltage direct current (VSC HVDC). This chapter will examine some additional system-level functions that VSC HVDC can provide and discuss controls. Some unusual HVDC installations are also investigated in order to illustrate the scope for possible future applications for VSC HVDC.

Normally HVDC will be operated to deliver the required active power between two AC busses and to supply reactive power at each AC bus independently. The additional control functions will be implemented as a master layer of controls, which modulate the power reference signal in the control diagram in Figure 17.11. Voltage-source converter HVDC can be employed to assist connecting AC systems in the following ways:

- *HVDC contribution to small signal stability (power oscillation damping).* Power oscillations typically occur as a consequence of electromechanical interactions between a large synchronous machine and a moderate-size AC grid. They are a 'regional' phenomenon (one machine versus the AC grid) and occur in frequencies above 0.1 Hz and up to 10 Hz. The subsynchronous resonance phenomenon occurs at slightly higher frequencies (20–50 Hz) and could also be included in this group. The HVDC can be made to damp power oscillations in this range by measuring the local variable (power, current or angle) that best observes the oscillatory mode. This signal is properly filtered and passed to a supplementary control circuit, which will modulate the HVDC power reference.
- *HVDC contribution to frequency stability.* Static frequency deviation will occur if there is an imbalance of load and generation in an AC system. To alleviate these issues, HVDC control can be configured to demand power exchange with a remote AC system in response to frequency deviation. This is implemented in a typical frequency-droop feedback resembling the response of a conventional generator. Dynamic frequency instability is slightly different and manifests typically as an interregional phenomenon in large AC systems. Here, a group of machines oscillates in synchronism against another remote group of synchronous generators. It typically occurs in a time

High-Voltage Direct-Current Transmission: Converters, Systems and DC Grids, First Edition.
Dragan Jovcic and Khaled Ahmed.
© 2015 John Wiley & Sons, Ltd. Published 2015 by John Wiley & Sons, Ltd.

frame of over 10 s or in the frequency range 0.01–0.1 Hz. In this case, the frequency signal or machine-speed signal is supplied to the HVDC controls and, when properly processed for required magnitude and phase, it is fed to modulate the HVDC power reference.

- *HVDC inertia contribution.* In very small systems, which have rotating machines with low inertia, the frequency deviations in response to load change may be significant. In such cases the HVDC power reference can be modulated to make the HVDC respond as a generator with inertia. In practice, these HVDC controls will be similar to those in the above case of frequency stabilization.
- *HVDC contribution to voltage stability.* Voltage stability is the phenomenon that describes voltage deviation in response to system loading, and the textbook PV-curve or 'nose curve' illustrates the border between stable and unstable operating regions. In typical transmission systems, voltage is sensitive to reactive power flow and the reactive power exchange capability of VSC HVDC is crucial to voltage stability. This problem is closely linked with the short-circuit ratio and it has been discussed in section 18.3. It should also be emphasized that each of the VSC HVDC terminals can operate as a STATCOM (static compensator) and regulate reactive power when there is no HVDC active power flow.
- *HVDC as a firewall.* Cascading outages are very rare but when they do occur widespread blackouts will result. At the heart of the problem is the inherent response of AC lines and power loads to draw more current when frequency and voltage reduce, which in turn sequentially overloads one system after another. HVDC system, however, is controlled to a power reference and responds as firm power/current (within limits) regardless of the actual frequency or voltage. This response prevents spreading system collapse through HVDC and it is well known that HVDC was beneficial in preventing spreading of recent blackouts (like the 2003 North American blackout).
- *Black start.* Black start is the capability of a unit to energize a dead network/load. Normally, HVDC connects by synchronizing with an established AC voltage. However, VSC HVDC can also start a dead network, using an internal oscillator driven by a frequency reference (line-commutated converter (LCC) HVDC does not have this capability).
- *HVDC establishing AC frequency.* High-voltage direct current can internally generate a stable frequency reference signal, which provides firm AC voltage at the point of common coupling (PCC). It can supply a completely passive AC load and other machines can lock onto HVDC-generated voltage. The frequency value is not restricted to 50/60 Hz.
- *HVDC controlling machine speed in a wide frequency range.* HVDC can also operate as a large variable-speed drive. Here the HVDC inverter controls machine speed using a flux vector control or other method with a wide speed range. In such cases, the VSC HVDC converter operates at a variable frequency.

The above control functions are further discussed using some common HVDC application topologies.

21.2   HVDC Embedded inside an AC Grid

Figure 21.1 shows a single line diagram of a power system where HVDC is embedded inside a single AC system (of limited size). In this application, HVDC cannot help with power imbalance. It cannot provide inertia support or frequency stabilization either, because on both sides of HVDC the frequency would be identical and HVDC has no energy storage. However, HVDC can provide power oscillation damping, voltage support or black start. Depending on the actual topology and protection system, HVDC can assist in preventing spreading blackouts. In this application, HVDC can also prevent loop flows, which can contribute indirectly to power flow according to market arrangements.

21.3   HVDC Connecting Two Separate AC Grids

Figure 21.2 shows an HVDC link employed to connect two separate AC systems. The two systems are not synchronized and therefore HVDC can draw power from one system to assist a system with (frequency, small signal, voltage) stability problems. In this application, HVDC can contribute to all the functions from section 21.1.

21.4   HVDC in Parallel with AC

Figure 21.3 shows an HVDC link in parallel with an AC interconnector, where HVDC might be added to increase capacity, to enhance reliability, or for other reasons. In normal operating conditions HVDC

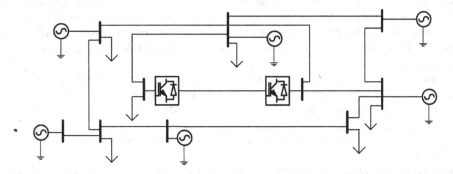

Figure 21.1   A VSC HVDC embedded in an AC grid.

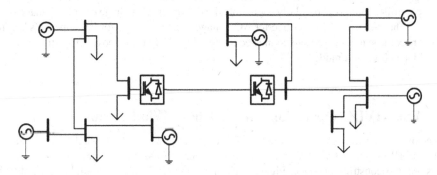

Figure 21.2   A VSC HVDC connecting two separate AC systems.

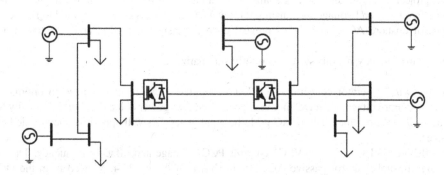

Figure 21.3   A VSC HVDC in parallel with an AC interconnector.

is embedded in a large AC system but it is likely that some operating restriction will apply. HVDC can typically operate as a generator on one terminal and as a constant power load on the other. The power transfer can also be scheduled to prevent overload of parallel AC line. If, however, the parallel AC line is tripped, then HVDC connects two unsynchronized AC systems, as in Figure 21.2. If one of the systems is weak or an island system, then HVDC controls can be configured to make HVDC stabilize frequency and respond as an infinite bus. An example installation is 300 MW, 350 kV, 2010 Caprivi VSC HVDC project, which has a similar topology.

21.5   Operation with a Passive AC System and Black Start Capability

A passive AC system is an AC network that has no inertia and has no voltage-controlling units. A VSC HVDC can supply power to such an AC system; however, special controls are required. If a VSC converter supplies a passive AC grid, as shown in Figure 21.4, the converter controller will internally set the required frequency ω_g and a voltage-controlled oscillator (VCO) is used to generate a reference sawtooth angle φ_{PLL}.

The converter will provide required active power in order to keep PCC voltage aligned with the coordinate frame ($V_{gq} = 0$), as shown in the control schematic in Figure 21.4. Another proportional integral (PI) controller will adjust the reactive power flow in order to keep the PCC voltage magnitude at the set value V_{gdref}.

A VSC HVDC has been used to supply power to (a passive) 292 km offshore AC system in 78 MW, 150 kV, 2011, Valhall project.

A VSC HVDC converter can also provide black-start capability, also called cold load pick-up, assuming that firm (controlled as an infinite bus) DC voltage is available. This is a similar condition as with a passive AC grid but it is only transient. The VSC converter AC voltage can be readily shaped from the controller and it is particularly useful that the converter AC voltage can be regulated from very low values. The frequency of the converter AC voltage is set internally for initial periods before large machines become synchronized and connected to the grid, and then the converter controller is switched to the grid-synchronized mode.

21.6   VSC HVDC Operation with Offshore Wind Farms

Voltage-source converter HVDC has been used to connect two offshore wind farms: the 400 MW ±150 kV, BorWin1 project in 2012, and the 800 MW ±320 kV DolWin1 project in 2014. Several other projects are in a construction stage. Figure 21.5 shows the schematic of an offshore wind farm connected to the onshore grid using VSC HVDC. Figure 21.6 shows an offshore platform for the BorWin1 HVDC project. A converter VSC1 is connected to a passive offshore system and it should maintain offshore frequency and voltage. A transmission-level DC chopper is commonly used to enable energy damping if onshore AC system faults prevent energy transfer. The chopper is inactive in normal operation.

The control strategy for this system is briefly summarized:

- The two wind-generator converters (5–10 MW) regulate local generator speed to optimal value, wind-generator voltage, local DC link voltage and reactive power at local AC grid (typically to zero value). Viewed from the offshore AC grid, the wind generators will appear as uncontrolled current sources.
- The offshore HVDC converter VSC1 controls PCC1 voltage and offshore frequency. The control topology is similar as for passive AC system, shown in Figure 21.4. Viewed from the DC line, the converter VSC1 will appear as uncontrolled DC current source.

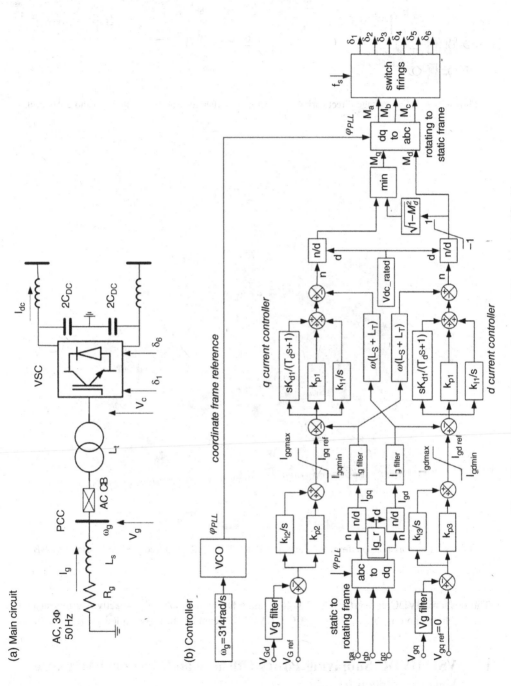

Figure 21.4   A VSC HVDC converter connected to a passive AC grid.

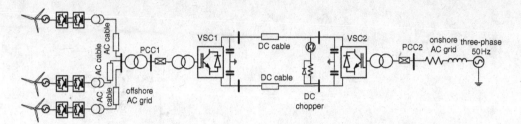

Figure 21.5   High-voltage direct current connecting an offshore wind farm with an onshore AC grid.

Figure 21.6   Offshore platform for BorWin1 HVDC (2012). Reproduced with permission of ABB.

- The onshore HVDC converter VSC2 will regulate DC voltage and also reactive current at PCC2. Viewed from the onshore system at PCC2, HVDC appears as an uncontrolled power injection.

21.7   VSC HVDC Supplying Power Offshore and Driving a MW-Size Variable-Speed Motor

Voltage-source converter HVDC has been employed to drive directly a MW-size variable speed motor loaded with a compressor in the 88 MW ±60 kV, 2005 Troll A project. In this project, HVDC supplies

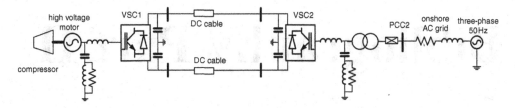

Figure 21.7   High-voltage direct current driving a MW-size variable-speed motor.

power 70 km offshore and directly controls motor speed. The offshore platform power is supplied from a local gas turbine.

Figure 21.7 shows the simplified schematic of the Troll A HVDC project. VSC 1 is directly coupled to a high-voltage motor and provides variable frequency (0–63 Hz) and variable voltage (0–56 kV) for motor control. No transformer is used in order to avoid core flux issues at low frequency and to reduce offshore space requirement. Instead, two special cable-wound high-voltage 40 MW motors are developed and directly connected to the VSC converter.

Bibliography Part II

Voltage Source Converter HVDC

Adam, G.P., Ahmed, K.H., Finney, S.J., and Williams, B.W. (2010) AC fault ride-through capability of VSC-HVDC transmission systems. *IEEE Energy Conversion Congress and Exposition (ECCE) Conference* 2010, pp. 3739–3745.

Adam, G.P., Ahmed, K.H., Finney, S.J., and Williams, B.W. (2011) H-Bridge modular multilevel converter for high-voltage applications. *The 21st International Conference and Exhibition Electricity Distribution (CIRED 2011)*.

Ahmed, K.H. and Adam, G.P. (2014) New modified staircase modulation and capacitor balancing strategy of 21-level modular multilevel converter for HVDC applications. *IEEE/IET PEMD Conference*, Apr. 2014.

Alsseid A.M., Jovcic, D., and Starkey, A. (2011) Small signal modelling and stability analysis of multiterminal VSC-HVDC. *Power Electronics and Applications, Proceedings of the 2011-14th European Conference on*, pp. 1–10.

Andersen, B.R., Xu, L., and Wong, K.T.G. (2001a) Topologies for VSC transmission. *Seventh International Conference on AC–DC Power Transmission IEE*, London, pp. 119–124.

Andersen, B.R., Xu L., and Wong, K.T.G. (2001b) Topologies for VSC transmission. *Seventh International Conference on AC-DC Power Transmission (IEE Conf. Publ. No.485)*. London, UK IEE, pp. 298–304.

Antonopoulos, A., Angquist, L., and Nee, H.P. (2009) On dynamics and volatge control of the modular multilevel converter. *13th European Conference on Power Electronics and Applications, EPE'09*, pp. 1–10.

Barnes, M. and Beddard, A. (2012) Voltage source converetr HVDC Links – the state of art and issues going forward. *Energy Procedia* **24**, 108–122.

CIGRE WG B4.37 (2005) VSC Transmission. CIGRE Brochure 269.

CIGRE WG B4.48 (2011) Components Testing of VSC System for HVDC Applications. CIGRE Brochure 447.

Du, C., Agneholm, E., and Olsson, G. (2008) Comparison of different frequency controllers for a VSC-HVDC supplied system. *IEEE Transactions on Power Delivery*, **23**, 4, 2224–2232.

Flourentzou, N., Agelidis, V.G., and Demetriades, G. D. (2009) VSC based HVDC power transmission systems: an overview. *IEEE Transactions on Power Systems*, **24**, 3, 592–602.

Friedrich, K. (2010) Modern HVDC PLUS application of VSC in Modular Multilevel Converter topology. *Industrial Electronics (ISIE), 2010 IEEE International Symposium on*, pp. 3807,3810, 4–7 July 2010.

Gnanarathna, U.N., Gole, A.M., and Jayasinghe, R.P. (Jan 2011) Efficient modeling of modular multilevel HVDC Converters (MMC) on electromagnetic transient simulation programs. *IEEE Transactions on Power Delivery*, **26**, 1, pp. 316–324.

Hammons, T.J., Woodford, D., Loughtan, J., Chamia, M., Donahoe, J., Povh, D., Bisewski, B., and Long, W. (Feb 2000) Role of HVDC transmission in future energy development. *Power Engineering Review, IEEE*, **20**, 2, 10,25.

Jaconson, B. Karlsson, P. Asplund, G. *et al.* (2010) vsc-hvdc transmission with cascaded two-level converters. *Proceeding of the Cigre Conference*, Paris, France, pp. 1–8.

High-Voltage Direct-Current Transmission: Converters, Systems and DC Grids, First Edition.
Dragan Jovcic and Khaled Ahmed.
© 2015 John Wiley & Sons, Ltd. Published 2015 by John Wiley & Sons, Ltd.

Jovcic, D. and Jamshidifar, A. (2015) Phasor model of modular multilevel converter with circulating current supression control. *IEEE Transactions on Power Delivery*. DOI: 10.1109/TPWRD.2014.2372780.

Jovcic, D., Lamont, L.A., and Xu, L. (2003) VSC Transmission model for analytical studies. *IEEE Power Engineering Society General Meeting, IEEE*, Vol. 3, pp. 1737–1742, 13–17 July 2003.

Kirby, N.M., Luckett, M.J., Xu, L., and Siepmann, W. (2001) HVDC transmission for large offshore wind farms. *IEE International Conference on AC-DC Power Transmission*, pp. 162–168.

Lamell, J.O., Trumbo, T., and Nestli, T.F. (2005) Offshore platform powered with new electrical motor drive system. *Petroleum and Chemical Industry Conference, 2005. Industry Applications Society 52nd Annual*, pp. 259,266, 12–14 Sept. 2005.

Lidong Zhang, Nee, H.-P. (2009) Multivariable feedback design of VSC-HVDC connected to weak ac systems. *Power-Tech, 2009 IEEE*, Bucharest, pp. 1, 8, June 28 2009–July 2 2009.

Modeer, T., Nee, H.-P., and Norrga, S. (2011) Loss comparison of different sub-module implementations for modular multilevel converters in HVDC applications. *Power Electronics and Applications (EPE 2011). Proceedings of the 2011-14th European Conference on*, pp. 1, 7, Aug. 30 2011–Sept. 1 2011.

Mohan, N., Undeland, T., and Robbins, W. (2003) *Power Electronics: Converters, Applications, and Design*. John Wiley & Sons, Inc., New York.

Papic, I. (2000) Mathematical Analysis of FACTS devices based on a voltage source converter. *Elsevier Electric Power Systems Research* **56**, 139–148.

Papic, I., Zunko, P., Povh, D., and Weinhold, M. (1997) Basic control of unified power flow controller. *IEEE Transactions on Power Systems*, **12**, 1734–1739.

Peralta, J., Saad, H., Dennetiere, S., Mahseredjian, J., and Nguefeu, S. (2012) Detailed and averaged models for a 401-level MMC–HVDC system. *IEEE Transactions on Power Delivery*, **27**, 3, 1501–1508.

Saad, H., Peralta, J., Dennetiere, S., and Mahseredjian, J. (2013) Dynamic averaged and simplified models for MMC-based HVDC transmission systems. *IEEE Transactions on Power Delivery*, **28**, 1723–1730.

Schauder, C. and Mehta, H. (1993) Vector analysis and control of advanced static VAr compensators. *IEE Proceedings C Generation, Transmission and Distribution*, **140**, 4, 299–306.

Svensson, J. (2001) Synchronisation methods for grid-connected voltage source converters, *IEE Proceedings – Generation, Transmission and Distribution*, **148**, 3, 229–235.

Welsby, V.G. (1960) *The Theory and Design of Inductance Coils*. MacDonald.

Yu, Q., Round, S.D., Norum, L.E., and Undeland, T.M. (1995) A new control strategy for a unified power flow controller. *European Conference on Power Electronics and Applications, EPE '95*, Seville, Sept. 1995, p. 2.901.

Zhang, L., Harnefors, L., and Nee, H.-P. (2010) Power-synchronization control of grid-connected voltage-source converters. *IEEE Transactions on Power Systems*, **25**, 2, 809–820.

Zhang, L., Harnefors, L., and Nee, H.P. (2011) Modeling and control of VSC-HVDC links connected to island systems. *IEEE Transactions on Power Systems* **26**, 2, 783–793.

Part III

DC Transmission Grids

22

Introduction to DC Grids

22.1 DC versus AC Transmission

Transferring electrical power between two locations some distance away can be achieved using direct or alternating current. Considering only power transfer at a single voltage level, DC power transmission is considerably better than AC, and the main advantages of DC include:

- Higher power transfer (up to 30%) for the same copper size and same insulation level. The AC cables must be insulated for peak AC voltage but power transfer is related to the RMS value of voltage.
- The losses of AC current (RMS value I_g) are equal to losses of the same magnitude of DC current ($I_{dc} = I_g$). However AC current will contain active current (which determines power transfer) and also reactive current. As it is impossible to eliminate reactive current circulation completely, the losses with AC transmission will always be larger. Therefore DC cables have smaller copper size and smaller insulation.
- The AC current is pushed towards the surface of the conductor and the skin depth at 50 Hz is around 9 mm. This increases losses and therefore the cross section of AC cables is limited while DC cables have no limitations in size except for the thermal management.

As a consequence of smaller wires/cables, the support towers (pylons) and transmission corridor (right of way) are smaller with DC transmission.

The reactive current circulation is the central issue with AC transmission. It depends on the voltages at the line ends but also on the line impedance. Transmission lines have dominant inductive reactance whereas cables have large capacitive reactance, which depends on the length. These reactances will, in either case, cause reactive current circulation if the AC current is transferred. The reactive current may become very large – even larger than the active current – and beyond a certain length AC power transmission may not be possible. In the case of long AC lines, the voltages at the line ends must be kept constant within a very narrow range to enable decent utilization of the line's thermal rating. The problem is particularly difficult with varying loading, in which case long AC cables may require power electronic systems (like Static VAR Compensators or static synchronous compensators (STATCOMs))

High-Voltage Direct-Current Transmission: Converters, Systems and DC Grids, First Edition.
Dragan Jovcic and Khaled Ahmed.
© 2015 John Wiley & Sons, Ltd. Published 2015 by John Wiley & Sons, Ltd.

to keep AC voltages constant, or power electronics controllers for line impedance like thyristor-controlled series capacitors (TCSCs).

The main disadvantages with DC transmission are related to fault isolation in networks. This is required for system reliability. Voltage stepping is also complicated with DC systems. Alternating current transmission has been the dominant technology for power transmission and distribution networks because of the simplicity of voltage stepping and fault isolation, which is achieved with inexpensive electro-mechanical components (transformers and circuit breakers (CBs)). In DC systems, it will be necessary to use power electronics for the components required for DC fault isolation and DC voltage stepping. While this implies higher costs, it also brings many performance benefits and operational flexibility.

22.2 Terminology

High voltage direct current (HVDC) transmission is the term used solely for a two-terminal DC system. It can assume monopolar or bipolar configuration but power is exchanged between only two AC terminals/systems. The power transfer is same at each terminal and it is coordinated.

Tapping on HVDC involves a small third terminal (or more terminals) along the HVDC lines. The tap terminal is very small compared with power level of the main HVDC line and the main HVDC needs no control modifications. No HVDC tap has been implemented but substantial demand exists and many studies have been completed. In particular, with very long HVDC lines passing close to communities with no access to power (Itaipu HVDC, Nelson River HVDC, etc.), HVDC tapping becomes an attractive option.

Multiterminal HVDC signifies a DC system of more than two terminals arranged along a single HVDC line. There are no meshed lines. If a DC line is in fault, there will be loss in capacity or loss in power-trading capability. Typically, a multiterminal HVDC will have a single protection zone, which implies tripping the whole system for any DC fault. However, DC line protection may be implemented on larger systems in order to isolate the smallest faulted segment in case of DC faults, thus allowing the remaining part to operate normally. There is a single DC voltage level. Control systems will be different from HVDC controls because power balancing within the system is required. Normally, there is no possibility for adding further terminals, unless substantial redesign is undertaken. Multiterminal HVDC has been implemented both with line-commutated converter (LCC) and voltage-source converter (VSC) HVDC and new projects are in the planning/design stages.

DC grid denotes a DC transmission system of more than two terminals with at least one meshed DC line. With DC grids there are multiple power-flow paths between two grid terminals. Power flow between two DC grid terminals may not be affected (or partially affected) by tripping a single DC line. DC grids will require some protection technology in order to isolate faulted lines/units allowing remaining part of the grid to continue power transfer. Tripping whole DC grid will not be allowed. Multiple nominal DC voltage levels may exist. Normally, any number of new terminals can be added to an existing DC grid. The control system will be considerably more complex than with two-terminal HVDC.

Node in a DC grid is any station where two or more DC lines connect.

terminal is a DC grid station where power is exchanged between DC grid and an AC system or another DC system (for example, when a DC/DC converter connects two DC systems). A terminal is a node with a converter.

22.3 DC Grid Planning, Topology and Power-Transfer Security

The primary motivation for DC grid development is the need to interconnect multiple HVDC links located in close proximity, and to enable power trading between all DC terminals. This brings benefits of better utilization of assets, better reliability and security of power transfer, better efficiency, enhanced power trading and operating flexibility, and all the advantages of interconnected systems (reserve sharing, control, etc.). The principal application drivers arise from the initiatives like the North Sea DC grid, Medtech, Desertec, European overlay Super grid and Atlantic wind.

It is expected that future DC grids will have similar level of security, reliability and performance as the existing AC grids. Although DC grid codes do not exist at present, the underlying principles of power trading, market access, security in power transfer and performance will be based on AC grid codes. The common $n - 1$ criterion (unaffected power transfer for loss of any single component) with AC systems is a valid expectation but an $n - 2$ or higher criteria would require significant capital investments because of high costs of converters and DC grid components. Note that no redundancy has been used with traditional HVDC (except at valve and control levels). On the other hand if a bipolar DC grid is built with neutral return then there is an option to operate terminals at 50% power in case of a DC cable outage. This is different from AC systems where a line fault (any line in a three-phase system) is always a circuit outage. A bipole is also different from an AC double circuit and requires considerably lower investment.

It is important that a DC grid provides access to all market participants, including renewable energy sources and this complicates regulatory aspects and imposes some high technical demands on the grid. The technical requirements for new entrants will also include stringent specifications (likely more demanding than with AC grids) in order not to disturb the DC grid under all foreseeable scenarios.

The rating of DC grid terminals will be dictated by the cable capacity but also by the operating codes of the AC grid where the terminal connects. The maximum acceptable single-unit loss is defined by the secondary reserve, which is different for each country. As an example it is 1500 MW in France, 1800 MW in the United Kingdom and 1300 MW in Spain.

Ideally, terminal controls should be able to separate/isolate disturbances on AC side from those on DC grid. A VSC terminal will also be expected to contribute to system stabilistaion on DC and/or AC side, which will present a major technical challenge and requires careful planning of DC and AC grid codes.

22.4   Technical Challenges

Direct-current grids will require substantial technical advances from the existing HVDC technology. The main challenges include DC grid protection and DC grid control:

22.4.1   DC Grid Protection

Developing both: DC protection components and protection systems is much more challenging compared with AC systems. A DC fault current has no zero crossing and therefore DC CBs are more complex. The protection coordination at DC grid level will be very challenging because the absence of reactive impedance on DC cables introduces complexities in finding the location of DC faults. The protection trip decision time window should also be shorter than with AC systems. The overcurrent and overvoltage capability of DC grid components is lower than with AC grids.

22.4.2   DC Grid Control

Direct current grid dynamics are an order of magnitude faster than dynamics of AC systems, and some converters can respond two orders of magnitude faster than electromechanical AC components. There is no inertia or any energy storage in DC systems and therefore energy balancing must be fast. Contrary to a unique frequency value across the whole AC system, DC voltage will be different at each terminal and cannot be used as a definite power balance indicator. Voltage-source converters are quite sensitive to DC voltage excursions and very fast DC grid control is essential. A VSC converter must be tripped if local DC voltage drops below 80–85%.

Table 22.1 Estimated cost of DC grid components.

Component	Cost (M€/MW)	Cost for 1 GW
AC/DC VSC converter	0.11 (1 pu)	110 M€
Hybrid DC CB (unidirectional)	0.25–0.35 pu	27–38 M€
Mechanical DC CB	0.002–0.005 pu	0.23–0.55 M€
DC/DC converter (low stepping ratio)	1 pu	110 M€
DC/DC converter (high stepping ratio)	1.5–1.8 pu	165–200 M€
DC-hub (multiport IGBT-based DC/DC)	0.8–0.9 pu/port	88–100 M€/port
Offshore platform for VSC (wind farm)	1.2–1.5 pu	130–165 M€
DC cable (pair)	0.013 pu/km	140 M€ (for 100 km)

22.5 DC Grid Building by Multiple Manufacturers

High-voltage direct-current systems are commonly developed by a single vendor coordinating works, perhaps with local suppliers. The technology is optimized and various subsystems (protection, control, valve design, etc.) are interdependent within HVDC topology for each different HVDC supplier. In only a few cases of HVDC station refurbishment, a vendor has been different from the supplier of the original equipment. In several other cases, the two poles in a large bipole HVDC have been built by different vendors. These are exceptional cases while in general there is no interoperability between inner HVDC subsystems or units by different suppliers.

Development of DC grids will involve multiple manufacturers. This is an essential requirement to facilitate proper functioning of the new market, to ensure public and political support and to establish competition, which will lead to better performance. If a complete terminal is developed by a single manufacturer there is still a significant challenge to integrate such a terminal into an existing DC grid that might contain terminals by several other manufacturers. Considering interoperability at lower component levels, for example interchangeable DC CBs by various manufacturers, there is need for substantial further work on standardization.

The development of DC grid codes and, in particular, international codes is essential to enhance interest in connecting to DC grids by various new players, which in turn will underpin DC grid development in the power industry.

22.6 Economic Aspects

A DC grid must be economically viable. The initial capital investment is likely to be higher than AC grid options (where technically feasible) because of high converter capital costs. However, cost of DC lines/cables is generally lower than comparable AC lines. The lower DC grid losses should be considered in the long-term studies of return on investment. The control of power flow will be much better in DC grids, which will enable better utilization of lines (closer to thermal limits) and which can eliminate problems of loop flow or overloading/underutilization of lines as experienced in AC grids.

The topology of future DC grids is not certain and different protection/control approaches are favoured by the main manufacturers. The grid planning will be a complex process considering technical performance, market aspects, security, reliability, safety, environmental considerations, economics and other factors. As a rough initial guide, Table 22.1 provides cost comparison for the basic building blocks of DC grids. Only the AC/DC VSC converter cost (assumed 1 pu) can be regarded as reliable because it is based on numerous existing projects and it has been publically endorsed by the Conseil International des Grands Réseaux Électriques (CIGRE). The DC cable cost is also derived from the public data on the recent HVDC projects. The other costs are estimates without market experience, and based on relative comparison of components with VSC converters. Note that the additional costs of offshore installations for converters should also be considered, if applicable.

23

DC Grids with Line-Commutated Converters

23.1 Multiterminal HVDC

The multiterminal high-voltage direct current (HVDC) has been studied since the first HVDC installation in 1950s but very few systems have been implemented. Numerous incentives exist for connecting a third terminal to an existing HVDC system, in particular when the DC line is long or when it passes close to load/generation centres.

There are two possible approaches in developing a multiterminal line-commutated converter (LCC) HVDC system: parallel and series connection, as schematically illustrated in Figure 23.1.

Some basic properties with parallel multiterminal HVDC system are:

- All terminals have the same DC voltage and it is assumed that DC voltage is tightly controlled.
- DC voltage is not allowed to change polarity. Note that some small systems (three terminals) can be configured to change DC polarity simultaneously on all terminals.
- One terminal controls DC voltage and all the other terminals regulate local DC current. The current-controlling terminals should also have DC voltage support function through local droop feedback.
- As all terminals are rated for full DC voltage, the low power terminals will be costly.
- A DC voltage fault (or a converter fault) will imply power interruption on all terminals.
- A commutation failure at any terminal implies DC voltage collapse and power interruption on all terminals.
- All terminals will operate with larger (ignition or extinction) angles in order to provide adequate control margin. This implies larger reactive power consumption, harmonics and switching losses.
- Fast power reversal is not possible at any terminal (as this would require voltage polarity reversal). Special mechanical switches may provide slow offline power reversal. Alternatively, at higher costs, bidirectional thyristor valves can be employed for fast current polarity reversal.

High-Voltage Direct-Current Transmission: Converters, Systems and DC Grids, First Edition.
Dragan Jovcic and Khaled Ahmed.
© 2015 John Wiley & Sons, Ltd. Published 2015 by John Wiley & Sons, Ltd.

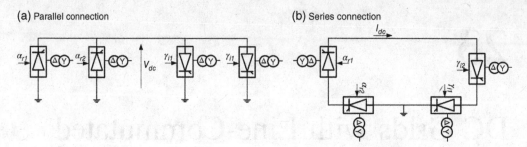

Figure 23.1 Four-terminal LCC HVDC in parallel and series connection.

The basic properties of series connected multiterminal HVDC are:

- All terminals have the same DC current. The DC current should be kept at a rated value at all times in order to provide full power control capability at all terminals. This implies large losses at partial loading.
- One (large) terminal controls the DC current and all the other terminals control local DC voltage. Small terminals will use low voltage rating, which has cost benefits.
- The DC voltage insulation will be unequal, dependent on operating conditions and complicated. It is likely that all segments will be insulated for highest DC voltage in order to provide operating flexibility when some terminals are out of service. This implies that costs will be similar as in a parallel connection.
- A commutation failure on any terminal implies only local power interruption, while all other terminals will operate normally. This is a major advantage of series connection.
- DC-side faults can be managed as partial power interruption.
- Power reversal on each terminal is readily achieved with independent reversal of local DC voltage, but restrictions may apply. No additional equipment is required.

It is generally accepted that in a typical high-power transmission application a parallel multiterminal connection is, overall, more attractive. A series connection might be suitable in case of small taps.

23.2 Italy–Corsica–Sardinia Multiterminal HVDC Link

The Italy–Corsica–Sardinia HVDC is the only multiterminal HVDC with long operating experience and it will be analysed in some depth. Note that the Quebec-New England HVDC has been designed as a multiterminal system but it does not have significant operating experience in multiterminal configuration. It is mentioned that the Indian four-terminal North-East Agra project should be commissioned in 2015.

The top-level schematic of the Italy–Corsica–Sardinia HVDC is shown in Figure 23.2. The Corsica converter was installed at a later stage with the intention not to disturb the operation of the main Italy–Sardinia link. The main operating mode of this link is the frequency control of Sardinia system. This implies that power transfer is highly variable (DC voltage is kept constant) and fast power reversal is feasible within 300 ms, which makes it difficult to add new terminals. Also, DC current on the main line is variable, and this has further eliminated the possibility of a series tap at Corsica.

The selected topology is shown in Figure 23.2 and the main characteristics are:

- The Italy station is a rectifier and controls the DC current.
- The Sardinia station is the main inverter, which controls DC voltage.

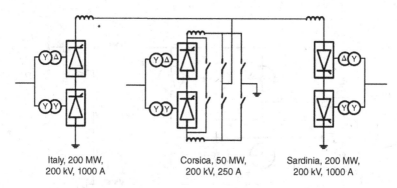

Figure 23.2 Italy–Sardinia HVDC with Corsica tap.

- The Corsica tap controls DC current (either as rectifier or inverter). Power reversal at Corsica is achieved using mechanical switches as main DC voltage polarity is assumed to be constant. This requires Corsica power interruption for a short period.
- If the main stations (Italy and Sardinia) change power direction by reversing DC voltage then the Corsica converter (in order to keep same local power direction) should change DC connections with mechanical contacts. This implies temporary Corsica power interruption.
- The Corsican converter is operating with a very large extinction/ignition angle of around 40°. This enables resilience to commutation failure for up to a 20% voltage drop on the local AC system voltage. On the downside, this solution implies large reactive power demand (equal to active power) and increased converter losses.

23.3 Connecting LCC Converter to a DC Grid

23.3.1 Power Reversal

This section assumes an established DC grid, which can be regarded as an infinite bus with AC systems. The DC grid has firm DC voltage of constant polarity, which is labelled $\pm V_{dc}$. It is capable of delivering or absorbing the DC current as an infinite DC bus, and the grid may contain LCC or voltage source converters (VSCs). The study is concerned with the methods of connecting a thyristor converter to DC grid.

Figure 23.3 shows the thyristor converter connection to a DC grid where mechanical switches are used to reverse power direction by changing current direction. The switches are low-power mechanical DC circuit breakers (CBs), which can break or make only limited current (essentially zero current). In order to change power direction the converter will bring current to zero and it will be blocked. The mechanical DC CBs will change over the connections and then the converter will be restarted in either rectification or inversion mode, depending on voltage polarity. The complete power reversal may take several seconds.

Figure 23.4 shows the thyristor converter with bidirectional thyristor valves connected to a DC grid. This converter is capable of rapidly changing DC current direction (100–200 ms). Although this converter has twice the number of semiconductors, compared with topology in Figure 23.3; the two antiparallel switches in a valve are not operated simultaneously and they will share many common components. Some estimates show that the station cost may not exceed 150% of the cost of the station in Figure 23.3.

Figure 23.5 shows another interesting integration option for a LCC converter. A DC/DC converter is used at the grid connection point. The DC/DC converter provides DC voltage stepping which can substantially reduce costs of DC cable and the new terminal in particular with low power terminals.

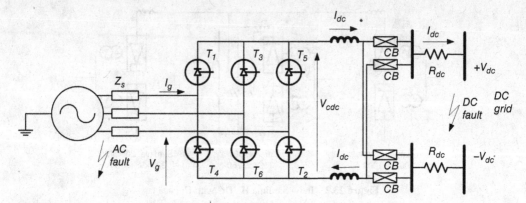

Figure 23.3 Connecting thyristor converter to a DC grid using mechanical switches for changing DC current direction.

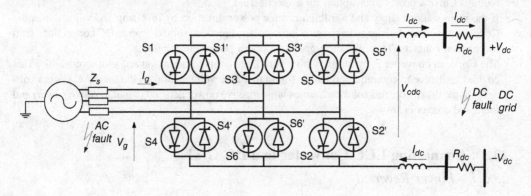

Figure 23.4 Connecting a thyristor converter with bidirectional valves to a DC grid.

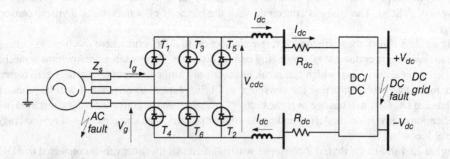

Figure 23.5 Connecting a thyristor converter to a DC grid using a DC/DC converter.

Some DC/DC topologies may provide voltage reversal on the side of LCC converter and current reversal on the DC grid side. A suitable DC/DC converter can provide power control and firewall against fault propagation. Although DC/DC topologies are not used with high-power DC networks; they are known from low-voltage DC applications and there are encouraging research results on DC/DC converters in DC grids.

23.3.2   DC Faults

If a DC fault happens in a DC grid, the LCC converter can rapidly control the fault current close to zero or to the rated converter output. There will therefore be no high current on the AC side and AC voltage/ grid will be unaffected. Note, however, that reactive power demand will transiently increase because of large firing angles.

Referring to Figure 23.3, if $V_{dc} = 0$, the LCC converter operating as a rectifier can reduce V_{cdc} in order to keep I_{dc} at rated values. If the LCC is operating as an inverter then the magnitude of V_{cdc} should be kept below V_{dc} in order to sustain the power transfer. This is feasible except in the case of very low impedance DC faults, which reduce $V_{dc} = 0$. An LCC converter, therefore, can readily operate through DC faults and it does not transfer DC faults to the AC side.

23.3.3   AC Faults

If an LCC operates as a rectifier, a small AC voltage depression will be compensated for by reducing the firing angle. For a nominal firing angle of 15°, the firing angle control can compensate for up to a 3.5% AC voltage drop. A more severe AC fault will imply reduction in power transfer. A fast voltage drop of over 20% will completely cease conduction in thyristors. This can have severe consequences if the LCC converter connects a large generation plant (loss of a generator). However such AC fault is not transferred to the DC side and DC voltage will not collapse. Note that slow AC voltage depression can be compensated with a transformer tap changer.

In the case of an LCC inverter, AC voltage depressions of over 5–10% will certainly cause commutation failure. Commutation failure is a short circuit on the DC side, which inevitably interrupts DC power transfer. Therefore an AC fault at a LCC inverter would cause power interruption and temporary DC voltage collapse in the entire DC grid. Typical recovery from DC commutation failure takes several cycles (50–100 ms) and, assuming that original AC fault is cleared, the power transfer is resumed in 200–500 ms. There are other ways to reduce commutation failure (like capacitor commutated converters or higher extinction angles) but they require further investments.

23.4   Control of LCC Converters in DC Grids

Figure 23.6 shows a simplified schematic for a LCC converter controller in a DC grid, which is very similar to a LCC converter controller in an HVDC and relates to both rectifier and inverter modes. If the converter is in power/current control mode it is necessary to add a stabilizing droop control loop. This control loop modifies the power order in response to DC voltage variation, as discussed further below.

If the converter is controlling DC voltage it would operate with an internal DC current control loop. This current loop limits DC currents in the case of DC faults.

A gamma controller is included as a safeguard from commutation failure and to aid postfault recoveries. A converter in the DC grid will not normally operate in gamma control mode and therefore the extinction angle will be larger than optimal.

23.5   Control of LCC DC Grids through DC Voltage Droop Feedback

The power balancing on a conventional, two-terminal HVDC is straightforward, as there is one rectifier supplying full power (into the DC line) as demanded by a single inverter. There is no possibility for power imbalance.

In the case of complex DC networks there will be many rectifiers supplying power to the DC grid and many inverters drawing power from the grid. Some converters will be uncontrollable by the DC grid

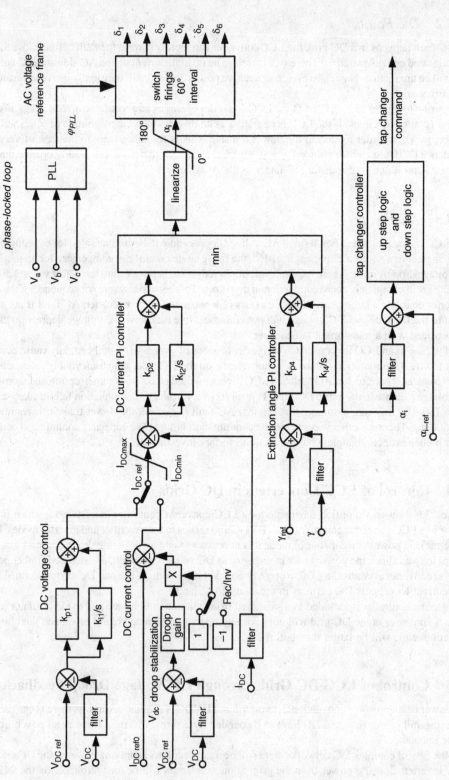

Figure 23.6 Control schematic for an LCC converter in DC grids.

operator (equivalent to load on conventional AC grids) and they draw power as required by the local control centre or by other factors (e.g. wind speed). The remaining converters, therefore, must instant-aneously balance power flow in the DC grid. A power imbalance may happen because of many other reasons like faults, converter tripping or other outages.

A power imbalance in DC grids will imply a DC voltage excursion. This mechanism is similar to frequency balancing on conventional AC grids.

In a large DC grid, one (large) rectifier station will have a proportional-integral DC voltage controller, which will be capable of maintaining DC voltage exactly at the reference value with zero error. Because of the rating limitation, this converter cannot balance DC grid power on its own. The remaining converters will operate in power control mode where the power reference will be scheduled from the grid control centre (dispatching and scheduling in conventional AC grids). It is also important that the converters in power-control mode respond to DC voltage variations for two reasons: (i) to ensure power balancing and (ii) to dynamically support DC grid stability and transient responses. This control response requires droop DC voltage feedback, just as droop frequency feedback is used with conventional AC grids. The droop gain $k_{DCdroop}$ adjusts the power reference in response to DC voltage variations, and it is calculated as:

$$k_{DCdroop} = \frac{2P_{DCmax}}{V_{DCmax} - V_{DCmin}} \qquad (23.1)$$

where V_{DCmax} and V_{DCmin} are the allowed limits for the DC voltage deviation and P_{DCmax} is the maximum power deviation allowed for system support (up to the converter's rated power). The droop feedback operates in such way that a depression on the DC voltage will lead to lower power demand from an inverter or increased power injection for a rectifier. Rectifiers will normally be required to par-ticipate in automatic power balancing, except, perhaps, if they connect to uncontrollable sources (wind farms or other renewable parks). Depending on the grid topology, inverters should also contribute to automatic power balancing, although some inverters may have a droop gain set to zero. The control scheme with the local DC voltage droop feedback will be able to balance DC grid power without communication between terminals.

If a dynamic DC grid support is also required (because of DC grid strength) the DC droop gain should be designed considering dynamic stability. The DC droop feedback can be developed using dynamic filters (frequency dependent transfer function) to enhance stability in a particular frequency range. This is likely to be an important requirement with small DC grids.

It is noted that the DC voltage magnitude will be different at different points in the DC grid because of the voltage drop along the DC lines. This implies that voltage limits in Eq. (23.1) will be different for different converters. The most important issue, however, is that the DC reference voltage V_{DCref} will be dependent on the DC node and on the operating condition, which will complicate grid control signifi-cantly. This issue is not seen when a frequency droop is used with AC grids because frequency is equivalent throughout the grid.

Figure 23.7 shows steady-state operating curves for a four-terminal DC grid shown in Figure 23.8. In this figure the rectifier 1 keeps DC voltage at a constant level (within its operating range). Each of the other three terminals regulates local current but the current reference depends on the DC voltage. For clarity of illustration, all currents are shown with positive sign, however inverter currents will have a negative sign and the sum of all currents will be zero at any instant.

23.6 Managing LCC DC Grid Faults

A DC grid with LCC converters will not have high DC fault currents as each LCC can regulate infeed from local AC grid. For permanent DC line or cable faults it is necessary to isolate the smal-lest segment of the grid (the faulted line) in order to enable normal operation of the remaining parts

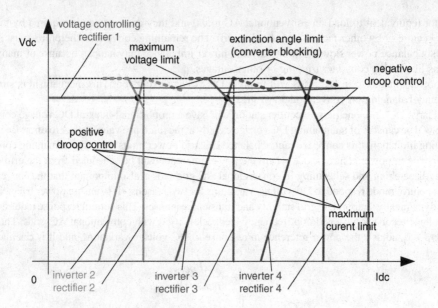

Figure 23.7 Steady-state VI characteristic of four-terminal LCC-based DC grid.

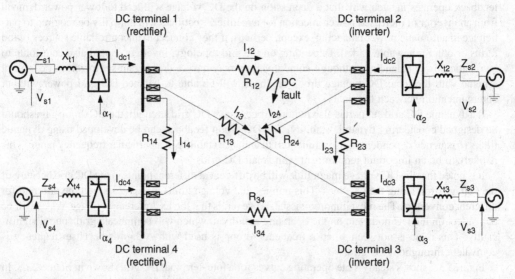

Figure 23.8 A four-terminal DC grid with LCC converters.

of the grid. This can be done using low-cost mechanical CBs considering that fault currents are low and clearance time is not critical.

If there is a fault on cable 12, as indicated in Figure 23.8, the two rectifier terminals will rapidly reduce local DC voltage in order to keep the local DC current below rated values. Within the controller time response, say 20–100 ms, the DC current on all rectifiers can be brought to zero. The inverter controllers will maintain the operating angles in the inversion range to avoid feeding the fault. The fast converter controllers can therefore reduce DC current in all DC lines to zero and this should clear all transient DC faults.

If a fault is permanent, the DC grid protection system can resolve the fault location and trip the two mechanical CBs at the ends of the affected line. Note that these CBs will be opening at zero or very low DC current. On isolating the fault, the grid DC voltage will be recovered and the grid assumes normal operation in a postfault topology.

23.7 Reactive Power Issues

An LCC converter in a DC grid may have more reactive power issues (compared with LCC in HVDC) for the following two reasons:

- An LCC converter will be operating with larger firing angles to enable an adequate control range, which requires more reactive power (both in rectifier or inverter modes).
- An LCC converter will operate with frequently varying firing angles. The angle variation is the result of active power changes and power balancing with the droop control. This implies requirement for variable reactive power support and frequent operation of CBs on capacitor/filter banks.

Some of these issues can be eliminated with DC/DC converters that provide variable DC voltage level, with the topology shown in Figure 23.5.

23.8 Large LCC Rectifier Stations in DC Grids

A DC grid may have a combination of VSC and LCC converters because each topology has advantages in certain applications. An LCC converter will cost less and has lower losses compared with a VSC converter.

A large rectifier station, perhaps connected to a nearby power station is a good candidate for LCC converter topology in DC grids. Assuming that the station is only operated as rectifier, the power direction change is not required and commutation failure is not an issue. The nearby generator should be able to supply fast-varying reactive power demand. A strong generator-based AC grid is also not likely to suffer AC voltage depressions and therefore no adverse impact on the DC grid is expected. Line-commutated converter technology may therefore have overall advantages for large rectifiers.

24

DC Grids with Voltage Source Converters and Power-Flow Model

24.1 Connecting a VSC Converter to a DC Grid

24.1.1 Power Reversal and Control

Figure 24.1 shows a two-level voltage-source converter (VSC) connected to a DC grid through a DC cable, R_{dc}. The DC grid is represented by an infinite DC bus, V_{dc}, and the power reversal is readily achieved by DC current direction reversal. The power exchange is controlled by adjusting the voltage difference across the resistor, R_{dc}.

24.1.2 DC Faults

DC faults are a serious problem with VSC converter topologies. A DC fault leads to a current increase through switches and converter protection will block insulated-gate bipolar transistors (IGBTs). When IGBTs are tripped, the VSC becomes an uncontrolled diode bridge that feeds the DC fault from the AC side. There is essentially a short circuit between the AC and DC sides and therefore the AC current will be high and the fault is transferred to the AC side. The AC voltage may also collapse. If another HVDC is connected nearby to the same AC system it would also be disturbed because of the low AC voltage. Such a fault is normally cleared with the AC-side circuit breaker (CB), AC CB in Figure 24.1, which has operating time of 20–100 ms. Once the converter AC CB is tripped, it may take considerably longer time to bring the converter back to operation.

The interfacing reactance, L_t, will help reduce the fault current magnitude, however there are other practical and performance limitations on its size. The inductor L_d reduces the fault current slope (derivative) but it cannot reduce the fault current magnitude as it is located on DC side.

24.1.3 AC Faults

A VSC converter readily controls AC voltage from zero to the rated value. This implies that the converter can remain operational and regulate the AC current for any AC system fault. A VSC controller

High-Voltage Direct-Current Transmission: Converters, Systems and DC Grids, First Edition.
Dragan Jovcic and Khaled Ahmed.
© 2015 John Wiley & Sons, Ltd. Published 2015 by John Wiley & Sons, Ltd.

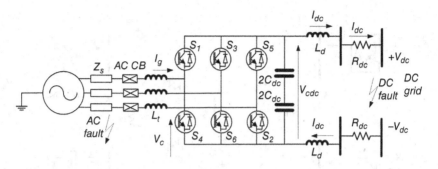

Figure 24.1 VSC converter connected to DC grid.

will have some time lag and, in the case of extreme AC faults, there will be a current overshoot in the first 10–20 ms. It is important that the first peak of the fault current is below the switch ratings and this is typically achieved by sizing the reactance, L_t, appropriately.

Except for the DC fault management, the VSC converters have clear advantages over the line-commutated converter (LCC) for employment in DC grids.

24.2 DC Grid Power Flow Model

This section presents the basic approaches for studying the power flow in DC transmission grids. In many aspects, the power-flow calculations in HVDC grids will be simpler than calculations in traditional HVAC grids. Firstly, only line resistances affect power flow and this implies that a single equation per DC line can be used. In HVAC grids a second equation is required to account for reactive power flow. Further, there is only one type of load in HVDC grids, which is a (VSC or LCC or DC/DC) converter. It is assumed that no other type of uncontrollable DC load exists. This simplifies power-flow studies significantly, considering that converters are fully controllable. In AC grids iterative calculations are needed because of voltage-dependent loads; this may not be required for DC loads.

In its simplest form, a VSC converter is represented as one of the following two components:

- *A constant DC current source.* The droop control can also be readily incorporated.
- *A constant DC voltage source.* One or more VSC converters in a grid can be configured to control the DC voltage by a zero-error integral controller.

Alternatively, the converter can be set to control the AC or DC power. If such a control is implemented, the power-flow calculations may require iterative procedures as the DC current injection depends on the DC voltage.

The DC grid will be connected to one or multiple HVAC grids. Under the above assumptions of controlled DC current/voltage sources, these HVAC grids have no influence on the HVDC grid power flow. The AC network will be disregarded in the remainder of the section and this enables effective studies of DC grid power flows and controls of any complexity. Otherwise, a combined AC/DC power-flow algorithm is required with separate power-flow solutions for each AC network.

This section develops a power-flow model for a fully meshed HVDC grid of any size, and with line impedances between any two DC nodes. Figure 24.2 shows a five-node (four-terminal) HVDC grid, which is used as the test system. Note that node 5 has no converter connection.

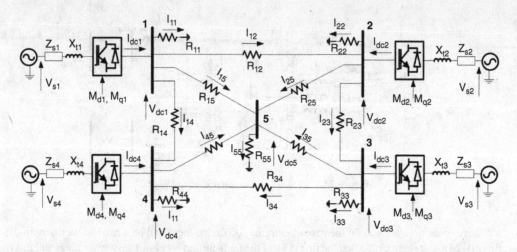

Figure 24.2 A five-node HVDC grid with VSC converters.

The first set of equations considers current flow in each DC line. For an n-node DC grid there is a maximum of $(n^2 - n)/2$ DC lines. For the test five-node system there are ten equations as follows:

$$V_1 - V_2 = I_{12}R_{12}$$

$$V_1 - V_3 = I_{13}R_{13}$$

$$V_1 - V_4 = I_{14}R_{14}$$

$$V_1 - V_5 = I_{15}R_{15}$$

$$V_2 - V_3 = I_{23}R_{23}$$

$$V_2 - V_4 = I_{24}R_{24} \qquad (24.1)$$

$$V_2 - V_5 = I_{25}R_{25}$$

$$V_3 - V_4 = I_{34}R_{34}$$

$$V_3 - V_5 = I_{35}R_{35}$$

$$V_4 - V_5 = I_{45}R_{45}$$

and for the system in Figure 24.2, $R_{13} = R_{24} = infinity$. The second set of equations considers the sum of currents at each DC bus/node. For an n-node DC grid there will be n equations and the test system has five equations:

$$I_{dc1} = I_{12} + I_{13} + I_{14} + I_{15} + V_1/R_{11}$$

$$I_{dc2} = -I_{12} + I_{23} + I_{24} + I_{25} + V_2/R_{22}$$

$$I_{dc3} = -I_{13} - I_{23} + I_{34} + I_{35} + V_3/R_{33} \qquad (24.2)$$

$$I_{dc4} = -I_{14} - I_{24} - I_{34} + I_{45} + V_4/R_{44}$$

$$I_{dc5} = -I_{15} - I_{25} - I_{35} - I_{45} + V_5/R_{55}$$

where all variables and parameters are defined referring to Figure 24.2. Further simplification is possible because, for the considered test system, $I_{dc5} = 0$. The local shunt impedance, R_{ii}, can represent portion of converter loss in normal operation and also can be conveniently used to study DC faults. The above set of equations should be expressed in the matrix form

$$x = Ax + Bu \tag{24.3}$$

where x is the state vector of the dimension $n + (n^2 - n)/2$ and u is the input vector of the dimension n. The state vector and input vectors will be different depending on the type of VSC converter control at each terminal. As an example, if terminal 1 is voltage controlling terminal and all other terminals control DC current then the state equations (the unknowns) include: terminal 1 DC current, all voltages except at terminal 1, and all cable currents, while inputs are voltage at terminal 1 and DC currents at all other terminals. This model in matrix form becomes:

$$
\begin{bmatrix} I_{dc1} \\ V_2 \\ V_3 \\ V_4 \\ V_5 \\ I_{12} \\ I_{13} \\ I_{14} \\ I_{15} \\ I_{23} \\ I_{24} \\ I_{25} \\ I_{34} \\ I_{35} \\ I_{45} \end{bmatrix} =
\begin{bmatrix}
0 & 0 & 0 & 0 & 0 & 1 & 1 & 1 & 1 & 0 & 0 & 0 & 0 & 0 & 0 \\
0 & 0 & 0 & 0 & 0 & R_{22} & 0 & 0 & 0 & -R_{22} & -R_{22} & -R_{22} & 0 & 0 & 0 \\
0 & 0 & 0 & 0 & 0 & 0 & R_{33} & 0 & 0 & R_{33} & 0 & 0 & -R_{33} & -R_{33} & 0 \\
0 & 0 & 0 & 0 & 0 & 0 & 0 & R_{44} & 0 & 0 & R_{44} & 0 & R_{44} & 0 & -R_{44} \\
0 & 0 & 0 & 0 & 0 & 0 & 0 & 0 & R_{55} & 0 & 0 & R_{55} & 0 & R_{55} & R_{55} \\
0 & -1/R_{12} & 0 & 0 & 0 & 0 & 0 & 0 & 0 & 0 & 0 & 0 & 0 & 0 & 0 \\
0 & 0 & -1/R_{13} & 0 & 0 & 0 & 0 & 0 & 0 & 0 & 0 & 0 & 0 & 0 & 0 \\
0 & 0 & 0 & -1/R_{14} & 0 & 0 & 0 & 0 & 0 & 0 & 0 & 0 & 0 & 0 & 0 \\
0 & 0 & 0 & 0 & -1/R_{15} & 0 & 0 & 0 & 0 & 0 & 0 & 0 & 0 & 0 & 0 \\
0 & 1/R_{23} & -1/R_{23} & 0 & 0 & 0 & 0 & 0 & 0 & 0 & 0 & 0 & 0 & 0 & 0 \\
0 & 1/R_{24} & 0 & -1/R_{24} & 0 & 0 & 0 & 0 & 0 & 0 & 0 & 0 & 0 & 0 & 0 \\
0 & 1/R_{25} & 0 & 0 & -1/R_{25} & 0 & 0 & 0 & 0 & 0 & 0 & 0 & 0 & 0 & 0 \\
0 & 0 & 1/R_{34} & -1/R_{34} & 0 & 0 & 0 & 0 & 0 & 0 & 0 & 0 & 0 & 0 & 0 \\
0 & 0 & 1/R_{35} & 0 & -1/R_{35} & 0 & 0 & 0 & 0 & 0 & 0 & 0 & 0 & 0 & 0 \\
0 & 0 & 0 & 1/R_{45} & -1/R_{45} & 0 & 0 & 0 & 0 & 0 & 0 & 0 & 0 & 0 & 0
\end{bmatrix}
\begin{bmatrix} I_{dc1} \\ V_2 \\ V_3 \\ V_4 \\ V_5 \\ I_{12} \\ I_{13} \\ I_{14} \\ I_{15} \\ I_{23} \\ I_{24} \\ I_{25} \\ I_{34} \\ I_{35} \\ I_{45} \end{bmatrix}
$$

$$
+ \begin{bmatrix}
1/R_{11} & 0 & 0 & 0 & 0 \\
0 & R_{22} & 0 & 0 & 0 \\
0 & 0 & R_{33} & 0 & 0 \\
0 & 0 & 0 & R_{44} & 0 \\
0 & 0 & 0 & 0 & R_{55} \\
1/R_{12} & 0 & 0 & 0 & 0 \\
1/R_{13} & 0 & 0 & 0 & 0 \\
1/R_{14} & 0 & 0 & 0 & 0 \\
1/R_{15} & 0 & 0 & 0 & 0 \\
0 & 0 & 0 & 0 & 0 \\
0 & 0 & 0 & 0 & 0 \\
0 & 0 & 0 & 0 & 0 \\
0 & 0 & 0 & 0 & 0 \\
0 & 0 & 0 & 0 & 0 \\
0 & 0 & 0 & 0 & 0
\end{bmatrix}
\begin{bmatrix} V_1 \\ I_{dc2} \\ I_{dc3} \\ I_{dc4} \\ I_{dc5} \end{bmatrix} \tag{24.4}
$$

The state variables can be obtained by solving:

$$x = [\mathbf{I} - \mathbf{A}]^{-1} \mathbf{B} u \qquad (24.5)$$

If terminals 2–5 use droop DC voltage feedback control, then Eq. (24.2) becomes:

$$I_{dc1} = I_{12} + I_{13} + I_{14} + I_{15} + V_1/R_{11}$$

$$I_{dc2} + k_{d2} I_{dc2\text{rated}}/V_{2\text{ref}}(V_2 - V_{2\text{ref}}) = -I_{12} + I_{23} + I_{24} + I_{25} + V_2/R_{22}$$

$$I_{dc3} + k_{d3} I_{dc3\text{rated}}/V_{3\text{ref}}(V_3 - V_{3\text{ref}}) = -I_{13} - I_{23} + I_{34} + I_{35} + V_3/R_{33} \qquad (24.6)$$

$$I_{dc4} + k_{d4} I_{dc4\text{rated}}/V_{4\text{ref}}(V_4 - V_{4\text{ref}}) = -I_{14} - I_{24} - I_{34} + I_{45} + V_4/R_{44}$$

$$I_{dc5} + k_{d5} I_{dc5\text{rated}}/V_{5\text{ref}}(V_5 - V_{5\text{ref}}) = -I_{15} - I_{25} - I_{35} - I_{45} + V_5/R_{55}$$

where k_{d2}–k_{d5} are droop gains (scaled relative to local converter power), $V_{2\text{ref}}$–$V_{5\text{ref}}$ are reference values of DC voltages and $I_{dc2\text{rated}}$–$I_{dc5\text{rated}}$ are nominal DC currents. The corresponding matrices can be readily derived.

In the case that all VSC converters control local DC voltage, Eq. (24.6) applies but the states are DC currents and cable currents, while inputs are all DC bus voltages (reference values). It is important to notice that the above system is linear and the power flow can be solved directly using Eq. (24.5).

Example 24.1

The four-terminal DC grid in Figure 24.2 uses terminal 1 to control the DC voltage to 400 kV. The remaining terminals control local DC current to the following set values: I_{dc2} = −650 A, I_{dc3} = 1200 A, I_{dc4} = 900 A. It is presumed, for simplicity, that droop feedback is not used. The resistance matrix for this grid is given as follows:

$$R\,(\Omega) = \begin{bmatrix} R_{11} & R_{12} & R_{13} & R_{14} & R_{15} \\ R_{21} & R_{22} & R_{23} & R_{24} & R_{25} \\ R_{31} & R_{32} & R_{33} & R_{34} & R_{35} \\ R_{41} & R_{42} & R_{43} & R_{44} & R_{45} \\ R_{51} & R_{52} & R_{53} & R_{54} & R_{55} \end{bmatrix} = \begin{bmatrix} 3e6 & 5 & \infty & 2 & 2 \\ 5 & 3e6 & 2 & \infty & 2 \\ \infty & 2 & 3e6 & 5 & 2 \\ 2 & \infty & 5 & 3e6 & 1.8 \\ 2 & 2 & 2 & 1.8 & 3e6 \end{bmatrix} \qquad (24.7)$$

Determine the DC current at terminal 1, voltages at all DC busses and currents in all DC cables.

Solution

Table 24.1 shows the currents and voltages at all terminals, while Table 24.2 shows the HVDC grid currents.

The current at terminal 1 (I_{dc1}) is equal to the sum of the connecting cable currents and it is equal to sum of currents at all other terminals plus the losses. The terminals that operate as inverters (terminals 2 and 3) have the lowest DC voltages. The terminals control local current, but the cable currents have widely different values, and in general a central dispatcher is required to optimise DC grid current flow.

Table 24.1 Terminal voltages and currents in Example 24.1.

Terminal	1	2	3	4	5
Vdc (kV)	400.00	398.09	397.75	399.90	398.96
Idc (A)	950.66	−650.00	−1200.00	900.00	0.00

Table 24.2 HVDC grid currents in Example 24.1.

Cable	I_{12} (A)	I_{13} (A)	I_{14} (A)	I_{15} (A)	I_{23} (A)	I_{24} (A)	I_{25} (A)	I_{34} (A)	I_{35} (A)	I_{45} (A)
Current	381.8	0	49.9	518.7	167.6	0	−435.9	−428.9	−603.6	520.8

24.3 DC Grid Power Flow under DC Faults

When a fault happens in a DC grid, the converter IGBTs are tripped and the VSC converter becomes the VSC diode bridge and therefore the converter model derived previously is no longer valid. The VSC diode bridge currents under low DC voltage at each terminal can be calculated as presented in section 20.5.1, using Eq. (20.25), which considers AC system parameters. However, this equation requires DC voltage as the input. On the other hand, the DC grid power-flow algorithm in Eq. (24.5) requires VSC converter currents as the external inputs. It is therefore required that an iterative algorithm be used as illustrated in Figure 24.3. Initially, DC voltages are assumed to be around 10% of

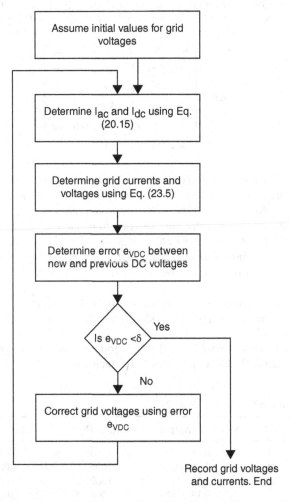

Figure 24.3 Iterative algorithm for calculating grid power flow under DC faults.

the rated values and the external currents are determined. These external currents are used as the inputs in the power-flow algorithm, which gives DC voltages at all nodes. If these voltages differ from the assumed values by more than a prespecified tolerance level, δ, then the DC voltages are corrected and the power-flow calculation is repeated. The process is repeated until the calculated DC voltages are close to the DC voltages from the previous step. Note that the power-flow algorithm in Eq. (24.2) is changed because all terminals are now operating as current injecting terminals (the first equation is modified).

Example 24.2

Consider the same four-terminal DC grid in Figure 24.2 (Example 24.1) and assume that there is a DC fault on cable 34 (fault resistance is 0.1 Ω) and the fault location is at 1/5 of the length from terminal 3. The AC system data for all terminals are given in Table 24.3. Determine DC currents in all VSC converters, in DC cables and all DC voltages.

Solution

It is recommended to introduce an artificial node 6 at the fault location (between terminals 3 and 4) to enable use of the power flow model. Since the fault is at 1/5th length towards terminal 3, the new cable resistances are $R_{36} = 1 \times 5/5 = 1\ \Omega$ and $R_{46} = 4 \times 5/5 = 4\ \Omega$. The new DC grid resistance matrix is:

$$R\ [\Omega] = \begin{bmatrix} 3e6 & 5 & \infty & 2 & 2 & \infty \\ 5 & 3e6 & 2 & \infty & 2 & \infty \\ \infty & 2 & 3e6 & \infty & 2 & 1 \\ 2 & \infty & \infty & 3e6 & 1.8 & 4 \\ 2 & 2 & 2 & 1.8 & 3e6 & \infty \\ \infty & \infty & 1 & 4 & \infty & 0.1 \end{bmatrix} \qquad (24.8)$$

Using the power-flow calculation method under faults described above, the grid-voltages and currents can be calculated as shown in Figure 24.4. All VSC converters have very large currents and IGBTs will be tripped immediately. The currents in the faulted cable can reach very large values (20 and 9 kA). Some DC cables, like cable 35, transfer fault currents from multiple AC grids, which is important for dimensioning DC CBs. Almost all the DC cables have a high fault current and this makes it more difficult to locate the fault. It is interesting that some DC cables will have DC fault current only marginally higher than nominal current (cable 12), but this conclusion depends on the fault location.

Also, the DC voltage has dropped from nominal 400 to 20–40 kV on most DC busses. At the faulted bus (node 6) the DC voltage has dropped to 3 kV for 0.1 Ω fault resistance.

It is seen in the above example that a DC fault brings grid-wide DC voltage collapse and high current infeed from all AC systems. The DC voltage under a DC fault is around 0.1 pu at all nodes. This implies

Table 24.3 AC system parameters in Example 24.2.

Terminal	SCR	X/R	$X_t = X_{t1} + X_l$ (pu)	$P_{dcnominal}$ (MW)	$V_{dcnominal}$ (kV)	$I_{dcnominal}$ (kA)
1	20	9	0.15	1000	±400	1.25
2	30	20	0.15	1000	±400	1.25
3	10	10	0.15	1000	±400	1.25
4	22	12	0.15	1000	±400	1.25

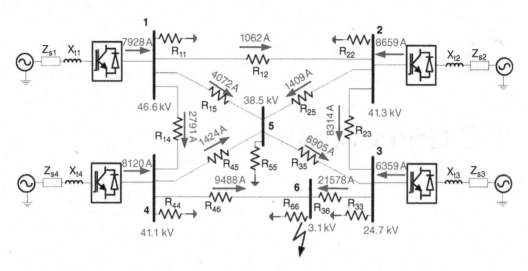

Figure 24.4   Power flow in five-node HVDC grid under DC fault at cable 34 in Example 24.2.

a loss of power on all terminals, likely AC voltage depression on all VSC terminals and therefore significant disruption in the overall power transfer. In case of large DC grids (supergrids), this will be a significant issue as AC systems experience multiple simultaneous faults and problems may occur with regional stability of large transmission systems. A fault at any point in DC grids leads to loss of power in the whole DC grid for the duration of the fault.

25

DC Grid Control

25.1 Introduction

If an AC/DC converter is installed with the main purpose of facilitating power trading between the local AC grid and the DC grid, the local AC system operator might primarily be interested in the exchange of active and reactive power through the converter. However, when connected in a DC grid, the voltage source converter (VSC) controllers will be required to contribute to DC grid stabilization in addition to local control. Because of the importance of DC grid integrity, the DC grid power balancing and stabilization may assume high priority in control systems at each terminal and some control functions may contradict local AC controls and may require tradeoffs in performance.

The DC grid control will involve local converter control at each terminal and a central dispatcher, as shown in Figure 25.1. The local controller at each terminal will receive local measurements to regulate local variables, but it will also receive commands (reference values, gain settings, etc.) from the dispatcher. Any communication with the dispatcher will involve delays (in each direction), depending on the grid size. The speed of the control signal propagation in optical cables is close to the speed of light and therefore the delay will be at least 1 ms per 300 km, plus processing delay, in each direction. It is also important to consider that communication equipment may not be reliable and therefore it should always be considered that dispatcher signals may not be available. The grid control must be developed to ensure integrity/stability for all foreseeable outages, with and without the dispatcher. The dispatcher can be best viewed as an 'optimizer' of DC grid performance.

25.2 Fast Local VSC Converter Control in DC Grids

Connecting to the DC grid will impose a range of demands on the VSC converter controller:

- The highest priority is the DC grid dynamic small-signal stability. Small-signal stability is the system's capability to return to an equilibrium state (all variables are bounded) following a disturbance. Aperiodic or oscillatory instability will imply the loss of a grid segment or loss of the whole grid. All

High-Voltage Direct-Current Transmission: Converters, Systems and DC Grids, First Edition.
Dragan Jovcic and Khaled Ahmed.
© 2015 John Wiley & Sons, Ltd. Published 2015 by John Wiley & Sons, Ltd.

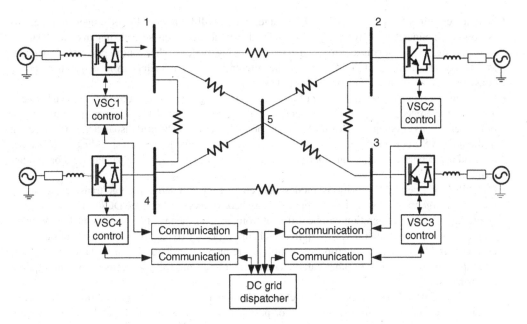

Figure 25.1 DC grid control structure.

the variables in a DC grid will be closely coupled dynamically because of low impedance of DC lines and low inherent inertia or energy storage (capacitances). Consequently, a growing oscillation on one variable will be observed widely across the whole DC grid. This is different from AC grids, where dynamic instability may develop locally and typically can be alleviated locally. DC grid dynamics will be very fast and the time constants will be of 1–2 orders of magnitude smaller than with AC grids. This implies considerably less time for corrective control action. It is therefore essential that a controller for each converter connecting to a DC grid has the capability of enhancing the stability of the DC grid. It will not be appropriate to develop converter controls using traditional assumptions of infinite DC bus and AC impedance. Steady-state approaches like V-I diagrams and droop gain study are the first steps towards developing controls for DC grid stability but they are not sufficient. In addition, the dynamics of the overall DC grid including converter-converter interactions should be considered and ideally the eigenvalue studies across all frequencies and transient time domain responses should be researched for all operating conditions.

- Maintaining DC voltage within the operating limits. The DC voltage operating limits will typically be within 5–10% around the DC grid nominal voltage. The upper DC voltage limit will be determined considering the component insulation requirements and it must be coordinated with surge arrester characteristics (insulation coordination) on the entire DC grid. As the cost of semiconductors is high, and much higher than cost of electromechanical components in the AC grids, the upper DC voltage limit will be tight and very important. Theoretically, all VSC converters can operate with high DC voltage (unless current ratings are exceeded) and therefore the upper limit is not crucial for functionality. However, depending on the designed control margin, a VSC converter loses control capability (control variables are in saturation) for a DC voltage drop of around 10–20%. If DC voltage falls further, the converter current increases and semiconductors will be tripped to prevent overheating.
- Receiving reference signals and control parameters from the dispatcher and reacting accordingly. Typically these signals will be changing slowly over a longer time period (secondary and tertiary control). The exact signals from the dispatcher will be defined by the DC grid code but they will likely

include set points for power and/or DC voltage, which will be adjusted over longer time periods (seconds and minutes). It may be required that the dispatcher also requests a change in control parameters, like droop gain or control limits, change in operating mode or other control parameters, depending on the prevailing grid condition. All the references and parameters from dispatcher can be considered constant in the fast transient studies.

- Detecting and reporting on the quality of local control. A large DC voltage error (difference between actual and reference DC voltage) or power error will imply abnormal DC grid operation (unbalanced power references or outages). This information should be sent to a DC grid dispatcher. The saturation of the inner VSC converter current control will prevent the converter from responding to dispatcher demand in the normal manner and therefore should be relayed to the grid dispatcher (and/or to other terminals).
- Capability of operation in various modes like emergency, limited ramping, transient recovery. Typically, large transients (faults or tripping of large converters) will change DC-grid characteristics significantly and grid operation will be different from nominal steady-state conditions. During these transients, a VSC controller may be required to change topology (typically to introduce various limiters and overrides) or locally override set points (like reducing transiently DC-node voltage set point) or change controller gains, or other parameters, in order to ensure DC grid recovery within required performance indicators.
- Capability of islanded operation and black start. Islanded operation or segmentation of a DC terminal may occur during transient events, or permanent faults, when all the connecting lines are tripped by grid protection logic. Under these circumstances, the grid voltage control should still satisfy all the above requirements (stability, limits, etc.). It is therefore essential that the VSC controller has the capability to reduce power to zero in response to fast DC voltage variation. A black-start capability is desired on all DC grid terminals and can be readily incorporated with typical VSC converter technologies. Note that a DC-grid black-start function will be very different from an AC grid black start, which is commonly incorporated in VSC HVDC controls.

It is understood that some VSC converters will not be able to satisfy some or all of the above requirements. A typical example is a VSC converter connected to an isolated (offshore) wind farm. Such a VSC will be a controlling power according to the wind-farm requirements because a wind farm has no means to regulate power (it is dependent on wind speed). Another example is a VSC connected to a passive, weak or critical AC system (offshore platform perhaps). The priority of such VSC controller will be the stability of the AC system and it cannot be expected to contribute to DC grid stabilization.

25.3 · DC Grid Dispatcher with Remote Communication

A dispatcher will be required with DC grids and, in particular, this is important if international trade is involved. DC grids will be designed to operate safely without dispatcher intervention but, as with AC grids, a dispatcher will be assigned to coordinate, optimize and reconfigure the grid.

The main roles of the dispatcher include:

- Sends DC control references (power and/or voltage) to all terminals. The factors that determine power dispatching include the operating situation, the desired power trade at each terminal, DC grid constraints, dispatch merit order, loss minimization and other requirements. The power references should be balanced (the sum of powers at all terminals is zero); however, this will never ideally be achieved because grid losses cannot be accurately determined. Furthermore, the tripping of any converter, cable or substation can happen locally at any time and such an event would lead to a significant power reference unbalance in the grid. Such an unbalance would be temporary, lasting only until the dispatcher receives accurate information on the situation and the availability of all grid components. In the meantime, the DC grid should be able to operate in a stable manner with unbalanced power references. The balancing of DC grid powers will be simpler than with AC grids because most terminals (sources and loads) are controllable. Some terminals may have a high-priority local AC power demand, not responding to dispatcher power demands or only weakly responding to dispatcher demands for grid stabilization.

- Sends other operating parameters, like droop gains, DC voltage controller limits or operating mode requests, in order to optimize DC grid operation and ensure the required disturbance sharing among terminals for future events.
- Continuously monitors the DC grid-operating conditions and reschedules power flow if any safe operating limits are violated. The VSC converters can automatically and independently control only local variables like converter current and DC terminal voltage. The currents in DC grid cables and voltages on uncontrolled DC nodes (like bus 5 in Figure 24.2) cannot be directly controlled and they will be the dispatcher's responsibility. The large cables will have thermal time constants in the order of tens of minutes and therefore there is no need for urgent control action. All these constraints will be incorporated in the grid models at the dispatcher, which will be running continuously under inputs from the terminals. The dispatcher may use 'alert-state' labelling to indicate an operating condition that may not ensure adequate responses to future disturbances ($n - 0$ condition). In such a case, more severe actions are required from the dispatcher, which should transform the grid to $n - 1$ or $n - 2$ operating conditions.

25.4 Primary, Secondary and Tertiary DC Grid Control

At the fastest control level (within 10 ms), all the converters have a number of feedback control loops that ensure that valve currents and converter voltages are within rated values and numerous other functions (balancing, harmonic reduction, etc.). These control loops are concerned with converter safety and cannot be removed. Typically they receive current references from the higher control levels, including the primary response DC grid control. The interface between converter control and DC grid control will be at current reference as schematically shown in Figure 25.2.

The primary system response is commonly described as converter reaction to a DC grid disturbance within the initial time window, spanning perhaps up to 100 ms. In this period all power and DC voltage reference signals are constant and no new communication signals have arrived to converter stations from the dispatcher. Only local controls are active. It is essential that the primary response acts to dynamically stabilize the DC grid – that it acts towards balancing power in the DC grid and keeps DC voltages within rated limits. However, the primary response alone will not be able to establish a new optimal state for the DC grid.

The secondary response involves adjustments of DC power and/or DC voltage reference points at each terminal in order to ensure that DC voltages are within the required margins and that power flow is as close as possible to the desired values. Secondary control will typically involve some communication between the dispatcher centre and the terminals, and this will happen with at least a 5–10 ms delay but the entire transient may involve hundreds of milliseconds or longer. The secondary response will not involve direct human intervention. The control will be implemented as a DC power/voltage reference change, control override, control limit setting, control mode change or similar. In the event of limited-magnitude disturbances (wind power change, load adjustment or similar), primary and secondary control will be adequate.

The tertiary control may include human intervention and will happen within minutes and perhaps tens of minutes. It will be activated following a significant outage on the DC grid, like a large converter

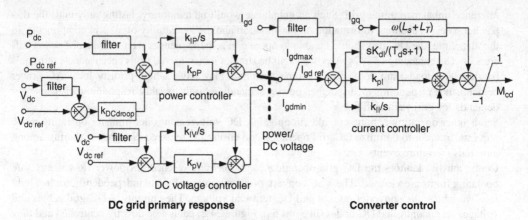

Figure 25.2   VSC converter controller with droop feedback for DC grid stabilization.

or line tripping. Tertiary control may include the following: significant changes in power references at terminals, tripping/connecting of terminals, changes in droop gains and others.

25.5   DC Voltage Droop Control for VSC Converters in DC Grids

The DC voltage is the single most important variable which determines integrity of the whole DC grid, the power flow and also the capability of all VSC converters to normally operate. DC voltage drop (even very short) of 10–20% will initiate tripping of VSC converters. Because of lack of any energy storage elements in DC grids, DC voltage will rapidly increase for a power flow unbalance (say inverter tripping). It is therefore essential that the DC voltage is robustly controlled throughout the DC grid for all operating conditions and outages, and DC voltage becomes bedrock of DC grid control. This implies that all converter stations (VSC terminals and DC/DC converters) should respond to DC voltage variations in a stabilizing manner. The DC voltage is an indicator of power balance in the DC grid, similarly as frequency is indicator of power balance in AC systems.

Figure 25.2 shows a simplified controller (only active power control, while reactive power control is unchanged) for a VSC converter in a DC grid. It is very similar to an ordinary VSC HVDC controller but a droop feedback loop is added to the power controller. This droop feedback constitutes the primary response function and it is required for the VSC converter to contribute to improving stability of DC grid. The converter detects DC voltage variation through V_{dc} droop feedback and adjusts the power exchange in such a way as to stabilize DC voltage.

The central challenge with the above method is what DC voltage reference V_{dcref} to use at each terminal and under particular operating conditions. As illustrated in Example 24.1, DC voltage is different at each terminal and it changes with the operating conditions. One possible solution is to use a single V_{dcref} on all terminals, which can be the rated grid DC voltage and which could be periodically updated by the central dispatcher. Inevitably there will be some error in power control on many terminals with this approach and the DC voltage limits may not be well defined in some operating conditions.

The second challenge with the droop feedback is the dynamic stability. The same droop gain is used for power balancing and also it will significantly influence dynamic responses. It is desired to have high droop gains in order to suppress steady-state DC voltage limits. However, high droop gains may have a negative impact on small-signal stability, which will be reflected in poorly damped oscillatory modes during transient responses.

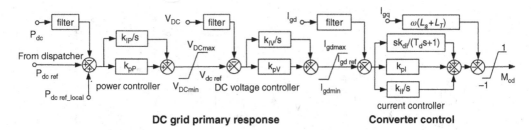

Figure 25.3   VSC converter controller with three levels.

25.6   Three-Level Control for VSC Converters with Dispatcher Droop

25.6.1   Three-Level Control for VSC Converters

Figure 25.3 shows the VSC converter control with three levels, which is a possible solution for converter control in DC grids. The controller includes:

- At the fastest level, there is inner current controller, which is identical to the controller described in section 17.6. The reference signal is the I_{gdref}, which is limited between $-I_{gdrated}$ and $I_{gdrated}$.
- A DC voltage controller is permanently operating at the middle level. The local DC voltage is controlled to a reference, which is derived from the power controller at the higher level. The voltage reference has hard limits, which are typically set to $-0.95V_{DCrated}$ and $1.05V_{DCrated}$. These limits are required to ensure DC grid integrity in case local power differs widely from the power reference (a power reference unbalance in the DC grid).
- At the highest level is the power controller. This controller is only active if the DC voltage is within $\pm 5\%$. The power reference signal will be received from a dispatcher. However, a provision is made for local power reference, which might be used as an example in case of local emergency situations. The two control loops (DC voltage and power) jointly constitute primary response.

The main advantages of the three-level controller are:

- DC voltage control has two gains (proportional and integral) at each terminal. This enables fast DC voltage control with excellent dynamic responses at each terminal. Integral gain ensures steady-state tracking with zero error, while proportional gain is used to adjust transient responses. Under all operating conditions, DC voltage control has priority and control should respond similarly in all conditions.
- There is a hard limit on V_{dcref} at each terminal (typically around $\pm 5\%$). As DC voltage is controlled at each terminal with a proportional-integral type controller, this method ensures that DC voltage will always stay within the hard limits at all terminals.
- The control method is very robust against unbalanced power orders and outages as any large demands by the power controller will be moderated by the hard limits on the DC voltage reference. Normally power references will be balanced by the dispatcher (sum of all power orders is 0). However, unbalanced power references may happen transiently if a communication link is broken or in the case of a terminal outage.

25.6.2   Dispatcher Controller

If three-level control is used at VSC terminals, then a dispatcher should be equipped with a droop-based power-reference calculation algorithm, as shown in Figure 25.4. This controller maintains the average

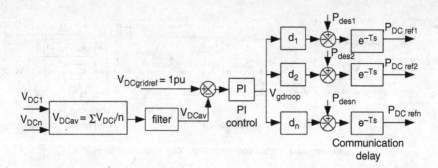

Figure 25.4 Dispatcher controller when VSC converters use three-level control.

DC voltage (for the whole DC grid) at the reference value of 1 pu, which is constant for all operating conditions. If average DC voltage deviations are detected then power orders are adjusted for all terminals according to the droop gains, d_1–d_n.

The advantage of this approach is that the global DC voltage average, which is a good indicator of power balance, is used for power adjustments.

Moreover, power balancing (which can be slow) is separated from the fast dynamic DC voltage stabilization at each terminal that must be fast.

25.7   Power Flow Algorithm When DC Powers are Regulated

This section will expand further the DC grid power-flow algorithm to include power regulation at each terminal, as this would correspond to power-controlling VSC converters in DC grids.

When DC power feedback is used (either proportional or integral) it is required to use an iterative algorithm in order to determine converter DC currents, as currents are now dependent on the grid DC voltages in the model in Eq. (24.6). An iterative algorithm, like Newton–Raphson, can be developed that has similar structure to the method for DC fault conditions, as shown in Figure 24.3.

Example 25.1
Consider the same four-terminal ±400 kV DC grid test system as in Example 24.1, where all terminals use a three-level controller from Figure 25.3 (there is no DC voltage controlling terminal). Assume that power references are given as: P_{ref1} = 760 MW, P_{ref2} = −520 MW, P_{ref3} = −960 MW, P_{ref4} = 720 MW. A central dispatcher has topology as shown in Figure 25.4, and it controls the average grid voltage to 400 kV. Assume, further, that DC voltage reference limits on all terminals are set to ±5%. Determine power flow, all DC voltages and currents:

a. *For the base case.*
b. *Assuming that the local operator at terminal 4 significantly changes the power reference to P_{4ref} = −100 MW because of an emergency situation on the local AC system.*
c. *Assuming that terminal 1 is tripped.*
d. *Assuming that cable 12 is tripped.*
e. *Assuming that cables 12 and 15 are tripped.*

Solution
a. *Table 25.1 shows the power flow for the base case. The average DC voltage is well controlled at 400 kV. The power references at each terminal are different from local power references as the dispatcher adjusts power to optimize DC voltage at 400 kV. The sum of all powers is not equal*

Table 25.1 Terminal voltages and currents in Example 25.1(a) (with balanced case).

Terminal	1	2	3	4	Node 5	Average/sum
Vdc (kV)	401.0	399.1	398.8	400.9	399.9	400.0
Pdes (MW)	760	−520	−960	720	0.0	−69
P, Pref (MW)	761.6	−518.3	−956.4	721.3	0.0	8.2
Idc (A)	949.5	−649.3	−1199.1	899.6	0.00	—

Cable	I_{12}	I_{13}	I_{14}	I_{15}	I_{23}	I_{24}	I_{25}	I_{34}	I_{35}	I_{45}
Current (A)	381.5	0	49.7	518.2	167.6	0	−435.5	−428.6	−603.1	520.5

Table 25.2 Terminal voltages and currents in Example 25.1(b) (when $P_{4ref} = -100$ MW).

Terminal	1	2	3	4	Node 5	Average/sum
Vdc (kV)	399.9	398.4	397.9	399.3	398.9	399.0
Pdes (MW)	760	−520	−960	−100	0.0	6.2
P, Pref (MW)	930.6	−302.9	−737.0	114.6	0.0	5.13
Idc (A)	1163.4	−380.3	−925.9	143.5	0.00	—

Cable	I_{12}	I_{13}	I_{14}	I_{15}	I_{23}	I_{24}	I_{25}	I_{34}	I_{35}	I_{45}
Current (A)	314.1	0	329.6	519.6	199.3	0	−265.6	−261.9	−464.9	211.1

Table 25.3 Terminal voltages and currents in Example 25.1(c) (when terminal 1 is tripped).

Terminal	1	2	3	4	Node 5	Average/sum
Vdc (kV)	399.6	398.7	398.5	400.2	399.3	399.2
Pdes (MW)	760	−520	−960	720	0.0	5.6
P, Pref (MW)	0.0	−266.1	−700.1	970.8	0.0	4.6
Idc (A)	0	−333.7	−878.4	1212.8	0.00	—

Cable	I_{12}	I_{13}	I_{14}	I_{15}	I_{23}	I_{24}	I_{25}	I_{34}	I_{35}	I_{45}
Current (A)	173.8	0	−326.7	152.8	121.8	0	−281.8	−353.2	−403.6	532.8

to zero because of grid losses. The actual distribution of the currents through the cables in the grid is also shown.

b. Table 25.2 shows the power flow for this case. The dispatcher controller successfully maintains average grid voltage at 400 kV. At all terminals the power levels are very different from the reference power. This situation occurs because reference powers are highly unbalanced and the dispatcher controller significantly adjusts power references on all terminals in order to maintain grid integrity. In this case, tertiary action would also be required (after a period of time) in order to adjust desired power to balanced values. This example examines only primary and secondary action.

c. A large terminal is suddenly lost and this introduces a significant unbalance in grid power references. Table 25.3 shows the power flow. It is seen that the unbalance is equally shared among all the remaining terminals because the actual power will be different from the reference at each terminal. The average DC voltage is maintained at 400 kV.

Table 25.4 Terminal voltages and currents in Example 25.1(d) (when cable 12 is tripped).

Terminal		1	2	3	4	Node 5	Average/sum
Vdc (kV)		401.6	398.7	398.7	401.2	400.1	400
Pdes (MW)		760	−520	−960	720	0.0	52.2
P, Pref (MW)		761.5	−516.8	−955.4	721.1	0.0	10.3
Idc (A)		948.1	−648.0	−1198.1	898.7	0.00	—

Cable	I_{12}	I_{13}	I_{14}	I_{15}	I_{23}	I_{24}	I_{25}	I_{34}	I_{35}	I_{45}
Current (A)	0	0	201.0	747.0	19.1	0	−667.2	−492.9	−686.3	606.7

Table 25.5 Terminal voltages and currents in Example 25.1(e) (when cables 12 and 15 are tripped).

Terminal		1	2	3	4	Node 5	Average/sum
Vdc (kV)		403.2	398.0	398.0	401.3	399.2	400
Pdes (MW)		760	−520	−960	720	0.0	54.1
P, Pref (MW)		762.7	−514.8	−952.9	721.4	0.0	16.3
Idc (A)		945.8	−646.8	−1197.1	898.7	0.00	—

Cable	I_{12}	I_{13}	I_{14}	I_{15}	I_{23}	I_{24}	I_{25}	I_{34}	I_{35}	I_{45}
Current (A)	0.0	0.0	945.6	0	−35.7	0	−611.2	−657.4	−575.5	1186.8

d. The power references are balanced but cable 12 is lost. Table 25.4 shows the power flow. The power orders are marginally adjusted to reflect a different power flow in the grid. The average DC voltage is well maintained at 400 kV. Cable 15 now takes significantly more power and cable 14 reverses power direction.

e. In this case two cables at terminal 1 are lost (cable 12 and cable 15) but the power references remain balanced and Table 25.5 shows the power flow. The average grid DC voltage is kept at 400 kV. The DC grid currents are distributed in a very different way than in the base case. Cables 14 and 45 take significantly more current and might approach thermal limits. In such a case tertiary action by dispatcher would be required to reduce some power orders or redistribute power flow.

Example 25.2

Consider the same four-terminal ±400 kV DC grid test system as in Example 25.1, The same power references and same DC voltage reference limits are used. Assume that all communication with the dispatcher is lost and therefore no updated signals are received from the dispatcher following the disturbances studied below. All DC grid terminals operate on their own, using only local signals.

Determine power flow, all DC voltages and currents:

a. For the base case,
b. Assuming that local operator at terminal 4 significantly changes power reference to $P_{4ref} = -100\,MW$, which is required because of emergency situation on the local AC system.
c. Assuming that terminal 1 is tripped.
d. Assuming that cable 12 is tripped.
e. Assuming that cables 12 and 15 are tripped.

Solution

a. *Table 25.6 shows the power-flow solution for the base case. Since the power references are balanced, the average DC voltage remains at 400 kV, without any central dispatcher control. The power flow is only marginally different from the case with the dispatcher active (in Example 25.1).*

b. *Table 25.7 shows the power flow for this case. As the dispatcher is inactive, the terminals retain the last good dispatcher order. However, the power reference change at terminal 4 implies that power references are unbalanced. There is more power demand than power injection and DC voltage drops in the whole grid. When the DC voltage falls to the lower ±5% limit at any terminal, that terminal becomes the DC voltage-controlling terminal. In this case the DC voltage-controlling terminals are terminals 2 and 3. The voltages at these terminals are controlled to 0.95 pu (with a small tolerance). At all other terminals the power becomes different from the reference value and the system will settle at a nearest stable point, which provides natural sharing of the unbalance between all terminals. It is seen that even without dispatcher the grid voltage is controlled and the power is balanced.*

c. *In this case, a large terminal is suddenly lost and this introduces a significant unbalance in grid power references, while the dispatcher is inactive. Table 25.8 shows the power flow. The voltage drops to the lower limit of 380 kV because of insufficient power input. The power unbalance is shared among all the remaining terminals and the final power flow is similar to the case with the dispatcher active (Example 25.1(c)). This is the result of natural balancing between terminals when multiple terminals cannot achieve the reference power value. Therefore, even when the dispatcher and a large terminal are lost, the powers will be naturally balanced between terminals, while DC voltage will not deviate beyond limits.*

d. *In this case the power references are balanced but cable 12 is lost. Table 25.9 shows the power flow. It is seen that the average DC voltage remains at 400 kV. The terminal power values are similar as in the base case, but the power flow in the DC grid is different. The final power flow is similar to the case when dispatcher is active (Example 25.1(d)).*

Table 25.6 Terminal voltages and currents in Example 25.2(a) (with balanced case).

Terminal	1	2	3	4	Node 5	Average/sum
Vdc (kV)	401.0	399.1	398.8	400.9	399.9	400.0
Pdes (MW)	760	−520	−960	720	0.0	−69
P, Pref (MW)	761.9	−518.6	−956.8	721.7	0.0	8.2
Idc (A)	950.1	−649.7	−1199.7	900.1	0.00	—

Cable	I_{12}	I_{13}	I_{14}	I_{15}	I_{23}	I_{24}	I_{25}	I_{34}	I_{35}	L_{45}
Current (A)	381.7	0	49.7	518.5	167.6	0	−435.8	−428.8	−603.4	520.8

Table 25.7 Terminal voltages and currents in Example 25.2(b) (when P_{4ref} = −100 MW).

Terminal	1	2	3	4	Node 5	Average/sum
Vdc (kV)	381.1	379.2	378.8	380.9	380.0	380.0
Pdes (MW)	760	−520	−960	−100	0.0	9.2
P, Pref (MW)	775.8	−495.3	−916.9	644.7	0.0	8.3
Idc (A)	1017.9	−653.2	−1210.3	846.2	0.00	—

Cable	I_{12}	I_{13}	I_{14}	I_{15}	I_{23}	I_{24}	I_{25}	I_{34}	I_{35}	L_{45}
Current (A)	390.9	0	85.8	540.9	173.9	0	−436.3	−426.2	−610.3	505.7

Table 25.8 Terminal voltages and currents in Example 25.2(c) (when terminal 1 is tripped).

Terminal	1	2	3	4	Node 5	Average/sum
Vdc (kV)	380.4	379.6	379.3	381.1	380.1	380
Pdes (MW)	760	−520	−960	720	0.0	5.6
P, Pref (MW)	0.0	−252.5	−668.9	926.1	0.0	4.6
Idc (A)	0	−332.6	−881.7	1215.0	0.00	—

Cable	I_{12}	I_{13}	I_{14}	I_{15}	I_{23}	I_{24}	I_{25}	I_{34}	I_{35}	I_{45}
Current (A)	173.9	0	−327.2	153.1	123.0	0	−281.8	−354.0	−404.8	533.6

Table 25.9 Terminal voltages and currents in Example 25.2(d) (when cable 12 is tripped).

Terminal	1	2	3	4	Node 5	Average/sum
Vdc (kV)	401.4	398.6	398.6	401.0	399.9	399.9
Pdes (MW)	760	−520	−960	720	0.0	52.2
P, Pref (MW)	762.9	−518.1	−956.4	722.0	0.0	10.3
Idc (A)	950.1	−649.8	−1199.8	900.2	0.00	—

Cable	I_{12}	I_{13}	I_{14}	I_{15}	I_{23}	I_{24}	I_{25}	I_{34}	I_{35}	I_{45}
Current (A)	0	0	201.51	748.5	18.7	0	−668.7	−493.8	−687.4	607.8

Table 25.10 Terminal voltages and currents in Example 25.2(e) (when cables 12 and 15 are tripped).

Terminal	1	2	3	4	Node 5	Average/sum
Vdc (kV)	403.0	397.7	397.8	401.1	398.9	399.7
Pdes (MW)	760	−520	−960	720	0.0	54.1
P, Pref (MW)	765.7	−516.8	−954.5	722.1	0.0	16.3
Idc (Å)	950.1	−649.8	−1200.0	900.0	0.00	—

Cable	I_{12}	I_{13}	I_{14}	I_{15}	I_{23}	I_{24}	I_{25}	I_{34}	I_{35}	I_{45}
Current (Å)	0.0	0.0	949.9	0	−36.5	0	−613.5	−659.4	−576.9	1190.6

e. *In this case two cables at terminal 1 are lost (cable 12 and cable 15) but the power references remain balanced and Table 25.10 shows the power flow. The average DC voltage stays at 400 kV, even though the dispatcher is inactive. The DC grid currents are very differently distributed than in the base case, but they are similar to the case with dispatcher active (Example 25.1(e)).*

25.8 Power Flow and Control Study of CIGRE DC Grid-Test System

25.8.1 CIGRE DC Grid Test System

The Conseil International des Grands Réseaux Électriques (CIGRE) DC grid-test system is developed jointly by the B4.57 and B4.58 working groups to enable more harmonized DC grid studies and bench-marking within CIGRE and in a wider professional community.

Figure 25.5 shows the CIGRE DC grid with a nominal power flow. The system consists of:

1. Two onshore AC systems: system A (busses: A_0 and A_1), and system B (busses: B_0, B_1, B_2 and B_3).
2. Four offshore AC systems: system C (busses: C_1 and C_2), system D (bus: D_1), system E (bus E_1) and system F (bus F_1). One offshore system (bus E) is a load (oil platform).
3. Three DC grids: DCS1 (±200 kV monopolar HVDC with busses A1 and C1), DCS2 (±200 kV monopolar five-terminal DC grid with busses B_2, B_3, B_5, E_1 and F_1) and DCS3 (±400 kV bipolar eight-terminal DC grid with busses A_1, C_2, D_1, E_1 B_1, B_4 and B_2).
4. Two DC/DC converters: Cd-E1, which enables a power exchange between 800 kV and 400 kV DC systems, and Cd-B1, which regulates power flow in the line between busses B1 and E1.

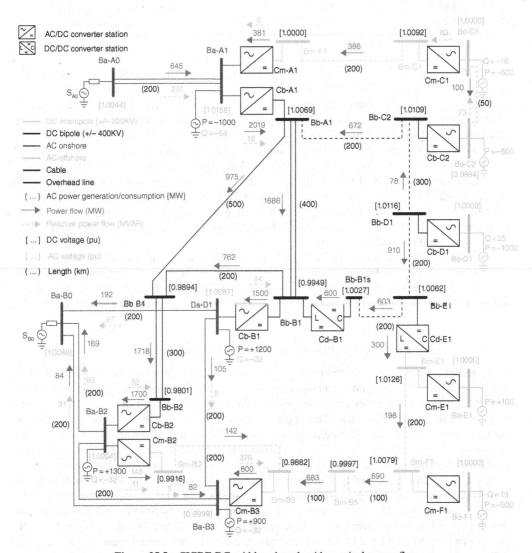

Figure 25.5   CIGRE DC grid benchmark with nominal power flow.

Table 25.11   Power reference values for the CIGRE DC grid.

Converter station	Power rating (MVA)	Power reference (MW)
Cm-A1	800	DC voltage control
Cm-C1	800	AC Slack (P = 500 MW)
Cm-B2	800	150
Cm-B3	1200	−900
Cm-E1	200	AC Slack (P = −100 MW)
Cm-F1	800	AC Slack (P = 500 MW)
Cb-A1	2 × 1200	2000
Cb-B1	2 × 1200	−1500
Cb-B2	2 × 1200	−1700
Cb-C2	2 × 400	500 MW
Cb-D1	2 × 800	AC Slack (P = 1000 MW)
Cd-E1	1000	300 (from Bb-E1 to Bm-E1)
Cd-B1	2000	600 (from Bb-E1 to Bb-B1)

The AC systems include infinite sources, fixed loads (or both) and transmission lines as indicated in the figure. Figure 25.5 also shows all the line lengths. A systematic labelling notation is adopted, where C stands for converter, B for bus, subscript *m* stands for monopolar, *b* for bipolar and *d* for DC/DC.

The proposed power set points are shown in Table 25.11. All the converters connected to onshore grids are equipped with the 3L (three-level) control from section 25.6.1. The DC/DC converter Cd-E1 also employs a similar three-level control and responds to DC voltage deviations on either side (DCS2 and DCS3). The offshore AC/DC converters regulate local power and do not respond to the dispatcher. There is a dispatcher at each of the two DC grids and they use average voltage controller as shown in Figure 25.4 with 1 pu reference voltages (800 kV and 400 kV) and a 20 ms delay is included on all signals.

The actual power flow is influenced by the droop setting and gives slightly different values from the reference powers for the onshore converters, as Figure 25.5 shows. All the voltages are shown scaled with respect to the nominal DC voltage.

The CIGRE test system power flow in Figure 25.5 is obtained using a nonlinear time-domain simulation where AC/DC and DC/DC converters use ABC frame nonlinear average models, while the controllers are modelled in detail, AC and DC cables are represented using distributed parameter models and loads are fixed power sources (using local active and reactive power feedback control).

25.8.2   Power Flow after Outage of the Largest Terminal

In this section, one of the outages, tripping of the largest converter (Cb-A1), is presented. Figure 25.6 shows the steady-state power flow after blocking Cb-A1, which results in 2 GW infeed loss in the DC grid DCS3. The grid establishes a balanced postfault power flow and that all DC voltages are close to rated values in a new steady state. The dispatcher controllers reschedule power references on the onshore converters while the offshore converters keep the power level unchanged. The DC/DC converter Cd-E1 reduces the power infeed to DCS2 (from 300 to 55 MW) in order to stabilize DCS3. This causes a disturbance to DCS2 and therefore the dispatcher on DCS2 also reschedules power.

Figure 25.7 examines the quality of transient responses on DC voltages. This significant disturbance causes an immediate DC voltage drop of almost 15% but the local DC voltage loops enable very fast DC voltage regulation and DC voltage is soon elevated to within ±5 limits. After around 300–500 ms the dispatcher action becomes visible and DC voltages return to 1 pu. Figure 25.8 shows the average DC voltages on the grid DCS2 and DCS3, which confirms that dispatcher control maintains both average voltages at 1 pu after this outage.

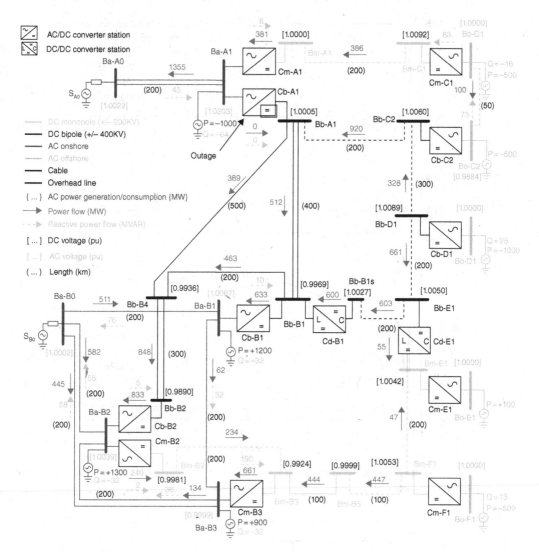

Figure 25.6 CIGRE DC grid benchmark after terminal Cb-A1 outage, assuming 3L converter control and dispatcher droop control.

If dispatcher communication is lost for the above outage (not shown in figures), the DC grid also establishes a stable new operating point but one of the DCS3 DC voltages settles at 0.95 pu limit and the average DC voltage will be below 1 pu.

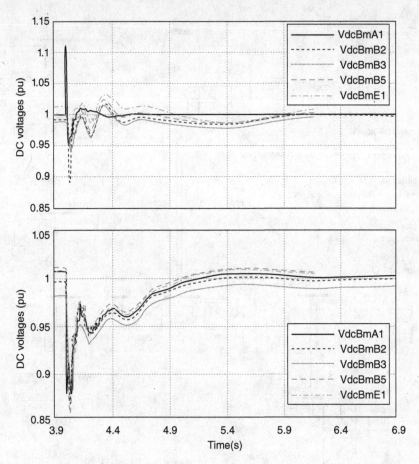

Figure 25.7 CIGRE DC grid benchmark DC voltages after terminal Cb-A1 outage, assuming 3L converter control and dispatcher droop control.

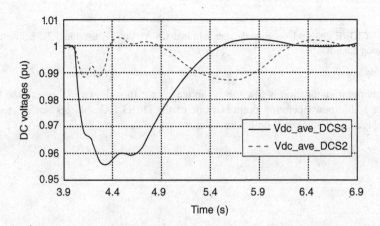

Figure 25.8 CIGRE DC grid benchmark average DC voltages after terminal Cb-A1 outage, assuming 3L converter control and dispatcher droop control.

26

DC Grid Fault Management and DC Circuit Breakers

26.1 Introduction

The protection systems for fault management in traditional AC systems have evolved into very reliable, flexible and cost-effective technologies. It is feasible to develop a protection system for any new AC system topology that designers and operators might demand. The management of faults in a DC grid, however, is considerably more difficult. The main challenges with DC grid protection include:

- *Direct-current fault current has no zero crossings.* This creates problems because all mechanical DC circuit breakers (CBs) exploit natural zero crossing to interrupt the current arc. Mechanical DC CBs can be developed based on AC CB technology but they become more complex and costly.
- *Direct-current line impedances are significantly lower.* In AC systems the impedances consist of resistive and reactive parts, where reactive component $(2\pi fL)$ may reach 10–20 times the resistive part. Zero frequency means that the DC grid impedances have only a resistive part, and the impedance magnitude is considerably smaller. This implies much larger fault current magnitudes and a low voltage level across the entire grid for a DC fault.
- Locating a fault in DC grids is more difficult because of low impedances.
- The semiconductor-based components in a DC grid – voltage-source converters (VSCs), DC/DC converters, DC CBs – have very small thermal constants and a very small overcurrent rating. Considering, too, that costs of semiconductor components are high, there is a strong requirement to clear DC faults in a short time and therefore the fast speed of protection system operation is very important.
- Voltage-source converters are blocked if DC voltage drops to around 80–90% of nominal value. This implies loss of a terminal and potentially loss of the whole DC grid for any DC fault. Converter reconnection involves the full startup sequence (charging, synchronization, etc.), which may take a long time (measured in seconds). In general, DC faults should be cleared within 3–5 ms to avoid tripping of VSC converters.

High-Voltage Direct-Current Transmission: Converters, Systems and DC Grids, First Edition.
Dragan Jovcic and Khaled Ahmed.
© 2015 John Wiley & Sons, Ltd. Published 2015 by John Wiley & Sons, Ltd.

- Many DC systems involve cables that have significant amount of shunt capacitive impedance and further capacitance is present with converter DC-side capacitors and DC-filters. Shunt capacitors will discharge into DC faults, increase fault currents and may cause oscillations. These capacitors only contribute to the fault current in the first few milliseconds but as DC protection requires a short operating time, the capacitances increase the requirement for current rating of DC circuit breakers.

The above challenges make development of DC protection systems and protection components considerably more challenging and costly than AC grids.

In a large DC grid the primary role of protection will be to ensure power transfer security and grid stability following DC faults. This is achieved by isolating the faulted segments in such way that the remaining part of the grid can continue to operate normally. In this aspect the protection should be:

- *Selective*. Only the smallest faulted segment should be isolated.
- *Sufficiently fast*. If the grid remains under a faulted condition for too long, the converters will be blocked, there will be large loss in capacity and AC machine stability may be endangered according to the angular stability criteria.
- *Reliable and robust*. In case of protection component failure a backup protection should isolate progressively larger grid segments.

An equally important role of the protection system is safety – both safety of personnel and prevention of component damage. The major components in DC grids that should be protected include AC/DC converters, DC capacitors, DC cables/lines and possibly DC/DC converters. Note that fault-protection equipment, that is DC CBs, cannot sustain fault conditions indefinitely and should also possess self-protection (in case of the failure of the main protection system). All semiconductor-based components have very short thermal time constants, small overcurrent capability and generally small I^2t energy store capability, which influences DC grid-protection speed requirements. DC cables, on the other hand, involve a significant mass of material and will not suffer a destructive temperature increase even under considerable I^2t energy over the rated values (time constant is in minutes).

26.2   Fault Current Components in DC Grids

A single AC/DC converter under DC faults is studied in Chapter 20. A complex DC grid may have numerous VSC converters and numerous DC lines. However, a DC grid will have only three energy sources that can feed a DC fault:

- the AC grids, which contribute to a steady-state fault current through AC/DC converters;
- the DC capacitors, which contribute to a transient fault current;
- the energy stored in DC lines, which contribute to a transient fault current.

This is an important conclusion because a steady-state fault current in a DC grid can only be supplied from the joining AC grids and therefore the fault level in connecting AC grids will ultimately determine the magnitude of DC fault currents. Under a fault condition in the DC grid only the back electromotive force (EMF) voltage on rotating machines in the AC grid can be assumed as constant voltage (the machine speed stays at the rated values for the fault duration). Therefore the total impedance in the fault path (including DC lines and AC lines) determines the fault current. A DC cable, depending on its location inside a DC grid, could have a fault current magnitude larger than any single VSC

Table 26.1 DC fault current contribution through AC/DC converter.

Converter type	LCC	Two and three level VSC	MMC VSC (half bridge)	MMC VSC (full bridge)
Uncontrolled DC fault current from AC system	No	Yes	Yes	No
Uncontrolled DC fault current from DC capacitors	No	Yes	No (assuming no DC filters)	No (assuming no DC filters)
Capability to reverse DC voltage to extinguish DC fault current	Yes	No	No	Yes
DC circuit continuity can be interrupted	Yes	No	No	Yes

converter because several AC systems could be contributing to the cable fault current. In the Example 24.2, the fault current in the cable 34 is over 21 kA.

Depending on the type of AC/DC converter, there may exist one or two sources of DC fault current on each DC terminal, as summarized in Table 26.1. The last two rows indicate the converter's ability to drain power from the DC cable by reversing voltage polarity, and to interrupt DC cable energy discharge, which are particularly important for recovering from transient DC faults (on overhead lines).

The shunt capacitors have been extensively used on AC systems and their impact on fault currents and protection requirements are well understood. The DC capacitors and lines, especially cables, will increase the initial peaks of transient fault currents but they do not contribute to the steady-state fault current as discussed in Section 20.4.

The typical VSC converters will have small DC side inductors. The purpose of these inductors is to limit the capacitor discharge current and also to limit the slope of increase of main fault current from AC grid. The presumption with this protection principle is that there will never be DC faults between the converter and the inductors (in the valve hall).

Direct-current CB costs are high, so it is desirable to have as low interrupting current as possible. One possible strategy to reduce DC CB duty is to operate DC CB in the short time before the fault current reaches high values. In order to reduce the rate of rise of the transient discharge current, and also to prolong the time for the current to reach steady-state values, additional inductors can be introduced at each end of the line. In practice, inductors will be placed in series with the DC breakers in the same way as inductors are placed on the DC side with VSC converters. This extra time will also enable the protection system to selectively make the trip decision to isolate only the faulted line. Introducing inductors will increase the overall losses in the system and increase the magnetic energy that needs to be cleared by the energy damps in DC breakers. A compromise needs to be found when installing coils: the lower derivative, the additional time and the subsequent gains in selectivity that can be gained by installing coils, must be weighed against the size and cost of the coils and the effect on the breaker requirements.

Example 26.1

Consider the six-node 640 kV DC grid as shown in Figure 26.1. Approximately, without detailed calculations, determine the worst case steady-state DC fault level at node 5.

Solution

Using the formula (20.31), the steady-state DC fault currents at each DC terminal can be found as shown in Table 26.2.

Table 26.2 DC fault current at each terminal in Example 26.1.

Terminal	1	2	3	4
Idcfpu (pu)	5.87	4.1	5.09	6.7
Idcf (A)	9 182	3 197	11 937	10 471

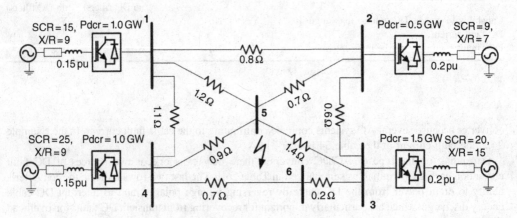

Figure 26.1   Six-node DC grid in Example 26.1.

Therefore, neglecting all DC resistances (they are small compared to AC side reactances) the maximum steady-state DC fault current at any point in the grid is the sum of the fault infeeds from the four terminals:

$$Idcft = Idcf1 + Idcf2 + Idcf3 + Idcf4 = 34\ 787\ A$$

Note that this approach is only valid if each terminal connects to a separate AC system.

26.3   DC System Protection Coordination with AC System Protection

The faults in DC grid should be cleared by using DC protection. The whole process of isolating a faulted DC cable should be completed before AC CBs (which will also see fault current) start tripping. In general DC CBs will have faster operating times than AC CBs, as will be discussed below. The VSC converter might be blocked under a DC fault and it should be possible to restart VSC converter after the DC fault is cleared. The whole process can happen while converter AC CB is closed.

The fast semiconductor-based DC CBs, which are discussed below, are capable of interrupting a fault current within few milliseconds and possibly even before VSCs are tripped. However in complex DC grids the process of finding the fault location and coordination between various DC CBs might take a longer time. The overview of protection operating timeframes for AC and DC sides is illustrated in Figure 26.2.

26.4   Mechanical DC Circuit Breaker

26.4.1   Operating Principles and Application

The mechanical kilovolt-range DC CBs are commercially available and they have been used in some industrial DC systems and as neutral switches with high voltage direct current (HVDC). The largest

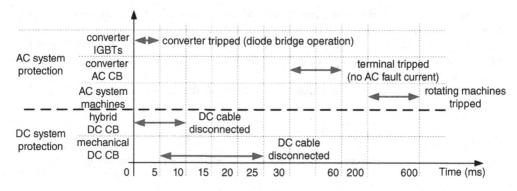

Figure 26.2 Operating timeframe for AC and DC system protection.

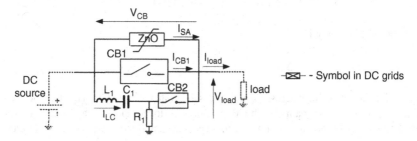

Figure 26.3 Mechanical DC circuit breaker.

commercial DC CB have a DC voltage rating of around 1–3 kV with nominal DC current of several kiloamperes and DC current interruption capability of 50–100 kA. In 1980s EPRI developed a prototype 250 kV mechanical DC CB but they have not been used commercially. With studies of DC grids in 2010–2015, there has been renewed interest in this technology and several manufacturers have already demonstrated prototype mechanical DC CB products with ratings of 50–100 kV, 5–10 kA.

Figure 26.3 shows the mechanical DC CB topology. It consists of the following components:

- Main CB1 and auxiliary CB2, which are similar to common AC CBs. The main circuit breaker CB1 is normally closed and it is rated for full DC voltage and peak interrupting current. The auxiliary CB2 is normally open and it has similar rating as CB1.
- Surge arrester Zno, for preventing overvoltage and for damping energy stored in the DC system.
- Resonant circuit (L_1 and C_1), which generates AC current when CB2 is closed. This AC current is superimposed on the main DC current in order to create current zero crossings in CB1.
- Charging resistor R1, which enables charging of C1 to the rated voltage.

The operating time of this DC CB depends on the speed of mechanical contacts, which must separate sufficiently to provide isolation distance for CB1 and fully close for CB2. This time is typically in the range 20–30 ms; however, the latest designs use very fast mechanical systems and multiple manufacturers have recently reported clearing times below 10 ms. The inner AC circuit resonant frequency will be in the order of kilohertz and this also helps achieving faster operating speed than 50 Hz AC CBs. In general, mechanical DC CBs will be able to operate faster than AC CBs, which is important for protection coordination. The mechanical DC CBs will be operating after VSC converters are tripped and therefore the diode bridge fault current will determine the DC CB interrupting current. Also, if mechanical DC CBs are used, the diode bridge must be overrated (to sustain DC fault for 10–20 ms) or some other means of fault current limiting will be used on all DC grid terminals. If very large series DC inductors are used it might be theoretically possible to keep VSC converters operating until the mechanical DC CBs are tripped.

On the positive side, the costs and losses of the above DC CB will be low.

26.4.2 Mathematical Model and Design Principles

Assuming that the initial current is zero, the basic equations for the LC circuit in Figure 26.3 are:

$$i_{LC} = \frac{v_{CB} - V_{c1o}}{z_o} \sin(\omega_o t) \tag{26.1}$$

$$v_{c1} = v_{CB} - (v_{CB} - V_{c1o}) \cos(\omega_o t) \tag{26.2}$$

where the natural frequency is:

$$\omega_o = 2\pi f_o = 1/\sqrt{L_1 C_1}, \ z_o = \sqrt{L_1/C_1}, \tag{26.3}$$

V_{c1o} is the initial value of voltage across C_1 and V_{CB} is the voltage across CB. The study here is concerned with the LC circuit when the main switch CB_1 is closed (prior to opening) and therefore $V_{CB} = 0$.

The peak current is particularly important for the design, and from Eq. (26.1) it is given by:

$$i_{LC\max} = \frac{V_{c1o}}{z_o} \tag{26.4}$$

The first derivative of the current and voltage are important for the practical design constraints, and from Eqs (26.1)–(26.2) they are given by:

$$\frac{di_{LC}}{dt} = \omega_o \frac{-V_{c1o}}{z_o} \cos(\omega_o t) \tag{26.5}$$

$$\frac{dv_{c1}}{dt} = \omega_o V_{c1o} \ \sin(\omega_o t) \tag{26.6}$$

The maximum values of the current derivative $(di_{LC}/dt)_{max}$ and voltage derivative $(dv_{C1}/dt)_{max}$ in Eqs (26.5) and (26.6) are:

$$\left(\frac{di_{LC}}{dt}\right)_{\max} = \omega_o \frac{V_{c1o}}{z_o} = \omega_o i_{LCp} \tag{26.7}$$

$$\left(\frac{dv_{c1}}{dt}\right)_{\max} = \omega_o V_{c1o} \tag{26.8}$$

Dividing Eq. (26.7) by Eq. (26.8) it is possible to obtain a relationship that links the peak values for current and voltage derivatives:

$$\left(\frac{dv_{c1}}{dt}\right)_{\max} = \left(\frac{di_{LC}}{dt}\right)_{\max} \frac{V_{c1o}}{i_{LCp}} \tag{26.9}$$

The above equation has practical significance as hardware limitations on mechanical switches (ability to provide isolation and prevent restriking) are linked to current and voltage derivatives.

The equations for L_1 and C_1 can be obtained by rearranging Eqs (26.1)–(26.3).

$$L_1 = \frac{V_{c1o}}{\left(\dfrac{di_{LC}}{dt}\right)_{max}} \tag{26.10}$$

$$C_1 = \frac{1}{L_1 \omega_o^2} \tag{26.11}$$

A test system is developed to illustrate calculation of the values for CB parameters assuming that DC voltage is $V_{CB} = V_{c10} = 400\,kV$. Two values for peak interrupting current are considered: $i_{LCp} = 3000\,A$ and $i_{LCp} = 9000\,A$. Figure 26.4 shows the design graphs. In this figure, the 'practical limit' curve shows maximum values for current and voltage derivatives to enable successful commutation (below the curve), which is drawn taking into account the prototype tests with EPRI's 250 kV, 9000 A DC CB, but might be very different with modern mechanical designs. It is seen that the larger interrupting current will imply a requirement for a lower resonant frequency for the LC circuit and it will demand a larger capacitor. The inductor value does not depend on the peak interrupting current, but it depends very much on the current derivative.

Figure 26.5 shows the simulation of DC fault clearing using a mechanical DC CB. A DC fault occurs at 0.5 s and the DC voltage drops. The DC fault current is kept approximately constant by a grid-connected LCC (line commutated converter) converter; however, there is small initial peak, which is the result of a grid/cable capacitor discharge and converter controller delays. A delay time of 20 ms is purposely introduced in the DC CB opening to simulate practical delays in the reaction of the mechanical system. At 0.52 s DC CB2 contacts fully close and DC CB1 contacts open. Closing of DC CB2 initiates oscillations (around 1 kHz) in the current that circulates through DC CB1. The peak of these oscillations must exceed the DC fault current in order to drive the summing current to zero in DC CB1. In the lower plot it is seen that the current in DC CB1 reaches zero crossing after approximately 1 ms (period of f_0) and the DC CB1 current extinguishes. DC CB2 opens soon after, enabling the charging of the capacitor in preparation for the next trip signal. DC CB1 may reclose after a period of time depending on the higher level protection strategy.

26.5 Semiconductor Based DC Circuit Breaker

Semiconductor DC CBs are developed to improve performance over the mechanical DC CBs. In particular the following properties are desired:

- *Faster operation.* The intention is to operate DC CB before VSC converter IGBTs are tripped under DC fault conditions.
- *Lower DC interrupting current.* Since DC CB operation is fast (few ms), the fault current will not reach full steady-state value, and therefore the rating of DC CB will be lower.
- Unlimited number of open-close operations.
- Theoretically, if fast semiconductor DC CBs are used through the DC grid, the AC/DC converters may not require diode overrating. However, the consequence of protection failure would be catastrophic for the AC/DC converters.

A semiconductor based DC CB is shown in Figure 26.6, which includes the following components:

- *A valve S_1 with integral antiparallel diode D_1.* It may consist of numerous high-power IGBTs or insulated-gate commutated thyristors (IGCT) connected in series.

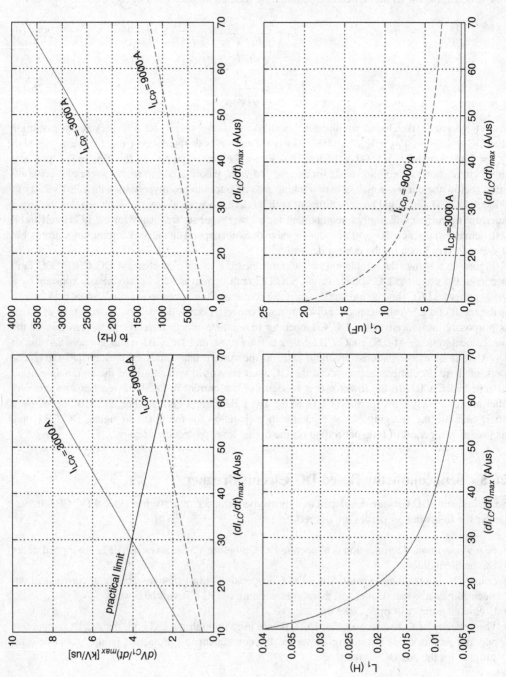

Figure 26.4 Selection of parameters for a mechanical DC CB, assuming $V_{CB} = 400$ kV and two peak interrupting currents $I_{LCp} = 3000$ A and $I_{LCp} = 9000$ A.

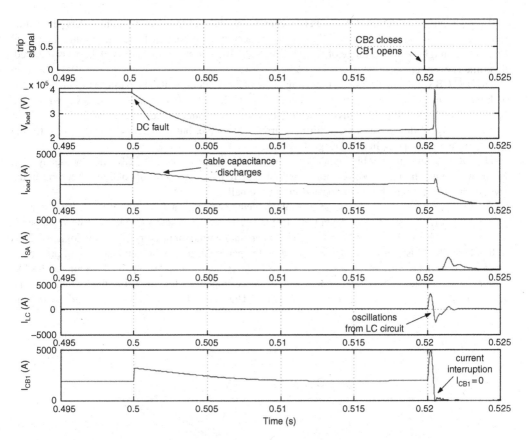

Figure 26.5 Mechanical DC CB simulation for a DC fault at 0.5 s, assuming 20 ms breaker opening delay.

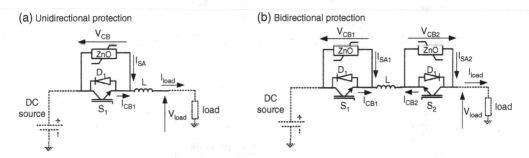

Figure 26.6 Semiconductor-based DC CB.

- An inductor, L, which limits the fault current derivative.
- A surge arrester, Zno, which dumps all fault energy in DC circuit.

The topology in Figure 26.6a is capable of interrupting fault current in only one direction, whereas diode D1 conducts uncontrollably in the opposite direction. If bidirectional protection is required then another valve is required, as shown in Figure 26.6b.

If a high current is detected the switch is capable of opening the circuit within a very short time (the microsecond range). However, the external protection coordination logic will require additional time, which might be in the range of few milliseconds. If communication across DC grid is required, this time is even longer.

This DC CB will be operating on the rising slope of DC fault current. The purpose of inductor, L, is to limit current derivative which constrains the DC fault current to an acceptable level within the protection operating time (2–5 ms). Note that this inductor cannot reduce the steady-state DC fault current. A large inductor can reduce further current derivative, but it increases energy rating for the surge arrester and it complicates DC grid dynamics.

This DC CB has considerably higher costs compared with mechanical DC CB. Also the steady-state losses will be much higher. Valve cooling will also be required (because of continuous current in the switch) and therefore it should be located in a valve hall.

Figure 26.7 shows the response of a semiconductor switch to a DC fault at 0.1 s. The nominal DC current is 1.2 kA and the trip level is set to 1.9 kA. The slope is set to 1.33 kA/ms, which depends on the inductor L and the line inductance. The total operating time for primary protection logic is 2.2 ms in this figure but this will be highly dependent on the particular installation and the type of protection system. The peak DC current in this case is 4 kA and the switch should be capable of opening at this DC current level. Typically this requirement will be more important for designing the switch than the requirement for nominal on-state current of 1.2 kA. The same DC CB may also be used for backup

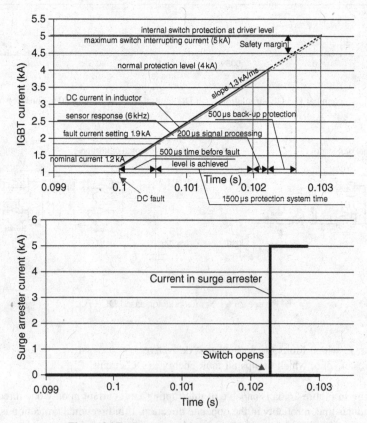

Figure 26.7   Response of a semiconductor DC CB.

protection in the event that another DC CB (on another line) fails to operate, and this further increases operating time and raises demand for peak interrupting current.

It is noted that a semiconductor switch will also have internal protection logic implemented as a hard-wired circuit at driver level. This is indiscriminative logic, which serves as the last line of protection in order to open switch before the current reaches levels that the switch can no longer interrupt. This is manufacturer-specified maximum interrupting current, which is presented as 5 kA limit in Figure 26.7.

Example 26.2

A 1.5 kA IGBT is used in the semiconductor DC circuit breaker on a DC line rated for 1.5 kA. This switch has the internal driver-level switch protection set at 5 kA with 20% tolerance. A 10% margin should be allowed between normal protection level and the internal switch protection. The normal protection time is 2 ms and backup protection is not used. Determine the size of DC CB reactor assuming that DC line is connected to an infinite DC bus of 400 kV voltage. Estimate the energy requirement for the surge arrester.

Solution
Considering the tolerance and margin, the normal DC current trip level will be

$$0.7 * 5\,kA = 3.5\,kA.$$

The required fault current slope is:

$$\Delta I / \Delta T = (3.5 - 1.5)\ kA/2\,ms = 1\,kA/ms.$$

The inductor value is determined as:

$$L = \Delta V / (\Delta I / \Delta T) = 400\,kV/1000\,kA/s = 0.4\,H.$$

This inductor size is large but may be acceptable with a LCC HVDC system. It is not likely to be acceptable with a VSC HVDC system. Larger switches should therefore be selected, capable of interrupting a higher DC current, which will facilitate the use of lower inductors. As an example, if a switch of 10 kA interrupting capability is used, the reactor size will be L = 0.145H, which might be acceptable with VSC systems.

The total energy requirement for SA is $E = 0.5I^2L = 0.5 * 3500^2 * 0.4 = 2.45\,MJ.$

26.6   Hybrid DC Circuit Breaker

A hybrid DC CB is developed to eliminate on-state losses with the semiconductor DC CB. This is beneficial in reducing direct energy loss, but it also simplifies auxiliary system requirements (like cooling systems). The basic schematic of one possible topology is shown in Figure 26.8. It consists of the following components:

- *Main semiconductor valve.* This valve is normally open but it does not take load current. It is rated for peak interrupting DC fault current and for surge arrester DC voltage.
- *Auxiliary semiconductor valve with surge arrester.* This valve is normally ON and it is rated for nominal DC current, for interrupting DC fault current and for voltage equal to the voltage drop across the main valve under fault current. This valve is commonly a single 3–4 kV IGBT.

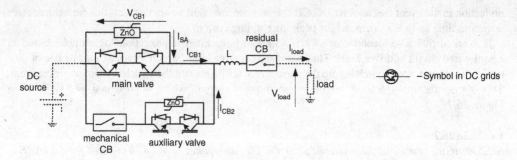

Figure 26.8 Hybrid DC circuit breaker for bidirectional protection.

- *Mechanical CB, which is normally closed and rated for nominal DC current and for full DC voltage.* It opens at zero DC current and can be called a fast disconnector. In some designs the auxiliary valve is not included and therefore the mechanical CB will have some interrupting DC current duty.
- *Surge arresters that will dissipate the fault energy and prevent overvoltages.* The rated DC voltage is a design tradeoff because too low voltage will cause a large leakage current.
- *The residual CB permanently isolates the unit to eliminate surge arrester leakage current:* It is rated for leakage surge arrester current and full DC voltage.

The nominal DC current runs through the mechanical DC CB and the auxiliary valve. This implies that the on-state loss is small as the auxiliary valve has low on-state resistance.

In case of a DC fault the following sequence of events occurs:

- The auxiliary valve and mechanical CB take full fault current. The main valve is closed, but it takes no current since it has large internal resistance (many IGBTs in series).
- The auxiliary valve opens redirecting the fault current to the main valve. At this stage the auxiliary valve blocks the voltage equal to the voltage drop across the main valve (few kV).
- The mechanical CB opens at zero current. This is necessary to provide isolation of the auxiliary valve. These first three steps can be executed proactively, even before final trip decision is received from higher protection levels.
- The main valve opens interrupting the full fault current. At this stage the element operates as a semi-conductor DC CB described in the section above. The fault current is interrupted and the fault energy is dumped in the surge arrester.
- After a certain delay, when the fault current drops to SA leakage level, if no reclosure is required, residual CB will open for long–term isolation.

The manufacturers have demonstrated high-power prototypes, which are capable of interrupting current in the range 7–16 kA within 2 ms. The interrupting current capability of hybrid DC CB will be much larger than that of semiconductor DC CB for the same main valve. This can be achieved since the main valve in hybrid DC CB is not conducting nominal current and therefore it is in a cold state before current interruption. The current interrupting capability from cold state is considerably higher than headline figures in manufacturers' sheets, which apply to a switch at operating temperature in a typical converter.

Another advantage of hybrid DC CB is that it may not require active cooling. The loss in the auxiliary switch is small and natural convection may suffice, which implies simpler and more reliable auxiliaries and valve hall (if needed).

The hybrid DC CBs further offer the possibility of actively limiting the fault current magnitude, which has been demonstrated on some prototypes. This feature might be attractive since such fault current limiting can be applied immediately on local detection of fault conditions, without danger of tripping the DC cable unnecessarily. DC CB tripping action must be coordinated with other DC CBs in the grid but fault current limiting action needs no coordination and this implies fast and reliable limiting of fault currents. The fault current limiting however requires higher rated DC CB components.

Example 26.3

A 400 kV, 750 A DC line employs a hybrid DC circuit breaker. The main valve uses 124 individual 6.5 kV, 750 A IGBTs in series. Each IGBT has a threshold voltage of $V_{TO} = 1 V$, and the on-state resistance of $r_{on} = 3.6 m\Omega$. If the normal DC fault interrupting current is 5 kA, determine the required rating for the auxiliary valve. Also estimate the percentage on-state loss for this DC CB.

Solution

At 5000 A the total voltage drop across the main valve will be:

$$V_{on} = n(V_{TO} + r_{on}I_T) = 124(1 + 0.0036 \times 5000) = 2356 \, V$$

Assuming that two times overrating is required, the auxiliary valve rated for 4710 V is required. Therefore a single switch that is used in the main valve will suffice. If current interrupting capability is inadequate, then parallel IGBTs will be employed.

The on-state loss consists solely of the loss in the auxiliary valve.

$$P_{on} = n(V_{TO}I_{nom} + r_{on}I_{nom}^2) = 1 \times (750 + 0.0036 \times 750^2) = 2775 \, W$$

$$P_{on\%} = P_{on}/(I_{nom} \times V_{nom}) = 2775/(750 \times 400000) = 0.001\%$$

This is very small loss and indicates the advantages of using hybrid DC CB.

26.7 DC Grid-Protection System Development

The DC grid-protection system must be capable of locating fault within reasonable time and sending trip signal only to DC CBs that isolate the faulted cable. This is not a simple task because, commonly, all DC cables will have high fault current for a DC fault at any location in the DC grid. This is clearly seen in the example in Figure 24.4, where only cable 34 should be tripped but all cables evidently have high current during the fault.

Figure 26.9 shows the DC CB location on a four-terminal meshed DC grid. Each DC cable has two DC CBs, one at each end. Protection from DC ground faults in the converters (if required) will demand an additional converter DC CB at the connection with the DC bus, as shown the figure. Converter faults are cleared by converter DC breakers together with the AC breaker. It is essential that the converter protection can distinguish between faults in the converter and faults on a line to prevent unnecessary AC CB tripping, which normally implies loss in capacity.

The central bus is split into two section connected through two CBs to provide bus bar and back-up protection. All the CBs are assumed to be unidirectional (they allow current in both directions but can interrupt in only one direction). The two DC CBs between bus-bar sections may be replaced with a single bidirectional DC CBs. However, unlike with AC CB, hybrid DC CB for bidirectional operation will have similar cost as two unidirectional devices.

The backup protection on DC grids may be required for high reliability systems. Taking, as an example, a fault on cable 34, then protection logic should trip CB34 3 and CB34 4. This will isolate the faulted cable and the grid can continue to operate without any loss in capacity (assuming that all the

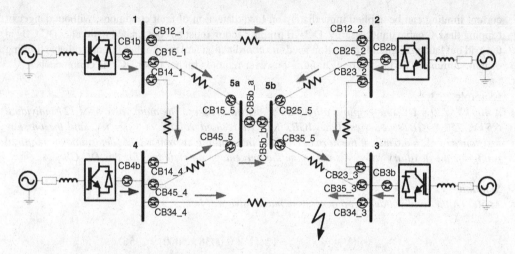

Figure 26.9 Protection system for a four-terminal meshed DC grid.

other cables have an adequate rating to take the redirected power flow). If a DC CB fails to trip, say CB34 3, then the backup logic will isolate the next wider area by tripping in this case: CB3b, CB23 2 and CB35 5, which would imply loss of one VSC. Backup protection will require additional overrating of components to allow for longer fault-clearing time.

26.8 DC Grid Selective Protection System Based on Current Derivative or Travelling Wave Identification

In large DC grids, as in AC systems, it is desired to use only local sensor signals to determine fault location. If controller (relay) at each DC CB determines location of the fault accurately then the decision on tripping can be made locally at each DC CB. No communication between DC CBs would be required which is significant benefit in terms of reliability and speed of protection operation.

Most AC system use distance protection relays where current rise versus time is compared with a specific preset curves. This method uses only current magnitude versus time signal for locating faults. It is suitable with AC systems as AC reactance provides sufficient discrimination between current-time signals at various locations. This method is much more difficult with DC systems because of very low DC line impedance.

Figure 26.10 shows the simulation of a DC fault at one end of a 600 km cable. A detailed frequency-dependent 2000 mm^2 three-layer DC cable model is used and four sensors are located at each 200 km points. It is seen that the four sensors read largely different signal in the first 15 ms after the fault. This is the consequence of a finite speed of travelling-wave propagation and different impedance between the sensor and the fault location. The speed of travelling wave propagation depends on the cable parameters. Theoretically therefore it is possible to preprogram the current-time curves or current derivative values for each distance and by comparing with the measured current-time waveforms it is possible to determine location of the fault. This method has the potential for fast fault location using only local measurements and adoption of AC grid protection practice. In practice, however, as cable length becomes shorter, the discrimination within 10 km becomes very difficult because of the very short time intervals, the steep current pulse on low-impedance DC cables and the noise content on the measured signals. An additional difficulty is the travelling wave reflections that will happen at each cable termination and discontinuity.

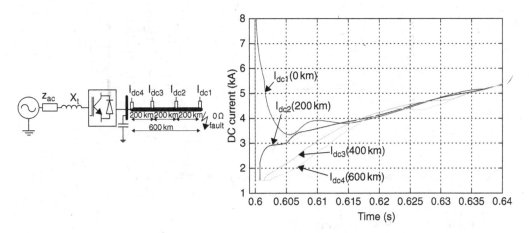

Figure 26.10   Fault currents in four sensors located at different places on a 600 km DC cable (number in brackets is distance from the fault).

26.9   Differential DC Grid Protection Strategy

If all current sensors are direction sensitive it is possible to develop robust protection system for DC grids. However, with large systems some communication between DC CBs will be required, as many DC CBs will have same direction of fault current. Consider CB at bus 2 (CB25 2) and CB at bus 3 (DC CB34 3) in Figure 26.9, which both see large positive fault current (towards the centre of the cable that they protect) for a fault on cable 34. These two CBs have no means to determine if the fault is on the local cable or on a remote cable. Only DC CB34 3 and DC CB34 4 should be tripped to clear this fault.

Differential protection using pilot wire is a well developed method for component protection in AC systems. It can be expanded for reliable protection of DC cables, as shown in Figure 26.11. The protection system consists of a direction sensitive sensor (S34), controller (C34) and a Circuit Breaker (CB34) at each cable end. Each controller communicates only with the controller at the opposite end of the same DC cable using a dedicated optical cable. No communication with DC CBs on other cables in the grid is required. The trip decision is made in at each controller if the sum of currents on the two sensors on a cable exceeds a threshold positive value for a specified period of time. This logic provides very reliable selectivity because for each DC fault location there will be only two controllers (those on the same cable) receiving positive differential signal.

The primary drawback of the differential protection is the delays associated with the geographical separation of sensors. Firstly, there will be a detection delay in the sensor caused by the travelling wave delay. In the example shown in Figure 26.10, if the threshold is set to 3 kA (2 pu), the sensor at the opposite end, 600 km away, would be able to detect fault current level after around 12 ms. On the other hand the local sensor can detect fault current level instantaneously.

Additionally, there will be a delay associated with the signal transfer through pilot wire between the controllers at the two cable ends. It is assumed that the speed of signal transfer in the optical communication cables will be close to the speed of light (300 000 km/s), not considering optical attenuation. This implies at least 1 ms delay for each 300 km of cable length. The processing delays in microcontrollers and A/D signal conditioners may account for additional delays.

Figure 26.12 illustrates the signal path in the differential protection system of cable 34 for a fault on cable 34 in the DC grid in Figure 26.9. Fault happens at time 0, as shown on the x-axis and it is located 100 km from the bus 3 as shown on the y-axis. Each controller needs to receive information from two sensors (local and remote) in order to make positive trip decision. The signal from the local

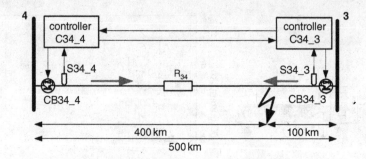

Figure 26.11 Differential protection system for the DC line 34.

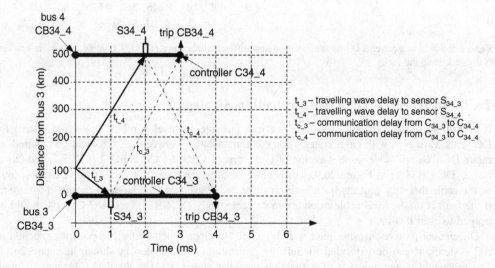

Figure 26.12 Communication and wave delays with DC line differential protection.

sensor will include only travelling wave delay (around 1 ms for controller C34 3 in Figure 26.12) but the signal from the opposite end will involve two delays: travelling wave delay and communication delay (around 4 ms for controller C34 3 in Figure 26.12). The travelling wave delays will be different for each sensor and will be dependent on the fault location, but the communication delays will stay constant regardless of the fault location. It is seen in the particular example that the total time to trip DC CB34 4 is around 3 ms while trip time for DC CB34 3 is around 4 ms. The DC CB at the cable end further away from the fault will always be first to receive the trip signal.

The fact that DC CBs at the two cable ends will not be tripped simultaneously may further complicate the situation in large DC grids. When the first DC CB is tripped, the system topology changes and now all the remote AC sources begin to feed the fault through the DC CB on the other end of the faulted cable and therefore this DC CB may take a larger trip current.

26.10 DC Grid Selective Protection System Based on Local Signals

It is very advantageous to have a DC fault-location logic that uses only local signals. Such logic would be very fast and reliable. Fast protection logic, in turn, implies that interrupting fault current will be lower and this has cost benefits for all components and protection units.

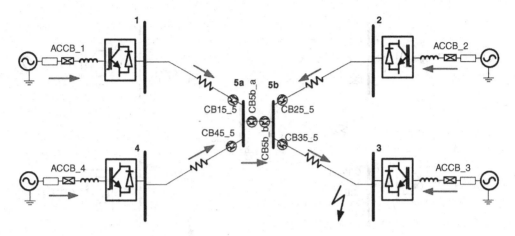

Figure 26.13 Protection system for a four-terminal star-connected DC grid.

A reliable protection logic based on local sensors is, however, feasible only with special star-grid topology, which has performance limitations. Figure 26.13 shows a four terminal DC grid based on star topology. All the DC CBs at the central bus use hybrid DC CB topology. Each VSC terminal includes an AC CB on the AC side of converter, as it would be normal case with HVDC converters.

A fault on cable 35 is shown and such fault would be cleared by DC CB35 5 and AC CB ACCB 3. The direction of fault currents is given in red arrows. It is seen that only one of the main DC CBs has fault current in the positive direction (CB35 5).

There is an independent protection system for each DC cable. It consists of a DC CB protection controller/relay and a local direction-sensitive current sensor. The protection logic is decentralized, uses only local signals and trip criterion is simple. A trip decision is made if the local current sensor detects current over a threshold and in positive direction. This implies that protection can operate as fast as hardware dynamics and processing speed will allow. As hybrid DC CBs are employed, the protection speed will be a few milliseconds.

A fault on any DC cable is cleared by one hybrid DC CB and one mechanical AC CB.

A split bus at terminal 5 is used to increase reliability. There are two CBs at bus 5, which serve as backup protection. Their protection logic is same as for other DC CBs but a delay is added in the control circuit. If, as an example, CB35_5 fails to operate for a specified time interval then CB5b a will operate and isolate bus 5b. Also ACCB 2 will isolate fault infeed from AC grid at terminal 2. The AC CBs will need a small delay in control logic if backup protection is used.

The above protection is applicable only to radial topologies and clearly cannot be used with meshed DC grids. The number of DC lines connecting to a single DC bus will also be limited by the DC CB rating as each DC CB in an n-terminal system will have fault current summing from $n - 1$ VSC converters.

26.11   DC Grids with DC Fault-Tolerant VSC Converters

26.11.1   Grid Topology and Strategy

The fault current in DC grids ultimately comes from AC systems through VSC converters. The study in this section aims to reduce magnitude of DC fault current through VSC converters, ideally to values comparable to the rated DC current. Such VSC converters would bring multiple benefits:

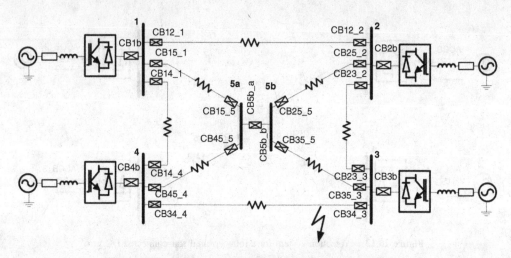

Figure 26.14   Four-terminal DC grid with DC fault tolerant VSCs.

- The IGBTs need not be tripped for DC faults. This means that control is retained through DC faults and postfault recovery is fast.
- The VSC converters would be able to indefinitely sustain DC fault situation, and therefore fault-clearing time can be extended to the range similar as with AC systems (20–100 ms). This implies that simpler and more reliable gridwide protection systems can be used.
- The reverse parallel diodes need not be overrated, implying cost savings in VSC converters.
- If all VSCs in a DC grid can limit the fault current, then the magnitude of fault currents in all cables inside the DC grid will be low. This implies less costly DC CBs.

Figure 26.14 shows a four-terminal DC grid, which employs four DC fault-tolerant VSCs. If each VSC limits the fault current to 1 pu then the worst-case fault current magnitude anywhere in the grid can reach 4 pu. There are 21 DC CBs in this simple grid and the potential cost saving in using lower specification DC CBs, like mechanical DC CBs, is significant. The total cost of 21 hybrid DC CBs would amount to around 7 pu (where 1 pu is VSC converter cost), according to data in Table 22.1, which is almost twice the total cost of all VSC AC/DC converters in this grid. The total cost of mechanical DC CBs may amount to 0.1 pu. Therefore further investment in making VSC converters fault tolerant should be carefully explored.

The sections below will review some options for limiting DC fault current in VSC converters.

26.11.2   VSC Converter with Increased AC Coupling Reactors

Figure 26.15 shows a VSC converter (a two-level, three-level or half-bridge modular multilevel converter (MMC) topology) connected to an AC grid through a reactor X_t, which can represent joint impedance of transformer and series reactor. It will be explored if an increased X_t can sufficiently limit the fault current and what drawbacks would occur.

The equation that gives the converter phasor current is:

$$jX_t\overline{I_{gdq}} = \overline{V_{gdq}} - \overline{V_{cdq}} \qquad (26.12)$$

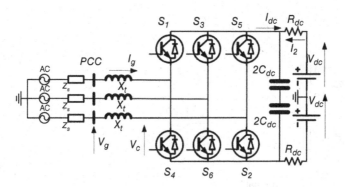

Figure 26.15 L VSC converter.

and when separated into real and imaginary components, assuming coordinate frame is linked with V_g:

$$I_{gd} = -\frac{V_{cq}}{X_t} \tag{26.13}$$

$$I_{gq} = \frac{V_g - V_{cd}}{X_t} \tag{26.14}$$

Assuming normal operation with $Q_g = 0$, the reactive current becomes $I_{gq} = 0$. From Eq. (26.14) it is seen that $V_{cd} = V_g$, and therefore this converter cannot achieve any stepping ratio V_g/V_c (transformers must be used if voltage stepping is required). Dividing Eq. (26.14) by Eq. (26.13), the current angle can be obtained and it can be concluded that this angle cannot be equal to the voltage V_c angle, and therefore this converter cannot have $Q_c = 0$ (reactive power at converter AC bus). Consequently converter voltage or current will be larger than the optimal value. Introducing the converter utilization ratio, U_r:

$$U_r = \frac{S_{grid}}{S_{conv}} \tag{26.15}$$

where S_{grid} is the required power at the grid-coupling point (S_1 in the above example) and S_{conv} is the required converter rating, enables study of the required converter rating. The ideal utilization ratio value is $U_r = 1$, which implies that converter rating (and therefore costs and losses) are optimal.

The fault current for the above converter can be determined as:

$$I_g^{fault} = \frac{V_g}{\sqrt{R_s^2 + (X_s + X_t)^2}} \tag{26.16}$$

Figure 26.16, in the top graph, shows the required converter DC voltage (in pu relative to optimal DC voltage) and the utilization ratio U_r, as the function of coupling reactance X_t. Larger reactance will imply a larger q component of AC voltage V_c and consequently larger DC voltage. As the DC voltage increases, the converter rating increases, the utilization ratio will be lower and, in practical terms, this implies higher capital costs and also higher operating losses. Normally, $0.1 < X_t < 0.2$ and it is seen that the DC voltage will be only 2–3% larger than optimal, plus perhaps few percentages for control room.

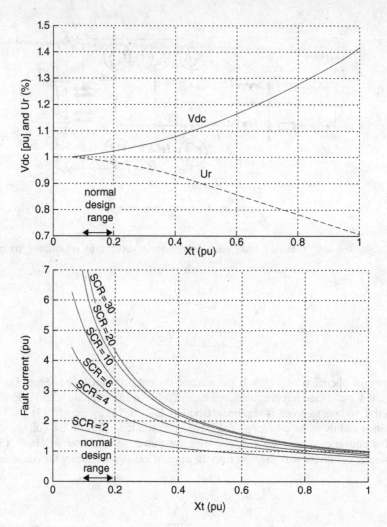

Figure 26.16   Converter utilization ratio and per-unit fault currents as the function of interface reactance.

Figure 26.16 on the lower graph shows the fault current relative to converter-rated current I_g^{fault}/I_g for a range of grid strengths and coupling reactance values. It is clear that it is possible to reduce the fault current significantly by increasing coupling reactance. For example, with a short-circuit ratio (SCR) = 6 and $X_t = 0.9$ pu the fault current will be 1 pu. Such converter would have around 35% larger costs and operating losses. On the positive side, such design will imply a range of benefits as discussed above (no requirement for additional antiparallel diodes, a lower rating for all DC CBs, simpler protection, etc.).

26.11.3   LCL VSC Converter

Figure 26.17 shows the LCL VSC converter. It includes an AC/DC converter, which can assume any topology (two-level, three-level or half-bridge MMC) and an LCL interface. Compared to the L interface, the LCL interface brings more design flexibility – for example, the possibility of voltage

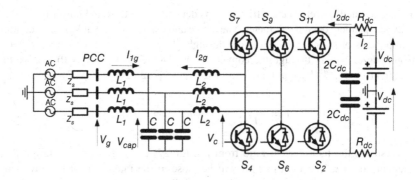

Figure 26.17 LCL VSC converter.

stepping, i.e. V_g/V_c, feasibility of achieving zero reactive power at both V_g and V_c, and reduced fault currents (the fault current is limited to a level close to the rated current).

The basic phasor equations for the LCL circuit in Figure 26.17 are:

$$j\omega L_1 \overline{I_{1gdq}} = \overline{V_{1gdq}} - \overline{V_{capdq}} \tag{26.17}$$

$$j\omega C \overline{V_{capdq}} = \overline{I_{1gdq}} + \overline{I_{2gdq}} \tag{26.18}$$

$$j\omega L_2 \overline{I_{2gdq}} = \overline{V_{cdq}} - \overline{V_{capdq}} \tag{26.19}$$

The above equations can be more conveniently written using parameters k_1–k_3 as:

$$\overline{I_{1gdq}} = \frac{V_g k_1 - \overline{V_{cdq}}}{j\omega \times k_3} \tag{26.20}$$

$$\overline{V_{capdq}} = \frac{V_g(k_1 - 1) + (k_2 - 1)\overline{V_{cdq}}}{k_2 k_1 - 1} \tag{26.21}$$

$$\overline{I_{2gdq}} = \frac{\overline{V_{cdq}} k_2 - V_g}{j\omega \times k_3} \tag{26.22}$$

where:

$$k_1 = 1 - \omega^2 L_2 C \tag{26.23}$$

$$k_2 = 1 - \omega^2 L_1 C \tag{26.24}$$

$$k_3 = L_1 + L_2 - \omega^2 L_1 L_2 C \tag{26.25}$$

$$k_1 < 1, \quad k_2 < 1, \quad k_3 < L_1 + L_2 \tag{26.26}$$

The above parameters are introduced to enable the study of fault currents. In Eq. (26.20), the parameter k_1 is the gain between V_g and I_{1g} and, similarly, in Eq. (26.22), parameter k_2 is the gain between V_c and I_{2g}. The final design parameters are L$_1$, L$_2$ and C, which can be obtained from Eqs (26.23) to (26.25) once k_1, k_2 and k_3 are finalized.

The converter losses will be minimal when reactive power $Q_c = 0$, implying $\angle V_c = \angle I_{2g}$, and it can be shown that this is achieved if $k_2 = k_1 s^2$, where $s = V_g/V_c$. The LCL VSC converter can achieve a 100%

utilization ratio at rated power. The coefficient k_3 is dependent on the rated power only, and therefore k_1 is the only independent design parameter. The active and reactive power at V_g can be controlled using the DQ components of the converter voltage (through M_d and M_q) as with common L converters.

The magnitude of current through the converter for a DC fault ($V_{dc} = V_c = 0$) relative to the rated converter current $r = I_{2g}^{fault}/I_{2g}$, can be determined from Eqs (26.20) to (26.22) as:

$$r^2 = \frac{1}{-k_1^2 s^2 + 1} \quad where \quad s = V_g/V_c \tag{26.27}$$

Figure 26.18 shows the fault current ratio from Eq. (26.27) as the function of design parameter k_1 and the stepping ratio s. The fault current will be close to the rated current for any stepping ratio and for a wide range of k_1. In general it is always possible to design the converter to keep the fault current to within 1.2 pu. In practice this result also implies that the converter will be able to operate under reduced DC voltage.

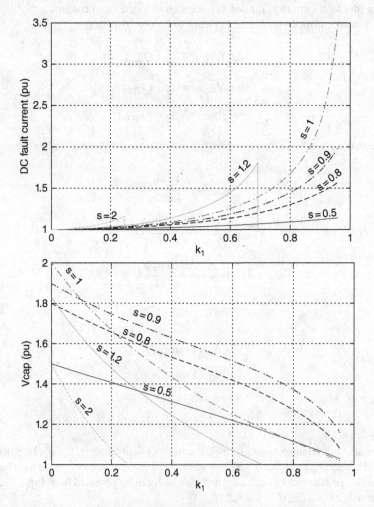

Figure 26.18 Converter DC fault current (pu) and capacitor voltage (pu) (base is largest of V_g and V_c) as the function of design parameter k_1, and stepping ratio $s = V_g/V_c$.

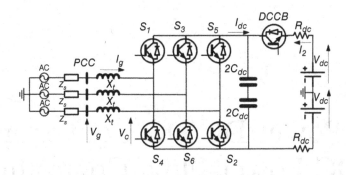

Figure 26.19 VSC converter with a DC CB operating as a fault-current limiter.

On the downside, as shown in the lower graph of Figure 26.18, the capacitor voltage V_c will be typically 1.2–1.4 pu (relative to the larger of V_g and V_c), which implies some cost penalties. Another drawback of this topology is that, at a partial load, it will not be possible to maintain $Q_c = 0$, which implies some more conduction losses. This drawback can be resolved if capacitor size is reduced at partial load, which can be achieved with switchable capacitor banks.

If DC fault reduction is a high priority, it is likely that an LCL converter will have overall advantages over a common L converter.

26.11.4 Full Bridge MMC Converters

Full-bridge MMC converters can operate with low DC voltage, they can limit the DC fault current, or bring DC fault current to zero, as has been discussed in section 20.8. *Therefore if all VSCs use FB MMC topology in a DC grid, like that in Figure 26.14, then the protection system would require low rated DC CBs.*

The FB MMC VSC topology can reduce a DC fault current to zero and this opens the possibility of using fast DC disconnectors (zero current devices) on DC cables. Nevertheless, many further challenges would arise if the protection precondition is that current is brought to zero in all cables in a complex DC grid. The time frame for the protection operation might be increased. It could be beneficial not to reduce all VSC currents to zero immediately under a DC fault, in order to allow *sufficient time for* protection logic to accurately determine fault location.

On the downside, FB MMC topology has higher costs and operating losses.

26.11.5 VSC Converter with Fault Current Limiter

It has been is explained in section 26.6 that semiconductor or hybrid DC CBs can be used as fault current limiting devices. In such cases, the DC CB active switch is operated periodically, not unlike the DC chopper, under an active feedback current control loop. The current can be controlled to low values, even under the worst case DC faults, but the duration of this operating mode is limited by the energy capability of switches, surge arresters and the cooling equipment.

Figure 26.19 shows the VSC converter with the DC CB acting as a fault current limiter. Jointly with the DC CB, this converter can limit the fault current like any of the topologies in the previous sections. Some manufacturers are planning to standardize a seven-switch VSC converter, as in Figure 26.19, which has capability of DC fault-current limiting and interrupting.

27

High Power DC/DC Converters and DC Power-Flow Controlling Devices

27.1 Introduction

It is expected that DC grids will eventually evolve into large meshed networks, which will inevitably have multiple DC voltage levels. A DC/DC converter will be needed in order to connect two DC grids operating at different DC voltage levels. One evident DC/DC application is to connect DC cables (which have DC voltage up to 600 kV) with overhead DC lines, which may have a higher DC voltage. The existing HVDC (high-voltage direct-current) links have wide range of highly optimized DC voltage levels and their possible integration into the DC grid will require DC/DC converters. It is also expected that medium-voltage DC grids, either distribution or collection systems (like those with offshore wind farms) will rapidly develop following acceptance of DC transmission grids, and their connection to DC transmission will require high-stepping ratio DC/DC converters. This role is similar to a transformer function in traditional AC systems.

Nevertheless, even in a DC grid with a single nominal DC voltage there might be a need for DC/DC devices in order to regulate the power flow in some cables or DC voltage level at some nodes. These devices may have low stepping ratio and perform a similar function to tap-changing transformers and phase-shifting transformers in AC systems. The power flow in DC grids will be primarily controlled using AC/DC converters located at grid terminals (connecting points with external AC grids). Assuming an n-terminal DC grid, it is theoretically possible to control power flow in up to n DC cables, with an allowance for inaccuracy because of DC voltage droop feedback. If the number of DC cables is greater than n then the power flow cannot be controlled in each cable.

A complex DC grid may have large number of meshed DC lines, since in a n-terminal DC grid it is possible to connect $(n^2 - n)/2$ cables. It is also beneficial to include some uncontrolled DC substations and in such case it is not possible to precisely control power in each DC cable. In order to avoid DC line overloading or underutilization there is a need for special DC power controlling elements. These devices will be installed in series with DC cables in order to adjust (increase) the cable resistance or to regulate DC voltage. Figure 27.1a shows a five-node, four-terminal (four AC/DC converters) DC grid with seven DC lines. All busses have the same nominal DC voltage level. The DC cable

High-Voltage Direct-Current Transmission: Converters, Systems and DC Grids, First Edition.
Dragan Jovcic and Khaled Ahmed.
© 2015 John Wiley & Sons, Ltd. Published 2015 by John Wiley & Sons, Ltd.

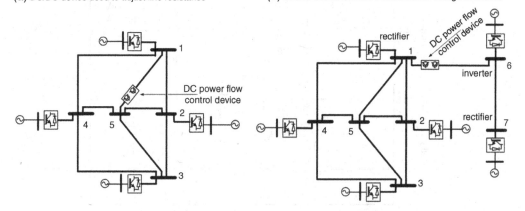

Figure 27.1 Applications of power-flow controlling device in DC grids.

1–5 may be very short (with low resistance) and if it connects to a large rectifier (terminal 1) it will have high current.

Another possible need for DC power flow device will be if there is an intention to interconnect two established DC systems. The DC voltage difference of several kilovolts might be too large for direct connection using a DC cable. DC power flows only from higher DC voltage to lower DC voltage, which might not always be consistent with the desired power flow direction. Figure 27.1b shows an established DC grid (terminals 1–5) and a new DC grid (terminals 6–7) in the vicinity, which have the same nominal DC voltage. If terminal 1 is a large rectifier and terminal 6 is an inverter connected to a long DC cable 6–7, the DC voltage difference between busses 1 and 6 may be of the order of 5–10 kV and it may be difficult to directly connect these busses with a short DC cable.

In this case two options are available:

- reduce operating voltage on the DC grid 1–5 with the associated reduction in capacity and increased losses;
- incorporate a DC/DC converter with low stepping ratio in the new DC line.

27.2 Power Flow Control Using Series Resistors

A simple solution for DC power-flow control is by using resistors that can be inserted in series with DC lines. The resistance value should be comparable to the DC cable resistance. A typical 1600 mm^2 submarine DC cable has resistance of around 0.01 Ω/km, and therefore several ohms are the maximum required resistance. Assuming typical ampacity of 1400 A for a 320 kV DC cable there will be loss of 2 MW for each additional 1 Ω of resistors. This heat dissipation is direct revenue loss but also large cooling equipment implications should be considered. Consequently, this method might be attractive only if a little additional resistance is required. Figure 27.2 shows the total loss in the resistors dependent on the DC voltage stepping ratio $d = V_{dc2}/V_{dc1}$, where the base power is assumed as P_{dc} = 320 kV × 1.4 kA = 450 MW. The loss becomes substantial if the device is used for DC voltage matching over 1%.

Figure 27.3a shows the DC power-flow controlling device based on series resistors with mechanical switches, where n identical series units are assumed. Each unit can be independently controlled in or out of the current path. The mechanical switches are DC circuit breakers (CBs), which must have the capability of breaking and making the rated DC current. The DC CB voltage rating will be equal to the

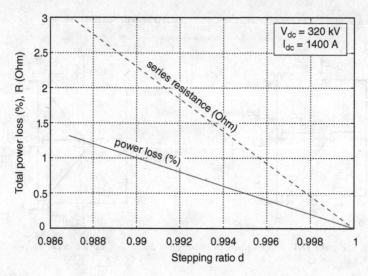

Figure 27.2 The series resistance value and total resistor loss depending on the stepping ratio V_{dc2}/V_{dc1}.

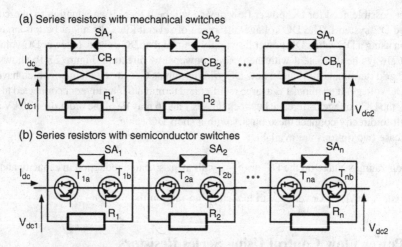

Figure 27.3 Power flow controlling device based on n series resistors.

voltage drop across resistors. The current flows either through the CB (CB closed) providing low resistance, or through the resistor (CB open) providing higher resistance. The surge arresters are required because voltage across resistors will be large in case of DC line faults or resistor open-circuit failure.

Considering that the device is based on mechanical contacts it will have negligible switch losses and the costs will be reasonable. On the downside, the wearing of contacts may become an issue if many switch operations are required. In the power-flow control application, it is expected that the device will be operated frequently. Some CB manufacturers specify up to 20 000 maximum operations for DC CBs under rated conditions. The additional disadvantage of this device is the speed of operation. The commercial units have the total operating speed in tens of milliseconds. Clearly this limits operating flexibility if the device is used for dynamic power-flow control.

Figure 27.3b shows the DC power-flow controlling device based on series resistors with semiconductor switches. Two insulated gate bipolar transistors (IGBTs) or (insulated-gate commutated thyristors) (IGCTs) are assumed in each unit considering two possible current directions. Each unit can be controlled separately by simultaneously setting both transistors a and b ON or OFF. Similarly to the mechanical device, the current flows either through the switches (switches T_a and T_b ON), providing low resistance, or through the resistor (switches OFF) providing higher resistance. The switches may not require additional snubbers, depending on the parasitic inductance of the resistors. With some more complex pulse-width modulation (PWM) control and additional components the device could be configured to make continuous adjustment of resistance. The number of ON/OFF operations is virtually unlimited and the speed of operation is excellent. On the downside, this device will require cooling systems for semiconductor switches as they will continuously be in an ON state, with the requirement for a valve hall. The on-state losses are the primary disadvantage compared with the mechanical devices.

Example 27.1

A 1.6 kA DC line requires series mechanical power flow-controlling device with 15 control positions and total resistance of 1.2 Ω. Determine rating of DC CBs. Calculate the voltage stress across each DC CB in case that maximal DC fault current reaches 30 kA.

Solution

Each resistor will have resistance of: $R_1 = R_{tot}/n = 1.2/15 = 0.08 \Omega$
  The voltage stress across each DC CB will be $V_1 = R_1 \times I_r = 0.08 \Omega \times 1600 A = 128 V$.
  Allowing two times the voltage margin, each DC CB should be rated for $V_{1r} = 256 V$.
  In the event of a 30 kA fault current the voltage across a DC CB (which is off) will be $V_{1f} = R_1 \times I_f = 0.08 \Omega \times 30\,000 A = 2400 V$. Therefore a surge arrester should be appropriately designed to limit the voltage stress.

Example 27.2

A semiconductor-based series power flow-control device uses seven resistor units of 0.3 Ω each. The IGBT switch data are show in Table 27.1. The nominal DC current in this 500 kV DC line is 1400 A. Determine total losses in semiconductors when all the resistors are bypassed.

Solution

The device consists of seven units and therefore there are seven IGBTs and seven diodes in the current path.

$$Ploss = 7 \times 2.4 \times 1400 + 7 \times 1.6 \times 1400 = 39.2 \ kW$$

The percentage loss is:

$$Ploss\% = Ploss/Pr = 39.2 \ kW/(500\,000 \times 1400) \times 100 = 0.0056\%$$

Table 27.1 Basic data for the IGBT switch in Example 27.2.

V_{CE} (V)	I_C (A)	IGBT on state voltage V_{CE} (V) (1400 A)	Diode on-state voltage V_F (V) (1400 A)
1700	1600	2.4	1.6

27.3   Low Stepping-Ratio DC/DC Converters

27.3.1   Converter Topology

A bipolar, low stepping ratio DC/DC converter is shown in Figure 27.4. It connects two bipolar DC grids that nominally have same DC voltage levels but because of power-flow conditions their DC voltages will differ. The topology is similar to buck-boost converters used widely in low power applications. The switch S_1 is switched ON/OFF at a constant operating frequency but with varying duty ratio $d = T_{on}/T$ as shown in Figure 27.4 while S_2 switch is operated complimentary. The converter is bidirectional because current can reverse direction. Note that $V_{dc1} < V_{dc2}$ (the inductor always connects to the lower DC voltage) must be satisfied and voltage cannot change polarity. If voltage V_{dc1} is expected to become larger than V_{dc2} in some operating conditions, then a full-bridge topology is needed (another converter leg is added) and costs would double. The converter is bipolar and each pole will have independent controller and can operate independently.

In case of DC faults on DC grid 1 (on the lower voltage side) the current through the converter will be increasing while the rate is limited by the inductor. This converter is capable of interrupting DC fault current by opening switch S_1 (either S_{1p} for positive pole faults or S_{1n} for faults on negative pole). However, the converter cannot interrupt faults on DC grid 2 as diodes D_1 would uncontrollably transfer fault currents. Note also that DC fault interruption depends on the controller speed and there will be some transient overcurrent.

This DC/DC converter can perform the following functions in a DC grid:

1. matching DC voltages;
2. regulating power flow;
3. interrupting DC fault current (for faults on V_{dc1} only);
4. balancing of loading between DC poles.

27.3.2   Converter Controller

The controller for the above converter is shown in Figure 27.5. It has an independent controller for each pole and each controller has two cascaded stages. In the inner loop, the DC current is regulated in order to prevent high currents in the switches and to improve response of the outer loops. There is a wide range of possible control options for the outer loop and the choice will depend on the particular application and the operating scenario. The DC voltage is controlled in the outer loop in Figure 27.5. This control method might be appropriate for a DC/DC converter inside a large DC grid and at a remote location, far from other terminals. It can operate indefinitely in DC voltage mode without any communication. If desired DC voltage cannot be achieved, the inner control loop will saturate and the converter will operate on maximum current. An alternative control strategy might be constant current/power in the cable, or a constant stepping ratio. In this case, this converter would be required to respond to DC voltage variation, using droop feedback, in similar manner as with AC/DC converters.

The current balancing control is used at the highest level to modulate the DC voltage reference. It can balance an unbalanced LV (low-voltage)-side DC poles. When it is enabled, the controller can eliminate LV ground current but this will generally result in some HV (high voltage)-side unbalance.

Example 27.3

A DC/DC converter is required to connect a 370 kV DC bus with a 400 kV DC system. The converter rating is 1800 A and assume that DC cable resistance on 400 kV side is 2 Ω while on 370 kV side it is 1 Ω. It is required to keep DC current deviation to 6%, and the switching frequency is limited to 1500 Hz in order to reduce switching losses.

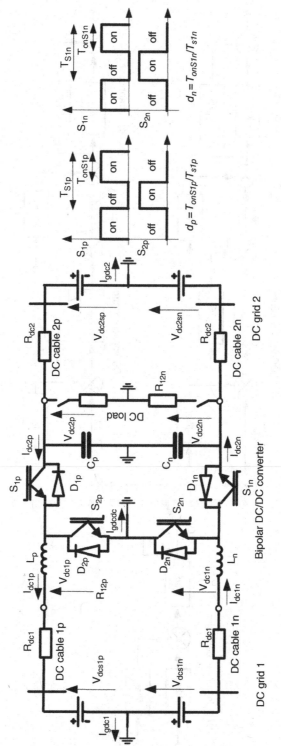

Figure 27.4 Bipolar DC/DC converter connecting DC grid 1 and DC grid 2.

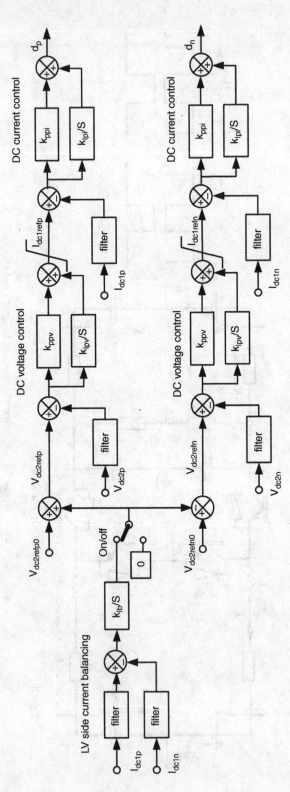

Figure 27.5 Controller for bipolar low stepping ratio DC/DC converter.

a. *Determine values for the converter inductor L.*
b. *Determine converter utilization ratio.*
c. *Design feedback controller and simulate converter responses to V_{dc2} drop to 395 kV.*
d. *Simulate converter response to 300 Ω unbalanced shunt load on negative pole of grid 2.*

Solution
a. *The converter duty ratio d is:*

$$d = \frac{V_{dc1}}{V_{dc2}} = 0.925 \tag{27.1}$$

The inductor value is calculated as:

$$L\frac{\Delta I_{dc1}}{T_{on}} = \Delta V$$

$$L = \frac{\Delta V\, d}{f\, 0.06 I_{dc1}} = \frac{30000 \times 0.925}{1500 \times 0.06 \times 1800} = 0.2H \tag{27.2}$$

b. *The converter utilization ratio is:*

$$UR = \frac{V_{dc1}}{V_{dc2}} = 0.925 \tag{27.3}$$

The DC voltage difference is 30 kV and, if the two grids are directly connected, the DC current would be 10 kA implying DC cable overload. Because of large DC voltage difference it would not be cost-effective to use series resistors. At 1500 A converter current (83% load) the converter regulated voltage is:

$$V_{dc2} = 400 - 1500 \times 2 = 397 \ kV.$$

c. *The converter testing results are shown in Figure 27.6. The following intervals are seen:*

Before 1 s the converter is regulating V_{dc2} = 397 kV and the power transfer is 600 MW towards grid 1 (step down operation).
   At 1 s, there is a significant drop in V_{dc2s} to 395 kV. The converter controller maintains V_{dc2} = 397 kV, which is achieved by power reversal through the DC/DC converter, which now feeds 400 MW to grid 2. This example illustrates DC/DC converter capability to support DC grid stability, and to transfer power to the DC grid with the higher DC voltage.
d. *At 1.5 s there is an asymmetrical load on grid 2 while DC current balancing is disabled. The converter manages to maintain balanced voltages V_{dc2p} = V_{dc2n} = 397 kV and there is no ground/neutral current in grid 2. The power required to balance grid 2 is drawn from grid 1 requiring a large ground current I_{g1}.*
   At 2 s, the current balancing control is enabled and grid 1 ground current is eliminated. However, DC grid 2 becomes unbalanced as a consequence.

27.3.3 DC/DC Converter Average Value Model

An average value model for a low stepping-ratio DC/DC converter is shown in Figure 27.7. It consists of a controlled voltage source on the LV side and a controlled current source on the HV side. Note that the LV side is interfaced with an inductor but the HV side is connected using a capacitor. The

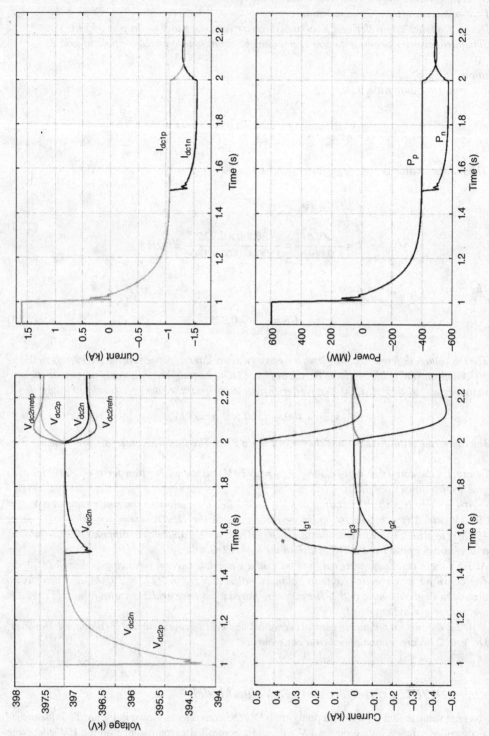

Figure 27.6 Demonstration of bipolar DC/DC converter operation in V_{dc2} control mode: at 1 s there is significant voltage drop in grid 2, at 1.5 s unbalanced load is added in grid 2, at 2.0 s DC current (I_{dc1}) balancing controller is enabled.

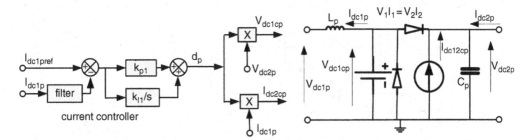

Figure 27.7 Averaged continuous model of bidirectional DC/DC converter.

power-balance equation $V_{dc1}I_{dc1} = V_{dc2}I_{dc2}$ (assuming lossless converter) is used as a simple approximation to link the two sides. This model can be used for power-flow studies in DC grids. It also has good dynamic representation in the frequency range of control loop dynamics. However the simplified model will not accurately represent harmonics or internal faults.

27.3.4 Fullbridge DC/DC Converter

Figure 27.8 shows one pole of a full bridge DC/DC converter. It resembles the topology in Figure 27.4 but it includes another two-valve converter leg, and therefore the number of switches is doubled. This converter can accommodate operation with $V_{dc1} < V_{dc2}$ or $V_{dc1} > V_{dc2}$. It can also interrupt DC faults on either DC grid 1 or DC grid 2. Like the converter in Figure 27.4, this converter can be used only for small voltage stepping ratio (say 0.8–1.2). It is not suitable for larger stepping ratios because extreme duty ratios cause significant reverse recovery loses in diodes and low utilization ratio.

27.4 DC/DC Converters with DC Polarity Reversal

In order to integrate thyristor line-commutated HVDC converters into DC grids with bidirectional power flow, as shown in Figure 23.5, a special DC/DC converter would be required. Figure 27.9 shows the three-phase thyristor-based DC/DC converter connecting the low-voltage DC grid V_1 with the HV DC grid V_2. This converter enables power-direction reversal by changing DC current polarity at HV side while changing DC voltage polarity at LV side.

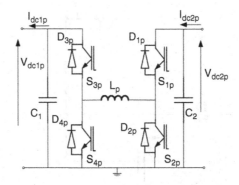

Figure 27.8 Full-bridge monopolar DC/DC chopper.

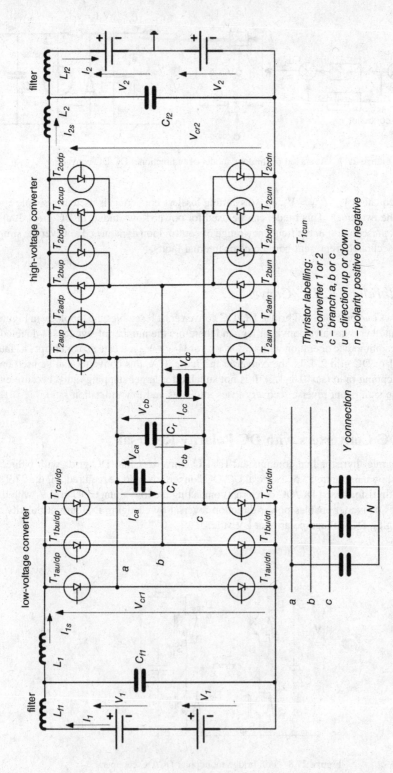

Figure 27.9 DC/DC converter enabling DC voltage polarity reversal at LV side and current polarity reversal at HV side.

27.5   High Stepping Ratio Isolated DC/DC Converter

Figure 27.10 shows the DC/DC converter employing an internal AC transformer, and two AC/DC converters front to front. The AC/DC converters can take single-phase or multiphase topologies. In addition to voltage stepping, the internal AC transformer provides electrical isolation between the LV and HV circuits. This has the advantage that the LV AC/DC converter is stressed for low voltage under all operating conditions. Similarly, the HV side is stressed for low current. Different grounding arrangements can also be selected for each side. The utilization ratio of both converters will be good, however both converters cannot simultaneously have an ideal utilization ratio. As discussed with fault tolerant voltage source converter (VSC) converters in section 26.11, it is not possible to have zero reactive power on both sides of a reactor.

The main shortcoming of this topology is the power loss in the magnetic circuit of internal transformer. The losses are directly proportional to the operating frequency and the practical aspects will limit the frequency to below 500 Hz for multi-MW size converters.

A fault current can always be interrupted by one of the two AC/DC converters that faces the fault on its AC side. This type of fault protection requires appropriately sized AC inductance and therefore this topology may require an additional inductor beside the internal AC transformer (shown as L_s in Figure 27.10). The transformer saturation under fault conditions may also bring design challenges.

27.6   High Stepping Ratio LCL DC/DC Converter

Figure 27.11 shows a DC/DC converter employing an internal LCL circuit. The appropriately designed LCL circuit (L_1, C and L_2) provides the following functions:

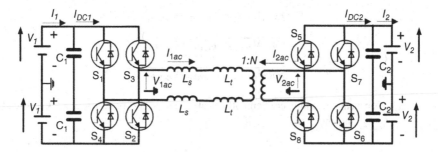

Figure 27.10   Dual active bridge DC/DC converter with internal AC transformer.

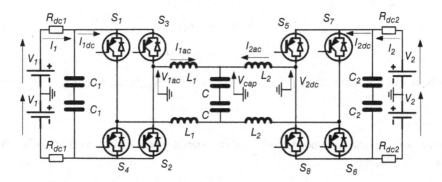

Figure 27.11   Symmetrical monopole, LCL IGBT-based DC/DC converter.

- it enables AC voltage and current stepping, similarly as conventional AC transformer;
- it provides inherent fault current regulation;
- it facilitates zero reactive power circulation (I_{ac1} is in phase with V_{ac1} and I_{ac2} is in phase with V_{ac2}).

The basic equations are the same as with LCL VSC converter Eqs (26.17)–(26.19); however, both AC voltages are fully controllable:

$$\overline{V_{ac1dq}} = V_{ac1} \angle \alpha_1 = V_{ac1d} + jV_{ac1q} \tag{27.4}$$

$$\overline{V_{ac2dq}} = V_{ac2} \angle \alpha_2 = V_{ac2d} + jV_{ac2q} \tag{27.5}$$

The voltage components will be controllable using the converter control signals, M_{1d}, M_{1q} and M_{2d}, M_{2q}:

$$V_{ac1d} = \frac{1}{\sqrt{2}} M_{1d} k_c V_1, \; V_{ac1q} = \frac{1}{\sqrt{2}} M_{1q} k_c V_1 \tag{27.6}$$

$$V_{ac2d} = \frac{1}{\sqrt{2}} M_{2d} k_c V_2, \; V_{ac2q} = \frac{1}{\sqrt{2}} M_{2q} k_c V_2 \tag{27.7}$$

where k_c is the gain between AC voltage and DC voltage, which depends on the converter topology and the modulation method (for sinusoidal PWM with two-level converters $k_c = 1$). The voltage angles are defined with respect to a common coordinate frame, which is linked with the central capacitor voltage Vc. The phasor domain internal circuit equations are similar as those with LCL VSC in section 26.11.3:

$$j\omega L_1 \overline{I_{ac1dq}} = \overline{V_{ac1dq}} - \overline{V_{capdq}} \tag{27.8}$$

$$j\omega C \overline{V_{capdq}} = \overline{I_{ac1dq}} + \overline{I_{ac2dq}} \tag{27.9}$$

$$j\omega L_2 \overline{I_{ac2dq}} = \overline{V_{ac2dq}} - \overline{V_{capdq}} \tag{27.10}$$

The above equations can be more conveniently written using the parameters k_1–k_3 instead of the parameters L_1, C, L_2, and the parameter link is given as:

$$k_1 = 1 - \omega^2 L_2 C \tag{27.11}$$

$$k_2 = 1 - \omega^2 L_1 C \tag{27.12}$$

$$k_3 = L_1 + L_2 - \omega^2 L_1 L_2 C \tag{27.13}$$

$$k_1 < 1, k_2 < 1, k_3 < L_1 + L_2 \tag{27.14}$$

It can be shown that zero reactive power condition at AC 1 ($V_{1acd}/V_{1acq} = I_{1acd}/I_{1acq}$) is achieved if

$$M_{2d} = sk_1 \tag{27.15}$$

where the stepping ratio is defined as $s = \frac{V_1}{V_2}$, and a zero reactive power condition at AC 2 ($V_{2acd}/V_{2acq} = I_{2acd}/I_{2acq}$) is achieved if

$$k_2 = k_1 s^2 \tag{27.16}$$

If Eq. (27.16) is satisfied, the converter will be operated with constant angle difference:

$$\tan(\alpha_2 - \alpha_1) = \sqrt{\frac{1}{k_1^2 s^2} - 1} \qquad (27.17)$$

The parameter k_3 is fully defined by the required power rating and there is therefore only one free design parameter, k_1. The design parameter k_1 can be selected in the range $-1/s < k_1 < 1$, depending on other requirements like fault current magnitude and the position of the natural resonant frequency with respect to the operating frequency ω.

Because of internal voltage stepping by the LCL circuit, the voltage stress on switches is low and similar to the converter with the internal transformer in Figure 27.10. Both AC/DC converters in the LCL DC/DC converter will operate with close to the maximum utilization ratio. One further advantage is that this converter uses inductors that can assume air-core design and therefore losses will be moderate even with increased operating frequency. Increased operating frequency implies smaller passive components.

The DC fault response of LCL DC/DC converter is excellent. As Figure 26.18 demonstrates, LCL circuit will keep the currents at low values for a short circuit at any of the terminals. This implies that VSC converters need not be overrated or tripped for DC faults. There will be no DC fault transfer through the DC/DC converter and the unfaulted side will only see a load rejection. A permanent fault isolation is possible by blocking IGBTs.

On the downside, this converter has no electrical isolation. It steps DC voltage symmetrically around the central point and therefore it cannot be used with asymmetric monopole HVDC topologies.

27.7 Building DC Grids with DC/DC Converters

A DC/DC converter is a very versatile device, which can take a range of functions in DC grids. It is very important that DC/DC converter can prevent DC fault propagation and enable the isolation of DC faults. Because of costs, it cannot be used instead of DC CBs but it can be strategically located in DC grids to provide protection zone separation.

Figure 27.12 shows an established 1.0 GW VSC HVDC (terminals 1–2), with two new 0.2 GW terminals (terminals 3 and 4) connected using a DC/DC converter. Connecting the third and fourth terminal without using a DC/DC converter would require both new terminals and DC cables 23 and 24 to be rated for high DC voltage (±320 kV) and additionally some DC CB arrangement would be necessary. The use of DC/DC brings the following benefits:

- The DC/DC converter crucially prevents propagation of faults from the small terminals to the main HVDC. A fault on DC cable 23 or VSC3 would not affect operation of terminals 1 and 2. However terminals 1, 2 and cable 12 still remain in a single protection zone and any fault in this zone implies that terminals 3 and 4 cannot operate.
- Cables 23 and 24 have lower DC voltage, implying cost savings. All equipment at VSC3 and VSC4 stations has lower voltage.
- DC/DC introduces one additional control channel, which enables DC voltage control in cable 23 and cable 24. It is assumed that VSC3 and VSC4 control local wind-farm power.

It is noted that some DC/DC converters enable DC voltage polarity reversal. This implies that the third terminal can have LCC converter technology.

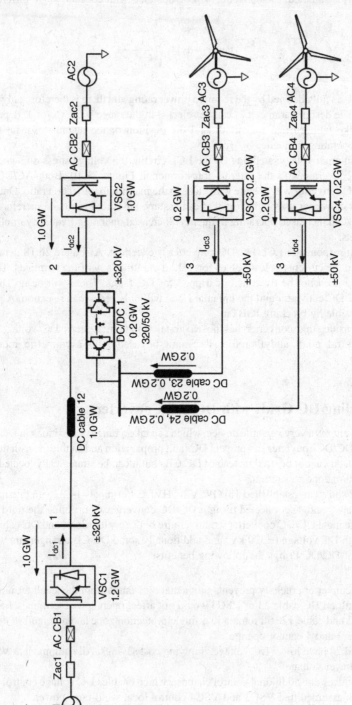

Figure 27.12 A VSC HVDC link with third terminal interfaced using a DC/DC converter.

Figure 27.13 shows an example of a 21-terminal DC grid, connecting nine onshore VSC terminals with 12 offshore wind farms. The grid is built using three star points, three DC/DC converters and 24 DC CBs. Some important properties of the grid are:

- each terminal has the opportunity to trade power with any other terminal;
- the grid can have three different DC voltage levels;
- DC faults are isolated using simple protection logic as discussed with star DC grids in section 26.10 (no need for communication between terminals);
- meshed topology (alternative power flow routing) exists only between star points.

The main principles for building large DC grids on the principle of star points and DC/DC converters are:

- Each new VSC terminal (either onshore of offshore) connects to the nearest star point, using a DC cable and a DC CB at the star connection point. Protection of the cable is simple because DC CB trips on a high positive signal from the local current sensor.
- Any two star points are interconnected using a DC cable with DC/DC at one end and a DC CB at the other end. A fault on the interconnecting cable is cleared by the DC CB and DC/DC. DC/DC prevents fault propagation from one star grid to another. If protection fails on one star grid, the remaining grids will not be affected. Each local star DC grid can have a different DC voltage and different protection arrangements (possibly a single vendor).
- The number of star points is determined considering required reliability, losses and protection methods. In general, more star points will provide better reliability but costs and losses will increase because of a larger number of DC/DC converters.

Redundancy can be incorporated as well as the $n - 1$ power security criterion. In order to enable full offshore power transfer to the shore, assuming that any single onshore VSC or DC cable is faulted, the onshore VSC and DC cables should be overrated. The rating for each VSC converter and cable should be $x/(x - 1)$ (9/8 for the grid in Figure 27.13) of the nominal power where x is the number of onshore terminals.

27.8 DC Hubs

The DC hub is a multiport DC/DC converter, which consists of an inner AC circuit and AC/DC converters interfacing each DC line. Figure 27.14 shows an N-port DC hub that uses only passive components in the inner two-phase AC circuit. The inner AC circuit may contain AC transformers and AC CBs for port isolation. The inner circuit will likely be operated at frequencies higher than 50 Hz in order to reduce component size. A multiphase topology will provide higher power ratings and increase reliability and possibly redundancy. The number of phases can be changed 'on the fly', which infers that a faulty inner phase can be substituted with a standby phase circuit. The DC hub will have complex inner control systems, which will manage power flow and stability of the isolated inner AC system. A different DC voltage is allowed on each connecting DC line.

The appropriate selection of inductor L_i and C_i for each port enables zero reactive power at each port under full active power.

The hub in Figure 27.14 has excellent DC fault responses. It can operate through DC faults on any port without any control action. Permanent DC faults can be isolated by opening an inner AC CB on the affected port.

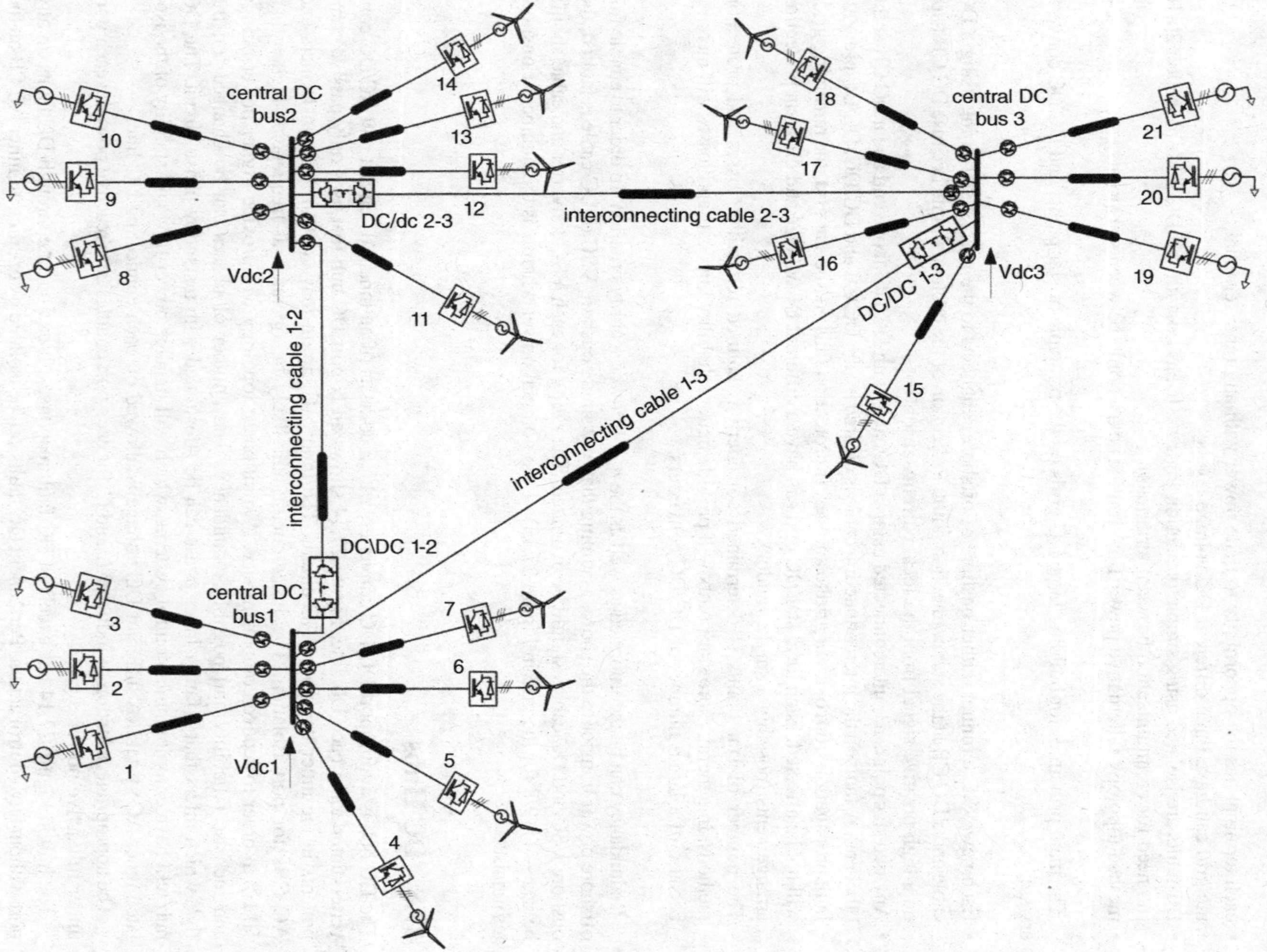

Figure 27.13 Twenty-one terminal DC grid with three star points and 3 DC/DC converters.

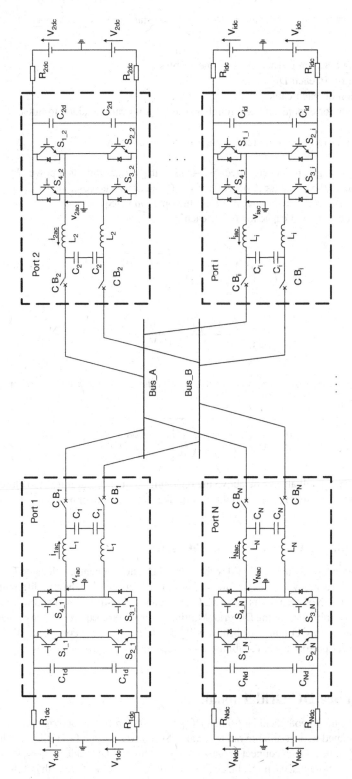

Figure 27.14 N-port, two-phase DC hub.

The DC hub can be viewed as an electronic DC substation in DC grids, which assumes these functions:

- Voltage stepping between connecting DC lines/cables.
- Power flow control in each DC cable.
- DC fault isolation on each cable.
- Connection points for expansion. Any number of new ports can be added to hubs.

The N-port hub design is summarized in the following steps:

1. For each port i, the given data are rated AC RMS voltage V_{acir} and rated power P_{ir}.
2. Select the value for the capacitor RMS voltage V_{cr}. Capacitor voltage will be 10–50% larger than the largest AC voltage in the inner circuit. Select the operating frequency ω.
3. Determine inductor L_i and capacitor C_i for each port i:

$$L_i = \frac{M_{ir}V_{acir}\sqrt{(V_{cr})^2 - (M_{ir})^2(V_{acir})^2}}{P_{ir}\omega} \tag{27.18}$$

$$C_i = \frac{1}{\omega(V_{cr})^2}\frac{P_{ir}\sqrt{(V_{cr})^2 - (M_{ir})^2(V_{acir})^2}}{M_{ir}V_{acir}} \tag{27.19}$$

4. For a given operating frequency ω, and selected V_{cr}, determine the value for resonant frequency ω_n using:

$$\frac{\omega_n}{\omega} = \sqrt{\sum_{k=1}^{N}\frac{1}{L_k} \bigg/ \sum_{k=1}^{N}\left(\frac{1}{L_k}\left[1 - \frac{(V_{ackr})^2}{(V_{cr})^2}\right]\right)} > 1 \tag{27.20}$$

The resonant frequency will always be above the operating frequency but if it is too close to the operating frequency the capacitor voltage should be adjusted to avoid undamped oscillations.

27.9 Developing DC Grids Using DC Hubs

Figure 27.15 shows a ten-terminal DC grid with a central DC hub. A centrally located DC hub enables power exchange between any terminals in the DC grid. Each terminal can have a different DC voltage and DC power ratings. Note that faults on any DC cable can be readily isolated and will not disturb the operation of the remaining part of the DC grid. Multiple hubs will be required for large, geographically dispersed grids, but the exact number of hubs will be determined using complex optimization while considering technical performance, reliability, power security, costs and losses.

27.10 North Sea DC Grid Topologies

Figure 27.16 shows the hypothetical North Sea DC grid, in the proposed topology with four DC hubs. Here, 1 GW is assumed to be a suitable rating for all VSC converters and DC cables. Each country has multiple 1 GW DC links, which connect to one of the hubs through DC cables. There is no terminal-to-terminal connection. Note that each DC link can have a different DC voltage with a different converter

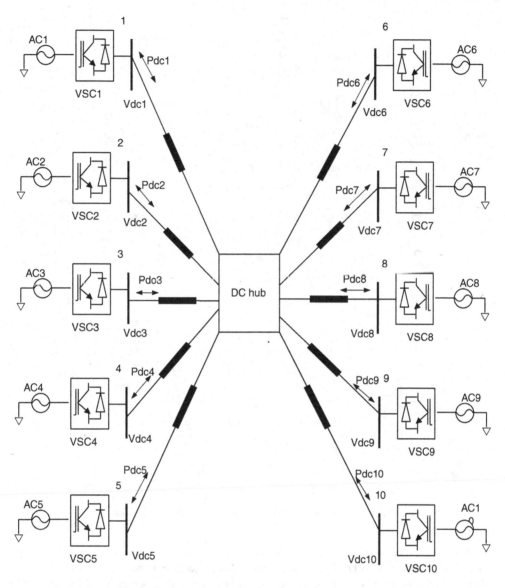

Figure 27.15 DC grid with ten terminals and one DC hub.

and protection topology. Expansion method is simple, since any number of future DC links can be connected to the hubs. No DC CBs are required. A hub of *n* ports will consist of *n* 1 GW VSC converters.

A similar DC grid can be developed using four DC busses instead of four DC hubs, by interconnecting regional star DC grids, as shown in Figure 27.17. The cables between two DC busses will have a DC/DC converter and a DC CB at the ends.

It is of interest to compare the two topologies in terms of total power electronics rating, which directly relates to the capital costs. Table 27.2 summarizes the converter rating for the grid with DC hubs and the DC grid with DC/DC converters, taking the basic rating assumptions from Table 22.1. It is seen that the amount of electronics in the inner grid (excluding VSC terminals) is 124 GW for DC hubs whereas it is

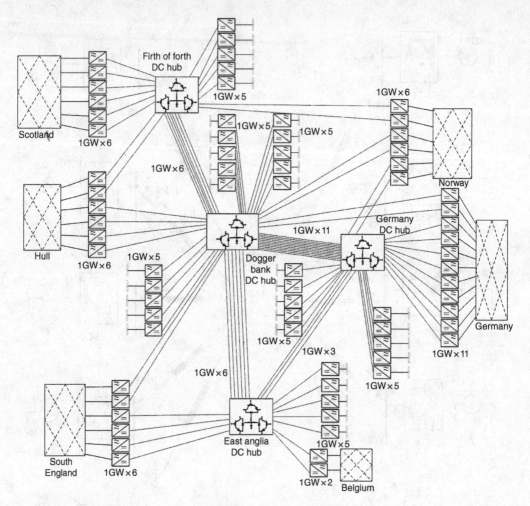

Figure 27.16 North sea DC grid schematic with four DC hubs.

93 GW for DC/DC converters. In terms of total DC grid costs, when VSC terminals and DC cables are included, the topology with DC hubs may require a few percentage higher investment.

However, the topology with DC hubs can use different DC voltage on each DC line and control/protection is considerably simpler. There are also other advantages with DC hubs like reliability and expansion flexibility.

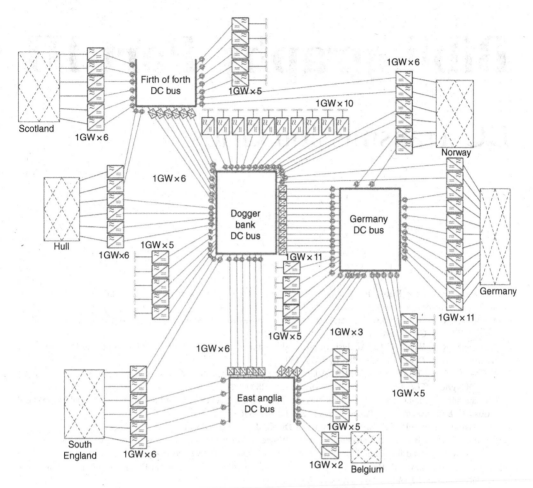

Figure 27.17 North sea DC grid schematic with four DC busses and DC/DC converters on all lines between busses.

Table 27.2 Comparison of North Sea DC grid with hubs and with DC/DC converters.

	Topology with 4 DC hubs	Topology with 4 DC busses
Number of 1 GW (1 pu) terminals	37 onshore + 35 offshore = 72	37 onshore + 35 offshore = 72
Number of 1 GW lines between hubs	26	26
Rating of grid converters (pu)	37 + 35 + 26 × 2 = 124	26 × 2 = 52
Rating of DC CBs (pu)	0	(37 + 35) × (1/3) + 26 × 2 × (1/3) = 41
Total converter rating (pu)	222	191

Bibliography Part III

DC Transmission Grids

B. Bachmann, G. Mauthe, Ruoss, E., *et al.* (1985) Development of a 500 kV airblast HVDC circuit breaker, *IEEE Transactions on Power Aparatus and Systems*, **PAS-104** (9), 2460–2466.

C.D. Barker, C.C. Davidson, D.R. Trainer, and R.S. Whitehouse (2012) Requirements of DC-DC converters to facilitate large DC Grids. Cigre Session 2012.

J. Beerten and R. Belmans (2013) Analysis of power sharing and volatge deviations in droop-controled DC grids, *IEEE Transactions on Power Systems*, **28** (4), 4588–4597.

N.R. Chaudhuri and B. Chaudhuri (2013) Adaptive droop control for effective power sharing in multi-terminal DC (MTDC) grids, *IEEE Transactions on Power Systems*, **28** (1), 21–22.

N.R. Chaudhuri, R. Majumder, and B. Chaudhuri (2013) System frequency support through multi-terminal DC (MTDC) grids, *IEEE Transactions on Power Systems*, **28** (1), 347–356.

CIGRE Working Group B4.52 (2013) Feasibility of DC Grids. CIGRE Brochure 533.

K. De Brabandere, B. Bolsens, J. Van den Keybus, A. Woyte, J. Driesen, and R. Belmans (2007) A voltage and frequency droop control method for parallel inverters, *IEEE Transactions on Power Electronics*, **22** (4), 1107–1115.

K. De Kerf, K. Srivastava, M. Reza, *et al.* (2011) Wavelet-based protection strategy for DC faults in multi-terminal VSC HVDC systems, *IET Generation, Transmission & Distribution*, **5** (4), 496–503.

J. Descloux, B. Raison, and J.-B. Curis (2013) Protection strategy for undersea MTDC grids. *Proceedings PowerTech, 2013 IEEE*, Grenoble, June 2013, pp. 1–6.

T. Eriksson, M. Backman, and S. Halen (2014) A Low Loss Mechanical HVDC Breaker for HVDC Grid Applications. B4-303, CIGRE, Paris.

J. Hafner and B. Jacobson (2011) Proactive hybrid HVDC breakers – A key innovation for reliable HVDC grids. *Proceeding of the 2011 CIGRE Bologna Conference*, pp. 1–8.

M. Hajian, L. Zhang, and D. Jovcic (2015) DC Transmission grid with low speed protection using mechanical DC circuit breakers, *IEEE Transactions on Power Delivery*. DOI: 10.1109/TPWRD.2014.2371618.

D. Jovcic (2009) Bidirectional high power DC transformer, *IEEE Transactions on Power Delivery*, **24** (4), 2276–2283.

D. Jovcic and A. Jamshidifar (2015) 3-Level cascaded voltage source converters converter controller with dispatcher droop feedback for direct current transmission grids Generation, *Transmission & Distribution, IET*, **9** (6), 571.

D. Jovcic and W. Lin (2014) Multiport high power LCL DC hub for use in DC transmission grids, *IEEE Transactions on Power Delivery*, **29** (2), 760–768.

D. Jovcic and B.T Ooi (2011) Theoretical aspects of fault isolation on DC lines using resonant DC/DC converters, *IET Generation Transmission and Distribution*, **5** (2), 153–160.

High-Voltage Direct-Current Transmission: Converters, Systems and DC Grids, First Edition.
Dragan Jovcic and Khaled Ahmed.
© 2015 John Wiley & Sons, Ltd. Published 2015 by John Wiley & Sons, Ltd.

D. Jovcic and L. Zhang (2013) LCL DC/DC converter for DC grids, *IEEE Transactions on Power Delivery*, **28** (4), 2071–2079.

D. Jovcic, L. Zhang, and M. Hajian (2013) LCL VSC converter for High power applications, *IEEE Transactions on Power Delivery*, **28** (1), 137–145.

D. Jovcic, M. Taherbaneh, J.P. Taisne, and S. Nguefeu (2014) Developing regional, radial DC grids and their interconnection into large DC grids. *IEEE PES General Meeting*, Washington, DC, July 2014.

D. Jovcic, M. Taherbaneh, J.P. Taisne, and S. Nguefeu (2015) Offshore DC grids as an interconnection of radial systems: protection and control aspects. *Smart Grid, IEEE Transactions on*, **6** (2), 903, 910.

E. Prieto-Araujo, F.D. Bianchi, A. Junyent-Ferré, and O. Gomis-Bellmunt (2011) Methodology for droop control dynamic analysis of multiterminal VSC-HVDC grids for offshore wind farms. *Power Delivery, IEEE Transactions on*, **26** (4), 2476, 2485.

C.E. Sheridan, M.M.C. Merlin, and T.C. Green (2012) Assesment of DC/DC converetrs for use in DC nodes for offshore DC grids. IET AC-DC Power Transmission, Birmingham, UK, December 2012, pp. 31–37.

J. Sneath and A.D. Rajapakse (2014) Fault detection and interruption in an earthed HVDC grid using ROCOV and hybrid DC breakers. *Power Delivery, IEEE Transactions on*, 99, 1, 1.

S. Tokuyama, K. Arimatsu, Y. Yoshioka, Y. Kato, and K. Hirata (1985) Development and interrupting tests on 250KV 8KA HVDC circuit breaker, *IEEE Transactions on Power Aparatus and Systems*, **PAS-104** (9), 2453–2459.

D. Van Hertem and M. Ghandhari (2010) Multi-terminal VSC HVDC for the European supergrid: obstacles. *Renewable and Sustainable Energy Reviews*, **14** (9), 3156–3163.

D. Van Hertem, M. Ghandhari, J.B. Curis, *et al.* (2011) Protection requirements for a multiterminal meshed DC grid. *CIGRE Symposium*, Bologna, Italy, September 2011.

E. Veilleux and B.-T. Ooi (2012) Multiterminal HVDC with thyristor power-flow controller, *IEEE Transactions on Power Delivery*, **27** (3), 1205–1212.

T.K. Vrana, J. Beerten, R. Belmans, and O.B. Fosso (2013a), A classification of DC node voltage control methods for HVDC Grids, *Electric Power Systems Research*, **103**, 137–144.

T.K. Vrana, Y. Yang, D. Jovcic, *et al.* (2013b) The CIGRE B4 DC Grid Test System, ELECTRA issue 270, October 2013, pp. 10–19.

W. Wang and M. Barnes (2014) Power flow algorithms for multiterminal VSC-HVDC with droop control, *IEEE Transactions on Power Systems*, **29** (4), 1721–1730.

J. Yang, J. E. Fletcher, and J. O'Reilly (2010) Multiterminal DC wind farm collection grid internal fault analysis and protection design, *IEEE Transactions on Power Delivery*, **25** (4), 2308–2318.

Appendix A

Variable Notations

Table A.1 shows the notation for two variables (current and voltage) in all three modelling systems used in this book:

- the static ABC frame;
- the rotating DQ frame;
- phasor notation with RMS values.

High-Voltage Direct-Current Transmission: Converters, Systems and DC Grids, First Edition.
Dragan Jovcic and Khaled Ahmed.
© 2015 John Wiley & Sons, Ltd. Published 2015 by John Wiley & Sons, Ltd.

Table A.1 Notation systems for two variables.

Variable	Current	Voltage
Static ABC frame		
Individual variables in static ABC frame	$i_a = I\cos(\omega t + \varphi_I) + I_0$	$v_a = V\cos(\omega t + \varphi_V) + V_0$
	$i_b = I\cos\left(\omega t - \dfrac{2}{3}\pi + \varphi_I\right) + I_0$	$v_b = V\cos\left(\omega t - \dfrac{2}{3}\pi + \varphi_V\right) + V_0$
	$i_c = I\cos\left(\omega t + \dfrac{2}{3}\pi + \varphi_I\right) + I_0$	$v_c = V\cos\left(\omega t + \dfrac{2}{3}\pi + \varphi_V\right) + V_0$
Vector in static ABC frame	$i_{abc} = \begin{bmatrix} i_a \\ i_b \\ i_c \end{bmatrix}$	$v_{abc} = \begin{bmatrix} v_a \\ v_b \\ v_c \end{bmatrix}$
Rotating DQ frame		
Orthogonal components in DQ frame	$I_d = I\cos(\varphi_I)$	$V_d = V\cos(\varphi_V)$
	$I_q = I\sin(\varphi_I)$	$V_q = V\sin(\varphi_V)$
	$I_0 = I_0$	$V_0 = V_0$
Polar components in DQ frame	$I = \sqrt{I_d^2 + I_q^2}$	$V = \sqrt{V_d^2 + V_q^2}$
	$\varphi_I = \arctan\dfrac{I_q}{I_d}$	$\varphi_V = \arctan\dfrac{V_q}{V_d}$
Vector in DQ frame	$I_{dq} = \begin{bmatrix} I_d \\ I_q \\ I_0 \end{bmatrix}$	$V_{dq} = \begin{bmatrix} V_d \\ V_q \\ V_0 \end{bmatrix}$
Phasor notation (employs RMS values)		
RMS (line–neutral) components in DQ frame	$\mathsf{I} = \dfrac{I}{\sqrt{2}}$	$\mathsf{V} = \dfrac{V}{\sqrt{2}}$
	$\mathsf{I_d} = \dfrac{I_d}{\sqrt{2}}$	$\mathsf{V_d} = \dfrac{V_d}{\sqrt{2}}$
	$\mathsf{I_q} = \dfrac{I_q}{\sqrt{2}}$	$\mathsf{V_q} = \dfrac{V_q}{\sqrt{2}}$
	$\mathsf{I_0} = I_0$	$\mathsf{V_0} = V_0$
RMS (line–line) components in DQ frame	—	$\mathsf{V\_{ll}} = \sqrt{3}\,\mathsf{V} = \dfrac{\sqrt{3}}{\sqrt{2}}V$
Phasor vector	$\overline{\mathsf{I}_{dq}} = \mathsf{I_d} + j\mathsf{I_q}$	$\overline{\mathsf{V}_{dq}} = \mathsf{V_d} + j\mathsf{V_q}$

Appendix B

Analytical Background for Rotating DQ Frame

B.1 Transforming AC Variables to DQ Frame

The AC system variables will be oscillating at a fundamental frequency and any transient changes will be seen as magnitude and phase variations of the oscillating variables. These oscillating variables are difficult to process in control and filtering circuits and they are commonly transferred to a rotating coordinate frame at the first processing stage in microcontrollers. The AC system is also modelled in the rotating coordinate frame in order to connect with DC system variables and to enable model linearization.

It is assumed that the AC system is symmetrical and balanced and therefore any three-phase current vector i_{abc} of magnitude I and phase angle φ_I can be represented in the static frame as:

$$i_{abc} = \begin{bmatrix} i_a \\ i_b \\ i_c \end{bmatrix} = \begin{bmatrix} I\cos(\omega t + \varphi_I) + I_0 \\ I\cos\left(\omega t - \dfrac{2}{3}\pi + \varphi_I\right) + I_0 \\ I\cos\left(\omega t + \dfrac{2}{3}\pi + \varphi_I\right) + I_0 \end{bmatrix} \tag{B1}$$

This variable can also be represented using rectangular components (I_0, I_d and I_q):

$$\begin{aligned} i_a &= I_0 + I\cos(\omega t + \varphi_I) \\ i_a &= I_0 + I\cos(\varphi_I)\cos(\omega t) - I\sin(\varphi_I)\sin(\omega t) \end{aligned} \tag{B2}$$

High-Voltage Direct-Current Transmission: Converters, Systems and DC Grids, First Edition.
Dragan Jovcic and Khaled Ahmed.
© 2015 John Wiley & Sons, Ltd. Published 2015 by John Wiley & Sons, Ltd.

$$i_a = I_0 + I_d \cos(\omega t) - I_q \sin(\omega t) \tag{B3}$$

where:

$$\begin{aligned} I_d &= I\cos(\varphi_I) \\ I_q &= I\sin(\varphi_I) \\ I_0 &= I_0 \end{aligned} \tag{B4}$$

The above transformation to a DQ coordinate frame, which is rotating at speed ω, is mathematically represented as:

$$\begin{bmatrix} I_d \\ I_q \\ I_0 \end{bmatrix} = P \begin{bmatrix} i_a \\ i_b \\ i_c \end{bmatrix}, \tag{B5}$$

where P is the transformation matrix given by:

$$P = \frac{2}{3} \begin{bmatrix} \cos(\omega t) & \cos(\omega t - 2\pi/3) & \cos(\omega t + 2\pi/3) \\ -\sin(\omega t) & -\sin(\omega t - 2\pi/3) & -\sin(\omega t + 2\pi/3) \\ 1/2 & 1/2 & 1/2 \end{bmatrix} \tag{B6}$$

The DQ coordinate frame orientation is shown in Figure B.1. Therefore using Eqs (B1)–(B6), the current d and q components are obtained as:

$$\begin{bmatrix} I_d \\ I_q \\ I_0 \end{bmatrix} = \frac{2}{3} \begin{bmatrix} \cos(\omega t) & \cos(\omega t - 2\pi/3) & \cos(\omega t + 2\pi/3) \\ -\sin(\omega t) & -\sin(\omega t - 2\pi/3) & -\sin(\omega t + 2\pi/3) \\ 1/2 & 1/2 & 1/2 \end{bmatrix} \begin{bmatrix} I\cos(\omega t + \varphi_I) + I_0 \\ I\cos\left(\omega t - \dfrac{2}{3}\pi + \varphi_I\right) + I_0 \\ I\cos\left(\omega t + \dfrac{2}{3}\pi + \varphi_I\right) + I_0 \end{bmatrix} \tag{B7}$$

Considering:

$$\cos k(\omega t) + \cos k\left(\omega t - \frac{2\pi}{3}\right) + \cos k\left(\omega t + \frac{2\pi}{3}\right) = 0, \quad k = 1, 2, \ldots$$

$$\sin k(\omega t) + \sin k\left(\omega t - \frac{2\pi}{3}\right) + \sin k\left(\omega t + \frac{2\pi}{3}\right) = 0, \quad k = 1, 2, \ldots \tag{B8}$$

$$\cos^2 x = (1 + \cos 2x)/2, \quad \sin^2 x = (1 - \cos 2x)/2,$$

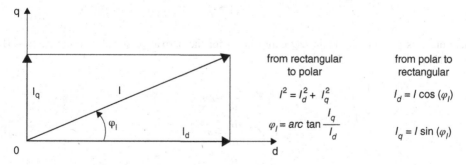

Figure B.1   AC current i in rotating DQ coordinate frame.

equation (B7) simplifies to:

$$I_{dq0} = \begin{bmatrix} I_d \\ I_q \\ I_0 \end{bmatrix} = \begin{bmatrix} I\cos(\varphi_I) \\ I\sin(\varphi_I) \\ I_0 \end{bmatrix} \tag{B9}$$

The vector components in Eq. (B9) are identical to those in Eq. (B4). The variables in Eq. (B9) are static (not oscillating), which illustrates the advantage of *DQ0* frame. The inverse transformation also exists:

$$\begin{bmatrix} i_a \\ i_b \\ i_c \end{bmatrix} = P^{-1} \begin{bmatrix} I_d \\ I_q \\ I_0 \end{bmatrix} \tag{B10}$$

where:

$$P^{-1} = \frac{3}{2} \begin{bmatrix} \cos(\omega t) & -\sin(\omega t) & 1 \\ \cos(\omega t - 2\pi/3) & -\sin(\omega t - 2\pi/3) & 1 \\ \cos(\omega t + 2\pi/3) & -\sin(\omega t + 2\pi/3) & 1 \end{bmatrix} \tag{B11}$$

B.2   Derivative of an Oscillating Signal in DQ Frame

The derivative of an oscillating signal, *i*, is a new oscillating signal, which can be labelled as *z*:

$$\frac{d}{dt}[i] = z$$
$$z = Z_o + Z_d\cos(\omega t) - Z_q\sin(\omega t) \tag{B12}$$

Taking derivative of Eq. (B3) the following is obtained:

$$\frac{d}{dt}[i] = \frac{d}{dt}\left[I_0 + I_d\cos(\omega t) - I_q\sin(\omega t)\right]$$

$$\frac{d}{dt}[i] = -\omega I_d\sin(\omega t) + \frac{dI_d}{dt}\cos(\omega t) - \omega I_q cos(\omega t) - \frac{dI_q}{dt}\sin(\omega t) \tag{B13}$$

$$\frac{d}{dt}[i] = \left[\frac{dI_d}{dt} - \omega I_q\right]_d \cos(\omega t) - \left[\frac{dI_q}{dt} + \omega I_d\right]_q \sin(\omega t)$$

Therefore it is possible to obtain *DQ* components for the derivative signal considering Eqs. (B12) and (B13):

$$Z_d = \frac{dI_d}{dt} - \omega I_q$$

$$Z_q = \frac{dI_q}{dt} + \omega I_d \tag{B14}$$

$$Z_o = \frac{dI_o}{dt}$$

If only steady-state values are of interest, by neglecting all derivatives in the above expression in Eq. (B14):

$$\begin{aligned} Z_d &= -\omega I_q \\ Z_q &= \omega I_d \\ Z_o &= 0 \end{aligned} \qquad \text{(B15)}$$

B.3   Transforming an AC System Dynamic Equation to DQ Frame

If AC dynamics are modelled in detail, the dynamic equations are first expressed in an ABC frame. Starting with a single first-order differential equation for phase a of a three-phase system in a Laplace domain:

$$sL_a i_a = -R_a i_a + v_a \qquad \text{(B16)}$$

where s is the Laplace operator $s = d/dt(\ )$ and where i_a and v_a are phase a oscillating variables. Under the assumption that the system is symmetrical and balanced, the following is valid:

$$R_a = R_b = R_c = R, \quad L_a = L_b = L_c = L \qquad \text{(B17)}$$

The three-phase system dynamic equation can now be represented as:

$$sLi_{abc} = -Ri_{abc} + v_{abc} \quad L = \begin{bmatrix} L & 0 & 0 \\ 0 & L & 0 \\ 0 & 0 & L \end{bmatrix}, \quad R = \begin{bmatrix} R & 0 & 0 \\ 0 & R & 0 \\ 0 & 0 & R \end{bmatrix} \qquad \text{(B18)}$$

The above model has all states as oscillating variables (50 Hz *sine* signals). It is required to transfer the model into rotating DQ frame using Park's transformation Eq. (B6).

Replacing all ABC variable with DQ variables in the model in Eq. (B18):

$$sLP^{-1}I_{dq0} = -RP^{-1}I_{dq0} + P^{-1}V_{dq0} \qquad \text{(B19)}$$

Multiply Eq. (B19) with P:

$$PsLP^{-1}I_{dq0} = -PRP^{-1}I_{dq0} + PP^{-1}V_{dq0} \qquad \text{(B20)}$$

Using the following properties of Park's matrix

$$PRP^{-1} = R$$
$$PP^{-1} = I = \begin{bmatrix} 1 & 0 & 0 \\ 0 & 1 & 0 \\ 0 & 0 & 1 \end{bmatrix}. \qquad \text{(B21)}$$

and expanding the dynamic term, Eq. (B20) is rearranged:

$$PsLP^{-1}I_{dq0} + LPLP^{-1}sI_{dq0} = -RI_{dq0} + BV_{dq0} \qquad \text{(B22)}$$

It can be shown that, for Park's matrix, the following holds:

$$PsP^{-1} = \omega \begin{bmatrix} 0 & -1 & 0 \\ 1 & 0 & 0 \\ 0 & 0 & 0 \end{bmatrix} = \omega W_0, \quad W_0 = \begin{bmatrix} 0 & -1 & 0 \\ 1 & 0 & 0 \\ 0 & 0 & 0 \end{bmatrix} \tag{B23}$$

therefore using Eq. (B23), Eq. (B22) is written as:

$$sI_{dq0} = -[RL^{-1} + \omega W_0]I_{dq0} + L^{-1}V_{dq0} \tag{B24}$$

The above equation is the DQ frame model for an AC system with a single dynamic equation (one dynamic element).

B.4   Transforming n-Order State Space AC System Model to DQ Frame

In case of a complex AC system, the dynamics are represented as an n-th order state-space model. A phase a model for the system with n states, m inputs and p outputs is represented as:

$$\begin{aligned} sx_a &= A_a x_a + B_a u_a \\ y_a &= C_a x_a + D_a u_a \quad A_a \in \mathfrak{R}^{n \times n}, B_a \in \mathfrak{R}^{n \times m}, C_a \in \mathfrak{R}^{p \times n}, D_a \in \mathfrak{R}^{p \times m} \end{aligned} \tag{B25}$$

$$x_a = \begin{bmatrix} x_1 \\ . \\ x_n \end{bmatrix}_a, \quad u_a = [u_1 . u_m]_a, \quad y_a = \begin{bmatrix} y_1 \\ . \\ y_p \end{bmatrix}_a$$

Then, the three-phase state-space system model in an ABC frame is:

$$\begin{aligned} sx_{abc} &= A_{abc} x_{abc} + B_{abc} u_{abc} \\ y_{abc} &= C_{abc} x_{abc} + D_{abc} u_{abc} \quad x_{abc} \end{aligned} = \begin{bmatrix} \begin{bmatrix} x_1 \\ . \\ x_n \end{bmatrix}_a \\ \begin{bmatrix} x_1 \\ . \\ x_n \end{bmatrix}_b \\ \begin{bmatrix} x_1 \\ . \\ x_n \end{bmatrix}_c \end{bmatrix}, \quad u_{abc} = \begin{bmatrix} [u_1 . u_m]_a \\ [u_1 . u_m]_b \\ [u_1 . u_m]_c \end{bmatrix}, \quad y_{abc} = \begin{bmatrix} \begin{bmatrix} y_1 \\ . \\ y_p \end{bmatrix}_a \\ \begin{bmatrix} y_1 \\ . \\ y_p \end{bmatrix}_b \\ \begin{bmatrix} y_1 \\ . \\ y_p \end{bmatrix}_c \end{bmatrix} \tag{B26}$$

$$A_{abc} = \begin{bmatrix} A_a & 0_{n \times n} & 0_{n \times n} \\ 0_{n \times n} & A_b & 0_{n \times n} \\ 0_{n \times n} & 0_{n \times n} & A_c \end{bmatrix}, \quad B_{abc} = \begin{bmatrix} B_a & 0_{n \times m} & 0_{n \times m} \\ 0_{n \times m} & B_b & 0_{n \times m} \\ 0_{n \times m} & 0_{n \times m} & B_c \end{bmatrix},$$

$$C_{abc} = \begin{bmatrix} C_a & 0_{p \times n} & 0_{p \times n} \\ 0_{pxn} & C_b & 0_{p \times n} \\ 0_{p \times n} & 0_{p \times n} & C_c \end{bmatrix}, \quad D_{abc} = \begin{bmatrix} D_a & 0_{p \times m} & 0_{p \times m} \\ 0_{p \times m} & D_b & 0_{p \times m} \\ 0_{p \times m} & 0_{p \times m} & D_c \end{bmatrix}$$

Using the above approach and assuming that the system is symmetrical and balanced, the state-space model (B26) is converted into rotating frame as:

$$
\begin{aligned}
sX_{dq0} &= A_{dq0}X_{dq0} + B_{dq0}U_{dq0} \\
Y_{dq0} &= C_{dq0}X_{dq0} + D_{dq0}U_{dq0}X_{dq0}
\end{aligned}
= \begin{bmatrix} \begin{bmatrix} x_1 \\ \cdot \\ x_n \end{bmatrix}_d \\ \begin{bmatrix} x_1 \\ \cdot \\ x_n \end{bmatrix}_q \\ \begin{bmatrix} x_1 \\ \cdot \\ x_n \end{bmatrix}_0 \end{bmatrix},\quad
U_{dq0} = \begin{bmatrix} [u_1 \; \cdot \; u_m]_d \\ [u_1 \; \cdot \; u_m]_q \\ [u_1 \; \cdot \; u_m]_0 \end{bmatrix},\quad
Y_{dq0} = \begin{bmatrix} \begin{bmatrix} y_1 \\ \cdot \\ y_p \end{bmatrix}_d \\ \begin{bmatrix} y_1 \\ \cdot \\ y_p \end{bmatrix}_q \\ \begin{bmatrix} y_1 \\ \cdot \\ y_p \end{bmatrix}_0 \end{bmatrix}
\tag{B27}
$$

$$
A_{dq0} = \begin{bmatrix} A_a & \omega I_{n\times n} & 0_{n\times n} \\ -\omega I_{n\times n} & A_u & 0_{n\times n} \\ 0_{n\times n} & 0_{n\times n} & A_a \end{bmatrix},\quad
B_{dq0} = \begin{bmatrix} B_a & 0_{n\times m} & 0_{n\times m} \\ 0_{n\times m} & B_a & 0_{n\times m} \\ 0_{n\times m} & 0_{n\times m} & B_a \end{bmatrix}
$$

$$
C_{dq0} = \begin{bmatrix} C_a & 0_{p\times n} & 0_{p\times n} \\ 0_{p\times n} & C_a & 0_{p\times n} \\ 0_{p\times n} & 0_{p\times n} & C_a \end{bmatrix},\quad
D_{dq0} = \begin{bmatrix} D_a & 0_{p\times m} & 0_{p\times m} \\ 0_{p\times m} & D_a & 0_{p\times m} \\ 0_{p\times m} & 0_{p\times m} & D_a \end{bmatrix}
$$

Note that, for a balanced and symmetrical system the set of zero sequence variables can be removed from the model. Equation (B27) shows that the system matrix has ω as a multiplier with off-diagonal terms. This implies that a frequency in the DQ frame is increased by f relative to the ABC frame. In general, a frequency component f_{abc} in the ABC frame is transferred to the DQ0 frame as f_{dq0} where:

$$
f_{dq0} = f_{abc} \pm f
\tag{B28}
$$

Once the A, B, C and D matrices are known in Eq. (B27) then the time domain solution can be determined as with any state-space system model. Alternatively, it is possible to converter model (B27) into transfer function representation:

$$
\begin{aligned}
Y_{dq0} &= W_{dq0}(s)U_{dq0} \\
W_{dq0}(s) &= C_{dq0}\left[sI - A_{dq0}\right]^{-1}B_{dq0} + D_{dq0}
\end{aligned}
\tag{B29}
$$

B.5 Static (Steady-State) Modelling in Rotating DQ Coordinate Frame

The above DQ model represents all the system dynamics and enables accurate transient responses and dynamic studies. If only the steady-state solution is of interest, the above model can be simplified by neglecting all dynamic terms, that is $s = 0$:

$$
\begin{aligned}
0 &= A_{dq0}X_{dq0} + B_{dq0}U_{dq0} \\
Y_{dq0} &= C_{dq0}X_{dq0} + D_{dq0}U_{dq0}
\end{aligned}
\tag{B30}
$$

Therefore the static model (B30) becomes:

$$X_{dq0} = -A_{dq0}^{-1} B_{dq0} U_{dq0}$$
$$Y_{dq0} = C_{dq0} X_{dq0} + D_{dq0} U_{dq0}$$

(B31)

The above model can be solved using standard matrix methods.

B.6 Representing the Product of Oscillating Signals in the DQ Frame

Assume that two single-phase AC signals x and y are given as:

$$x = X_0 + X_m \cos(\omega t + \varphi_x) + X_{m2} \cos(2\omega t + \varphi_x) = X_0 + X_d \cos\omega t - X_q \sin\omega t + X_{d2} \cos 2\omega t - X_{q2} \sin 2\omega t$$
$$y = Y_0 + Y_m \cos(\omega t + \varphi_y) = Y_0 + Y_d \cos\omega t - Y_q \sin\omega t$$

(B32)

where signal x also includes second harmonic for completeness. It is of interest to determine the DQ components of the product signal z:

$$z = xy$$

(B33)

$$z = Z_0 + Z_d \cos(\omega t) - Z_q \sin(\omega t) + Z_{d2} \cos(2\omega t) - Z_{q2} \sin(2\omega t) + Z_{d3} \cos(3\omega t) - Z_{q3} \sin(3\omega t)$$

(B34)

Replacing Eq. (B32) in Eq. (B33), and using the following trigonometric identities:

$$\cos\alpha\cos\beta = \frac{\cos(\alpha-\beta) + \cos(\alpha+\beta)}{2}, \quad \sin\alpha\sin\beta = \frac{\cos(\alpha-\beta) - \cos(\alpha+\beta)}{2}$$
$$\sin\alpha\cos\beta = \frac{\sin(\alpha+\beta) + \sin(\alpha-\beta)}{2}, \quad \cos\alpha\sin\beta = \frac{\sin(\alpha+\beta) - \sin(\alpha-\beta)}{2},$$

(B35)

Equation (B33) becomes:

$$z = \left(X_0 Y_0 + \frac{X_d Y_d}{2} + \frac{X_q Y_q}{2} \right)_0$$
$$+ \left(X_d Y_0 + X_0 Y_d + \frac{1}{2} X_{d2} Y_d + \frac{1}{2} X_{q2} Y_q \right)_d \cos(\omega t) - \left(X_q Y_0 + X_0 Y_q - \frac{1}{2} X_{d2} Y_q + \frac{1}{2} X_{q2} Y_d \right)_q \sin(\omega t)$$
$$+ \left(\frac{X_d Y_d}{2} - \frac{X_q Y_q}{2} + X_{d2} Y_0 \right)_{d2} \cos(2\omega t) - \left(\frac{X_q Y_d}{2} + \frac{X_d Y_q}{2} + X_{q2} Y_0 \right)_{q2} \sin(2\omega t)$$
$$+ \left(\frac{1}{2} X_{d2} Y_d - \frac{1}{2} X_{q2} Y_q \right)_{d3} \cos(3\omega t) - \left(\frac{1}{2} X_{d2} Y_q + \frac{1}{2} X_{q2} Y_d \right)_{q3} \sin(3\omega t)$$

(B36)

Therefore equating Eq. (B34) with Eq. (B36), the components of the product z signal are:

$$Z_0 = X_0 Y_0 + \frac{X_d Y_d}{2} + \frac{X_q Y_q}{2}$$

$$Z_d = X_d Y_0 + X_0 Y_d + \frac{1}{2}X_{d2}Y_d + \frac{1}{2}X_{q2}Y_q, \quad Z_q = X_q Y_0 + X_0 Y_q - \frac{1}{2}X_{d2}Y_q + \frac{1}{2}X_{q2}Y_0,$$

$$Z_{d2} = \frac{X_d Y_d}{2} - \frac{X_q Y_q}{2} + X_{d2}Y_0, \quad Z_{q2} = \frac{X_q Y_d}{2} + \frac{X_d Y_q}{2} + X_{q2}Y_0, \tag{B37}$$

$$Z_{d3} = \frac{1}{2}X_{d2}Y_d - \frac{1}{2}X_{q2}Y_q, \quad Z_{q3} = \frac{1}{2}X_{d2}Y_q + \frac{1}{2}X_{q2}Y_d,$$

If $X_{d2} = X_{q2} = 0$, as this is a common case in power engineering for symmetrical systems:

$$Z_0 = X_0 Y_0 + \frac{X_d Y_d}{2} + \frac{X_q Y_q}{2}$$

$$Z_d = X_d Y_0 + X_0 Y_d, \quad Z_q = X_q Y_0 + X_0 Y_q, \tag{B38}$$

$$Z_{d2} = \frac{X_d Y_d}{2} - \frac{X_q Y_q}{2}, \quad Z_{q2} = \frac{X_q Y_d}{2} + \frac{X_d Y_q}{2},$$

B.7 Representing Power in the DQ Frame

Assuming that the instantaneous voltage v and current i are given as:

$$i = I_0 + I\cos(\omega t + \varphi_i) = I_0 + I_d\cos(\omega t) - I_q\sin(\omega t) \tag{B39}$$

$$v = V_0 + V\cos(\omega t + \varphi_v) = V_0 + V_d\cos(\omega t) - V_q\sin(\omega t) \tag{B40}$$

The instantaneous electrical power per phase S_{1p} is defined as the product of instantaneous voltage and current:

$$S_{1p} = vi = V_0 I_0 + VI_0\cos(\omega t + \varphi_v) + V_0 I\cos(\omega t + \varphi_i) + VI\cos(\omega t + \varphi_v)\cos(\omega t + \varphi_i) \tag{B41}$$

Therefore when expanded:

$$S_{1p} = \frac{VI}{2}(\cos(\varphi_v - \varphi_i) + \cos(2\omega t + \varphi_v + \varphi_i)) + V_0 I_0 + VI_0\cos(\omega t + \varphi_v) + V_0 I\cos(\omega t + \varphi_i)$$

$$S_{1p} = \frac{VI}{2}(\cos(\varphi_v - \varphi_i) + \cos(2\omega t + 2\varphi_v + \varphi_i - \varphi_v)) + V_0 I_0 + VI_0\cos(\omega t + \varphi_v) + V_0 I\cos(\omega t + \varphi_i)$$

$$S_{1p} = \frac{VI}{2}\cos(\varphi_i - \varphi_v) + \frac{VI}{2}\cos(\varphi_i - \varphi_v)\cos(2\omega t + 2\varphi_v) - \frac{VI}{2}\sin(\varphi_i - \varphi_v)\sin(2\omega t + 2\varphi_v) \tag{B42}$$

$$+ V_0 I_0 + VI_0\cos(\omega t + \varphi_v) + V_0 I\cos(\omega t + \varphi_i)$$

$$S_{1p} = \frac{VI}{2}\cos(\varphi_i - \varphi_v)[1 + \cos(2\omega t + 2\varphi_v)] - \frac{VI}{2}\sin(\varphi_i - \varphi_v)\sin(2\omega t + 2\varphi_v)$$

$$+ V_0 I_0 + VI_0\cos(\omega t + \varphi_v) + V_0 I\cos(\omega t + \varphi_i)$$

Note that, with three-phase systems, the *cosine* term of second harmonic and also all first harmonics will cancel. By definition, active power is the average of the product of voltage and current in Eq. (B42):

$$P_{1p} = I_0 V_o + \frac{VI}{2}\cos(\varphi_i - \varphi_v) \tag{B43}$$

while the reactive power is the magnitude of the *sine* term of second harmonic:

$$Q_{1p} = -\frac{VI}{2}\sin(\varphi_i - \varphi_v) \tag{B44}$$

Replacing now Eqs (B39) and (B40) in Eq. (B43), the active power is obtained in terms of DQ components:

$$P_{1p} = V_0 I_0 + \frac{V_d I_d}{2} + \frac{V_q I_q}{2} \tag{B45}$$

Replacing Eqs (B39) and (B40) in Eq. (B44):

$$Q_{1p} = \frac{V_q I_d}{2} - \frac{V_d I_q}{2} \tag{B46}$$

If the components are given as RMS values, $I_d = \sqrt{2}I_d$ then Eqs (B45) and (B46) become:

$$P_{1p} = V_0 I_0 + V_d I_d + V_q I_q \tag{B47}$$

$$Q_{1p} = V_q I_d - V_d I_q \tag{B48}$$

Therefore, for a balanced and symmetrical system, the three-phase power is:

$$P = 3(V_0 I_0 + V_d I_d + V_q I_q) \tag{B49}$$

$$Q = 3(V_q I_d - V_d I_q) \tag{B50}$$

Example B.1
A simple three-phase system is shown in Figure B.2.

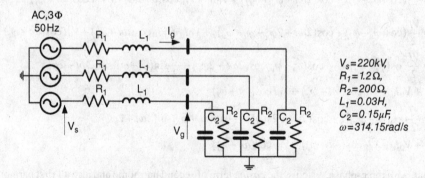

Figure B.2 Sample AC system for DQ frame analysis.

a. *Develop dynamic model in ABC frame.*
b. *Develop dynamic system model in rotating DQ frame.*
c. *Develop static model in DQ frame and calculate DQ components of current I_g and voltage V_g.*

Solution

1. The single phase state-space model in fixed ABC frame is:

$$s\begin{bmatrix} i_g \\ v_g \end{bmatrix} = \begin{bmatrix} \dfrac{-R_1}{L_1} & \dfrac{-1}{L_1} \\ \dfrac{1}{C_2} & \dfrac{-1}{C_2 R_2} \end{bmatrix} \begin{bmatrix} i_g \\ v_g \end{bmatrix} + \begin{bmatrix} \dfrac{1}{L_1} \\ 0 \end{bmatrix} v_s$$

$$\begin{bmatrix} i_g \\ v_g \end{bmatrix} = \begin{bmatrix} 1 & 0 \\ 0 & 1 \end{bmatrix} \begin{bmatrix} i_g \\ v_g \end{bmatrix} + \begin{bmatrix} 0 \\ 0 \end{bmatrix} v_s$$

(B51)

2. The DQ frame dynamic model is:

$$s\begin{bmatrix} I_{gd} \\ V_{gd} \\ I_{gq} \\ V_{gq} \end{bmatrix} = \begin{bmatrix} \dfrac{-R_1}{L_1} & \dfrac{-1}{L_1} & \omega & 0 \\ \dfrac{1}{C_2} & \dfrac{-1}{C_2 R_2} & 0 & \omega \\ -\omega & 0 & \dfrac{-R_1}{L_1} & \dfrac{-1}{L_1} \\ 0 & -\omega & \dfrac{1}{C_2} & \dfrac{-1}{C_2 R_2} \end{bmatrix} \begin{bmatrix} I_{gd} \\ V_{gd} \\ I_{gq} \\ V_{gq} \end{bmatrix} + \begin{bmatrix} \dfrac{1}{L_1} & 0 \\ 0 & 0 \\ 0 & \dfrac{1}{L_1} \\ 0 & 0 \end{bmatrix} \begin{bmatrix} V_{sd} \\ V_{sq} \end{bmatrix}$$

$$\begin{bmatrix} I_{gd} \\ V_{gd} \\ I_{gq} \\ V_{gq} \end{bmatrix} = \begin{bmatrix} 1 & 0 & 0 & 0 \\ 0 & 1 & 0 & 0 \\ 0 & 0 & 1 & 0 \\ 0 & 0 & 0 & 1 \end{bmatrix} \begin{bmatrix} I_{gd} \\ V_{gd} \\ I_{gq} \\ V_{gq} \end{bmatrix} + \begin{bmatrix} 0 & 0 \\ 0 & 0 \\ 0 & 0 \\ 0 & 0 \end{bmatrix} \begin{bmatrix} V_{sd} \\ V_{sq} \end{bmatrix}$$

(B52)

3. The static model is:

$$\begin{bmatrix} I_{gd} \\ V_{gd} \\ I_{gq} \\ V_{gq} \end{bmatrix} = -\begin{bmatrix} \dfrac{-R_1}{L_1} & \dfrac{-1}{L_1} & \omega & 0 \\ \dfrac{1}{C_2} & \dfrac{-1}{C_2 R_2} & 0 & \omega \\ -\omega & 0 & \dfrac{-R_1}{L_1} & \dfrac{-1}{L_1} \\ 0 & -\omega & \dfrac{1}{C_2} & \dfrac{-1}{C_2 R_2} \end{bmatrix}^{-1} \begin{bmatrix} \dfrac{1}{L_1} & 0 \\ 0 & 0 \\ 0 & \dfrac{1}{L_1} \\ 0 & 0 \end{bmatrix} \begin{bmatrix} V_{sd} \\ V_{sq} \end{bmatrix}$$

(B53)

Considering that $V_{sd} = 220e3/\sqrt{3}$, *and* $V_{sq} = 0$, *the currents are obtained:*

$$\begin{bmatrix} I_{gd} \\ V_{gd} \\ I_{gq} \\ V_{gq} \end{bmatrix} = \begin{bmatrix} 630.47\,A \\ 126.04\,kV \\ -23.63\,A \\ -5.9\,kV \end{bmatrix} \tag{B54}$$

The powers are calculated as:

$$P = 3\left(V_{gd}I_{gd} + V_{gq}I_{gq}\right) = 238.8\,MW \tag{B55}$$

$$Q = 3\left(V_{gq}I_{gd} - V_{gd}I_{gq}\right) = -2.25\,MVAr \tag{B56}$$

Appendix C

System Modelling Using Complex Numbers and Phasors

The phasor modelling of AC variables is adopted under the following assumptions:

- An AC system has three phases and is symmetrical and balanced. There are no zero sequence components.
- All variables are represented as RMS values assuming constant frequency ω.
- All components with energy storage (reactors and capacitors) are represented using reactances at fundamental frequency.

Using phasor notation, the voltage and current are represented as:

$$\overline{I_g} = I_{gd} + jI_{gq} \tag{C.1}$$

$$\overline{V_g} = V_{gd} + jV_{gq} \tag{C.2}$$

The system variables are solved using Kirchhoff's circuit rules with all the elements represented using impedance values at fundamental frequency. The basic impedances are as shown in Figure C.1.

Under the above assumptions, the complex power per phase is the product of a voltage and current conjugate complex:

$$\overline{S} = 3\overline{V_g}\left(\overline{I_g}\right)^* \tag{C.3}$$

where $(\ )^*$ notation stands for conjugate complex:

$$\begin{aligned}
\overline{S} &= 3\left(V_{gd} + jV_{gq}\right)\left(I_{gd} - jI_{gq}\right) \\
\overline{S} &= 3\left(V_{gd}I_{gd} + V_{gq}I_{gq} + j\left(V_{gq}I_{gd} - V_{gd}I_{gq}\right)\right)
\end{aligned} \tag{C.4}$$

High-Voltage Direct-Current Transmission: Converters, Systems and DC Grids, First Edition.
Dragan Jovcic and Khaled Ahmed.
© 2015 John Wiley & Sons, Ltd. Published 2015 by John Wiley & Sons, Ltd.

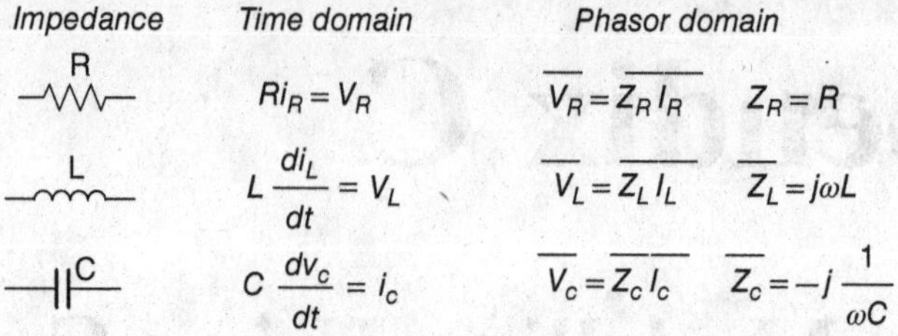

Impedance   Time domain   Phasor domain

$$R i_R = V_R \qquad \overline{V_R} = \overline{Z_R}\,\overline{I_R} \qquad Z_R = R$$

$$L \frac{di_L}{dt} = V_L \qquad \overline{V_L} = \overline{Z_L}\,\overline{I_L} \qquad \overline{Z_L} = j\omega L$$

$$C \frac{dv_c}{dt} = i_c \qquad \overline{V_c} = \overline{Z_c}\,\overline{I_c} \qquad \overline{Z_c} = -j\frac{1}{\omega C}$$

Figure C.1 Impedances in the time domain and in the phasor domain.

The active power is the real part whereas reactive power is the reactive part:

$$P = 3\left(V_{gd}I_{gd} + V_{gq}I_{gq}\right)$$
$$Q = 3\left(V_{gq}I_{gd} - V_{gd}I_{gq}\right) \tag{C.5}$$

Note that the above method with complex numbers cannot be used for systems with a zero-sequence component. The method is also not valid for general multiplication of two signals.

Example C.1
Consider the same system as in Example B.1. Determine system current I_g, voltage V_g and power at V_g using phasor methods.

Solution

$$\overline{z_1} = R_1 + j\omega L_1 = 1.2 + j9.425\,\Omega$$

$$\overline{z_2} = \frac{\dfrac{R_2}{j\omega C_2}}{R_2 + \dfrac{1}{j\omega C_2}} = \frac{R_2}{R_2 j\omega C_2 + 1} = 199.98 - j1.885\,\Omega \tag{C.6}$$

$$\overline{I_g} = \frac{\overline{V_s}}{\overline{z_1} + \overline{z_2}} = 630.5 - j23.63\,\text{A}$$

$$I_{gd} = 630.5\,\text{A} \tag{C.7}$$

$$I_{gq} = -23.63\,\text{A}$$

$$\overline{V_g} = \overline{I_g}\,\overline{z_2} = 126.03 - j5.913\,\text{kV}$$

$$V_{gd} = 126.03\,\text{kV} \tag{C.8}$$

$$V_{gq} = -5.913\,\text{kV}$$

The powers are calculated as:

$$S = 3\left(V_{gd} + jV_{gq}\right)\left(I_{gd} - jI_{gq}\right) = P_g + jQ_g \tag{C.9}$$

$$P = 3\left(V_{gd}I_{acd} + V_{gq}I_{acq}\right) = 238.8\,\text{MW} \tag{C.10}$$

$$Q = 3\left(V_{gq}I_{gd} - V_{gd}I_{gq}\right) = -2.25\,\text{MVAr} \tag{C.11}$$

The above results are consistent with those in Example B.1.

Appendix D

Simulink Examples

D.1 Chapter 3 Examples

Example D.1

Consider a rectifier connecting an AC source and a DC source as shown in Figure D.1

a. *Determine (by calculation or by simulation) the value of the firing angle in order to keep DC current at 1000 A.*
b. *Determine the maximum possible AC voltage drop that can be compensated for by firing angle variation, assuming that DC current is kept at 1000 A.*
c. *Determine the AC voltage that will reduce the DC current to zero.*
d. *Determine the maximum possible DC overvoltage that can be compensated for by firing angle variation, assuming that DC current is kept at 1000 A.*

Solution

A SIMULINK model is developed and the AC source magnitude is progressively reduced as shown in Figure D.2.

a. *The firing angle is around 16°.*
b. *The AC voltage is around 390 kV or 4.8% drop.*
c. *The AC voltage is 368 kV or 20% drop.*
d. *The DC overvoltage is same as AC voltage drop above (4.8%).*

High-Voltage Direct-Current Transmission: Converters, Systems and DC Grids, First Edition.
Dragan Jovcic and Khaled Ahmed.
© 2015 John Wiley & Sons, Ltd. Published 2015 by John Wiley & Sons, Ltd.

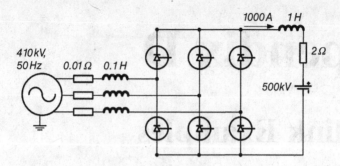

Figure D.1   Rectifier in Example D.1.

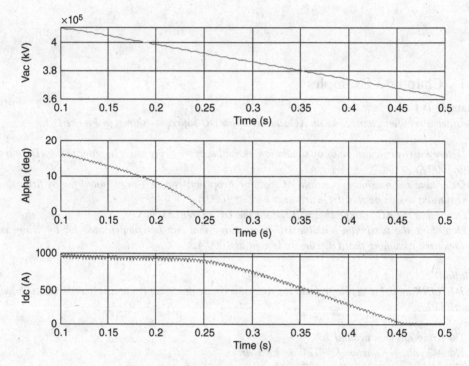

Figure D.2   Rectifier responses in Example D.1.

D.2   Chapter 5 Examples

Example D.2

Study the inverter connecting an AC and a DC source as shown in Figure D.3. Assume that this converter is operated in a closed-loop DC current control, maintaining $I_{dc} = 1000\,A$.

a. *Determine (by simulation or otherwise) the minimum symmetrical AC voltage drop that causes commutation failure.*
b. *Determine minimum DC current step increase that causes commutation failure.*
c. *Explain the impact of commutation failure on the AC system.*

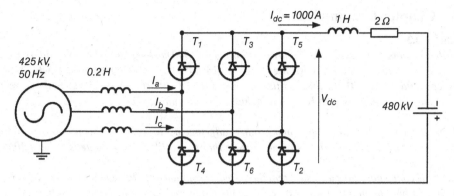

Figure D.3   Converter in Example D.2.

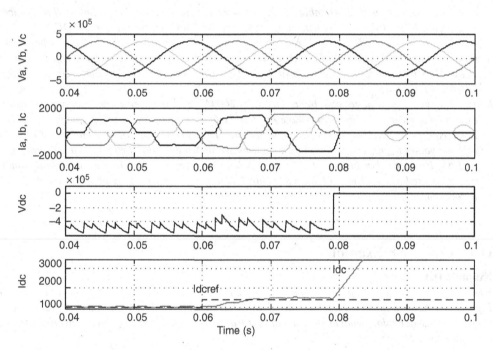

Figure D.4   Converter responses in Example D.2.

Solution

a. *By studying responses for reduced source voltage magnitude it can be determined that any voltage drop over 4% will cause commutation failure.*

b. *A DC current step to 1400 A causes commutation failure as seen by zero DC voltage.*

c. *Figure D.4 shows responses for a commutation failure caused by a DC current step to 1400 A. The AC voltage is not affected by the commutation failure. The AC current is also not affected, even though the DC current rises to very high values. Problems may arise in practical systems during subsequent recovery from commutation failures because of a large reactive power demand.*

D.3   Chapter 6 Examples

Example D.3

Consider the same test system as in Figure 6.6, however a 1000 km long DC cable is considered. Use the DC cable 'T' model as in the CIGRE high-voltage direct current (HVDC)benchmark, where the total cable capacitance is 240 μF, the rectifier/inverter side DC resistance is 13.5 Ω and the rectifier/inverter side inductance is 0.75 H (which includes DC smoothing reactors).

a. Tune the controller gains and make other adjustments to ensure system stability assuming inverter is in DC voltage-control mode. Ensure that the rectifier firing angle is at least 20° at full power.
b. Tune the controller gains to ensure that the system is dynamically stable in constant extinction angle mode. To operate in a constant extinction angle mode, reduce the inverter AC voltage.

Solution

a. If the same settings as in short-cable test system are used, the rectifier firing angle will be reduced to around 15° in order to compensate for the large DC voltage drop along the DC cable. It is required to reduce the DC voltage reference to around 480 kV. Alternatively, the rectifier transformer ratio can be adjusted to increase the converter-side voltage. The responses are shown in Figure D.5 for a current order step at 1.6 s. The transient response is slow because of large DC-cable impedance.
b. The inverter AC voltage can be reduced to 200 kV to test inverter operation in constant extinction angle mode (simulates remote AC voltage fault or line tripping). The system is unstable. It is required to reduce the rectifier and inverter DC current controller gains to 20% of the original values to ensure stability. The responses are shown in Figure D.6 for a current order step at 1.6 s. It is seen that inverter DC voltage is below reference value because gamma minimum control is used. Comparing with the Vdc control mode, in (a), the damping of the oscillatory mode is reduced.

D.4   Chapter 8 Examples

Example D.4

Study a 80% AC voltage drop on a rectifier AC system at 0.45 s for 100 ms for the same test system in Figure 6.6. Assume normal HVDC controls and that no voltage-dependent current order limiter (VDCOL) is used. Explain what mode transitions are occurring at the rectifier and the inverter. Discuss the impact on AC systems.

Solution

Figure D.7 shows the response to an 80% rectifier AC voltage drop.

At rectifier side the converter moves to alpha min mode but it still cannot maintain DC current. As the rectifier stays at a very low angle (around 2°) there is no increase in reactive power consumption.

At the inverter side, the DC current drops significantly and the inverter moves to current control. However this requires significantly lower firing-angle alpha (around 100°) in order to reduce DC voltage below the DC voltage at rectifier side. This implies very large reactive power consumption. At the inverter side, active power transfer is only 50 MW but reactive power consumption is over 500 MVAr and this may have negative consequences for the stability of the AC grid. For this reason VDCOL control logic is introduced, in order to reduce DC current during voltage depressions. A VDCOL helps fault recovery in particular with weak AC systems.

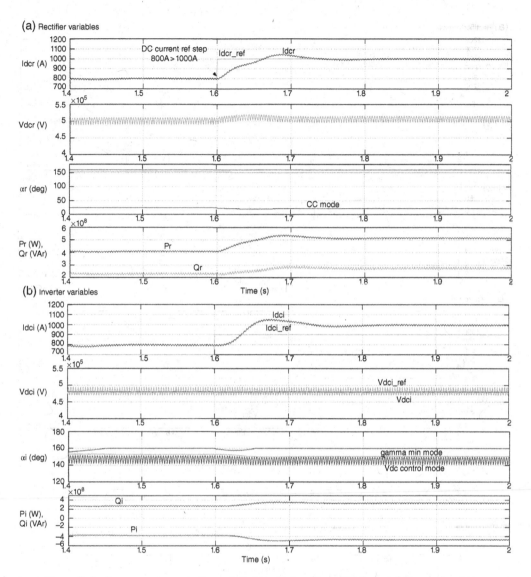

Figure D.5 System responses in Example D.3(a). (a) Rectifier variables and (b) inverter variables.

D.5   Chapter 14 Examples

Example D.5

Simulate a single-phase inverter, with a circuit as in Figure 14.1, and with $V_{DC} = 220\,V$, $f = 50\,Hz$.

a. *Assume a purely resistive load, $R = 50\,\Omega$. Sketch the load voltage and current.*
b. *Add a series inductor of 100 mH to the AC load. Determine the fundamental maximum AC voltage.*

Solution

a. *As can be seen from Figure D.8, voltage and current are in phase as expected with a resistive load.*
b. *When an inductor is added to the load, the current lags the voltage by the angle of the load power factor, as shown in Figure D.9. The maximum fundamental voltage is 280 V.*

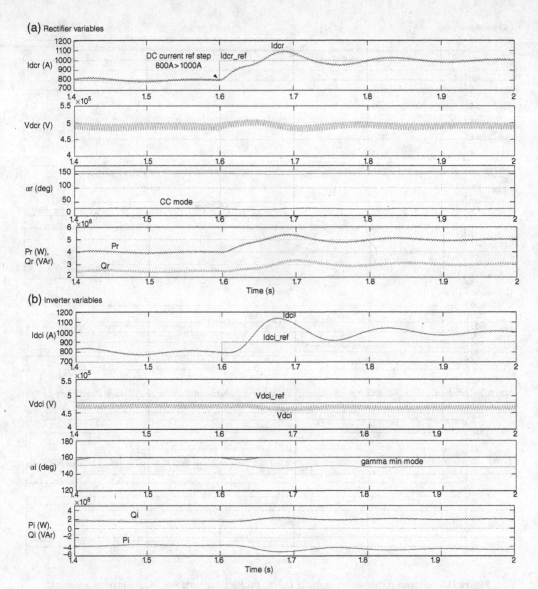

Figure D.6 System responses in Example D.3(b). (a) Rectifier variables and (b) inverter variables.

D.6 Chapter 16 Examples

Example D.6

Simulate a three-phase, three-level diode-clamped multilevel converter, with a circuit as in Figure 16.3, and with $V_{DC} = 1000\,V$, $f = 50\,Hz$, assuming switching frequency is $fs = 1.35\,kHz$.

a. *Assuming a resistive load of $R = 50\,\Omega$, determine the AC current of phase a and comment on the waveform.*

b. *Connect an inductor of 100 mH in series to the AC load and comment on the current waveform. Determine the output maximum phase voltage value.*

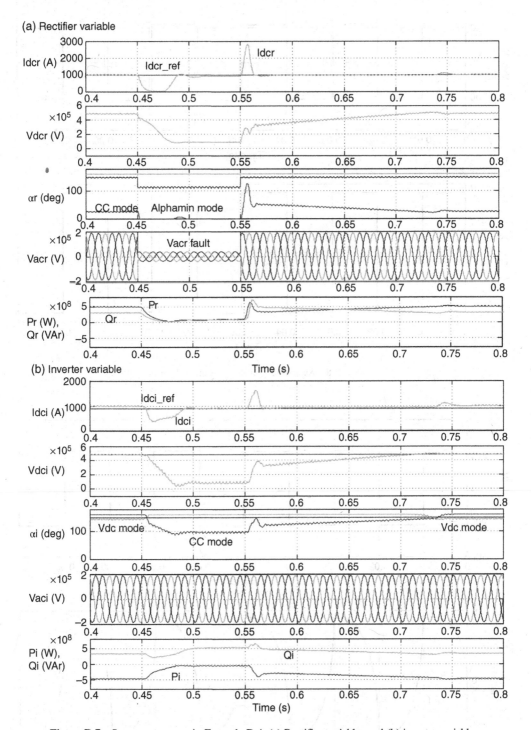

Figure D.7 System responses in Example D.4. (a) Rectifier variables and (b) inverter variables.

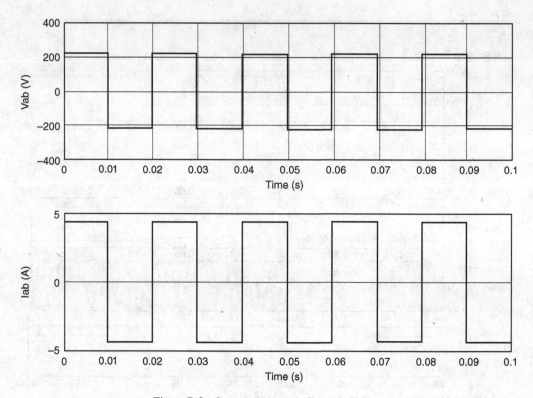

Figure D.8 System responses in Example D.5(a).

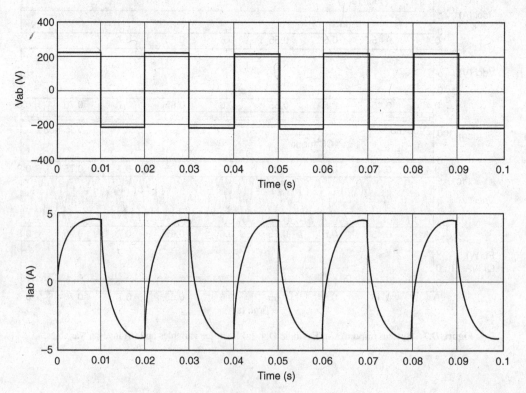

Figure D.9 System response with inductor in Example D.5(b).

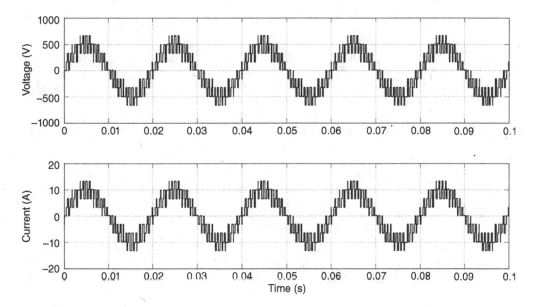

Figure D.10   System responses in Example D.6(a).

Solution

a. With resistive load, the voltage and current will be in phase, as can be seen in Figure D.10.
b. In the case of an inductive load, as seen in Figure D.11, the load acts as a low pass filter and the current is closer to ideal sinusoidal wave. The load-phase current is lagging the phase voltage by the load angle. The maximum phase fundamental voltage is 500 V.

Example D.7
Simulate a three-phase, nine-level half bridge modular converter, with a circuit as in Figure 16.7 and with $V_{DC} = 1000$ V, $f = 50$ Hz, presuming square wave modulation.

a. Assuming a resistive load of $R = 50 \, \Omega$, determine the AC current of phase A and comment on the waveform.
b. Connect an inductor of 100 mH in series to the AC load and comment on the current waveform. Determine the output maximum phase voltage value.

Solution

a. It can be seen in Figure D.12 that voltage and current are in phase.
b. In the case of an inductive load, as shown in Figure D.13, the load acts as a low pass filter and the current is closer to an ideal sinusoidal wave. The load current lags voltage by the load angle. The maximum phase fundamental voltage is 500 V.

D.7   Chapter 17 Examples

Example D.8
A symmetrical monopolar 500 MVA, 100 MVAr, 320 kV HVDC system is shown in Figure D.14. Converter station 1 is controlling the active power between the two AC grids, while converter station 2

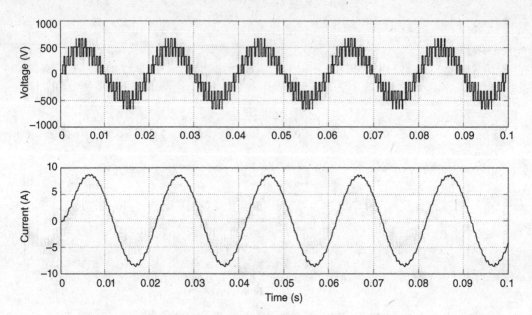

Figure D.11   System responses in Example D.6(b).

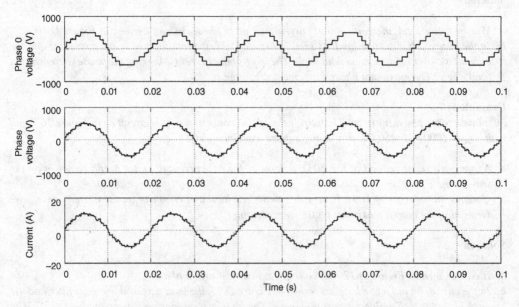

Figure D.12   System responses in Example D.7(a).

controller is regulating the DC voltage. Q channel at controller 2 regulates the AC voltage at PCC2 to support the AC grid. At station 1, the reactive current is controlled to zero. Sinusoidal pulse-width modulation (SPWM) switching is used with a triangular carrier wave with a frequency of 1.35 kHz. The copper DC cable is selected with a total distance of 100 km. The cable parameters are R = 0.0095 Ω/km, L = 2.11 mH/km, and C = 0.2 μF/km.

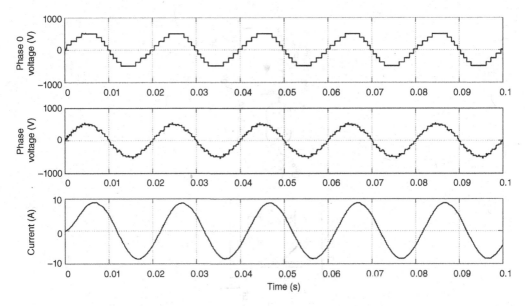

Figure D.13 System responses in Example D.7(b).

a. *Develop a SIMULINK model for this system and ensure that the system operates in stable manner for typical operating conditions.*

b. *Determine (by simulation or otherwise) the transformer ratios. It is desired that, at full power in both power transfer directions, the converters operate at around 0.9 modulation index.*

c. *Test transient responses for full-power reversal.*

d. *Assuming that the system is working at 50% power, test transient responses for a step to 80% power.*

e. *Apply a line-line DC fault at DC bus 2 and determine the DC and AC fault current at station 1. Compare the results with analytical values.*

f. *Run the system at full power but with station 1 as an inverter. Apply a solid three-phase fault at AC grid 1 and determine the peak overcurrent in VSC1.*

Solution

1. *The model responses are shown in Figure D.15.*
2. *Assume that the power flow is from station 1 to 2 at the rated value:*

Station 2: inverter mode

$$V_{dc2} = 320 \text{ kV}$$

$$V_{c2} = 0.9 \frac{V_{dc2}}{2\sqrt{2}} = 101.82 \text{ kV, } \textit{with 0.9 modulation index}$$

$$V_{c2\_LL} = \sqrt{3}V_{c2} = 176.36 \text{ kV,}$$

$$X_{tr2} = X_{tr2pu} \frac{V_{gll2}^2}{S_{t2}} = 0.1 \frac{230^2}{500} = 10.58 \text{ } \Omega$$

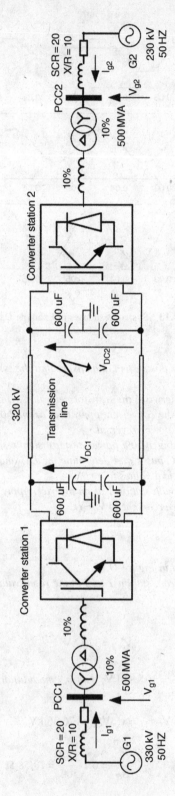

Figure D.14 Voltage source converter (VSC) HVDC system in Example D.8.

$$X_{r2} = X_{r2pu} \frac{V_{gll2}^2}{S_{t2}} = 0.1 \frac{230^2}{500} = 10.58\Omega$$

$$P_2 = \sqrt{S_2^2 - Q_2^2} = \sqrt{(500)^2 - (100)^2} = 489.9 \text{ MW}$$

$$SCR_2 = \frac{V_{sll2}^2}{z_{s2}P_{dc}} \Rightarrow z_{g2} = \frac{V_{sll2}^2}{SCR_2 \times P_{dc}} = \frac{230^2}{20 \times 489.9} = 5.4\Omega$$

$$R_{s2} = \frac{z_{s2}}{\sqrt{1+(X/R)^2}} = \frac{5.4}{\sqrt{1+(10)^2}} = 0.537\Omega, \quad X_{s2} = R_{s2} \times X/R = 0.537\Omega \times 10 = 5.37\Omega$$

Total impedance on the grid side:

$$z_{s2} = R_{s2} + j(X_{s2} + X_{tr2} + X_{r2})$$

$$z_{s2} = 0.537 + j(5.37 + 10.58 + 10.58) = 0.537 + j26.53\Omega$$

Grid current:

$$I_{g2} = \frac{S}{\sqrt{3} \times V_{s2\_ll}} = \frac{500 \text{ MVA}}{\sqrt{3} \times 230 \text{ kV}} = 1.255 \text{ kA}$$

Estimated point of common coupling (PCC) voltage:

$$V_{g2\_es} = V_{s2} + Z_{s2}I_{g2}$$

$$= 230/\sqrt{3} + 0.537 \times 1255 + j26.53 \times 1255$$

$$= 132.8 \text{ kV} + 0.674 + j33.295 \text{ kV} = 137.56 \text{ kV}$$

$$V_{g2\_est} = 238.26 \text{ kV}$$

Transformer turns ratio: 176.36/137.56 = 1.28 (transformer turns ratio with phase values)
  Winding voltages: 175/230
  Station 1: rectifier mode:

$$I_{dc1} = \frac{P_{dc}}{V_{dc2}} = 1.531 \text{ } kA$$

$$V_{dc1} = V_{dc2} + R_{dc}I_{dc} = 320\,000 + 2 \times 1531 \times 0.95 = 322.91 \text{ } kV$$

$$V_{c1} = 0.9\frac{V_{dc1}}{2\sqrt{2}} = 102.75 \text{ } kV, \text{ with 0.9 modulation index}$$

$$V_{c1\_ll} = \sqrt{3}V_{c1} = 178 \text{ } kV,$$

$$X_{tr1} = X_{tr1pu}\frac{V_{gll1}^2}{S_{t1}} = 0.1\frac{330^2}{500} = 21.78\Omega$$

$$X_{r1} = X_{r1pu} \frac{V_{gll1}^2}{S_{t1}} = 0.1 \frac{330^2}{500} = 21.78\Omega$$

$$SCR_1 = \frac{V_{sll1}^2}{z_{s1}P_{dc}} \Rightarrow z_{s1} = \frac{V_{sll1}^2}{SCR_1 \times P_{dc}} = \frac{330^2}{20 \times 489.9} = 11.11\Omega$$

$$R_{s1} = \frac{z_{s1}}{\sqrt{1+(X/R)^2}} = \frac{11.11}{\sqrt{1+(10)^2}} = 1.1\Omega, \quad X_{s1} = R_{s1} \times X/R = 1.1\Omega \times 10 = 11\Omega$$

Total impedance on the grid side:

$$z_{s1} = R_{s1} + j(X_{s1} + X_{tr1} + X_{r1})$$

$$z_{s1} = 1.1 + j(11 + 21.78 + 21.78) = 54.571\Omega\angle 88.7°$$

$$I_{g1} = \frac{S_1}{\sqrt{3} \times V_{s1\_ll}} = \frac{500 \text{ MVA}}{\sqrt{3} \times 330 \text{ kV}} = 874.7 A$$

Estimated PCC voltage:

$$V_{g1\_est} = V_{s1} - jz_{s1}I_{g1} = 330/\sqrt{3} - 1.1 \times 874.7 - j54.56 \times 874.7$$

$$= 190.5 \text{ kV} - 0.962 \text{ kV} - j47.7 \text{ kV}$$

$$= 189.5 \text{ kV} - j47.7 \text{ kV} = 195.44 \text{ kV}$$

$$V_{g1\_est} = 338.52 \text{ kV}$$

Transformer turns ratio: 178/195.44 = 0.91 (transformer turns ratio with phase values)
  Winding voltages: 175/330.

3. Figure D.15 shows the simulation of full power reversal.
4. Figure D.16 shows the simulation of power step from 50% to 80%.
5. DC fault
     Grid-side fault current

$$I_{gf} = \frac{V_{s1\_ll}/\sqrt{3}}{z_{s1}} = \frac{330000/\sqrt{3}}{54.571} = 3491.33 \text{ A}$$

Converter-side fault current

$$I_{gfc} = I_{gf} \frac{V_g}{V_c} = 3491.33 \times \frac{330}{175} = 6583.6 \text{ A}$$

$$I_{dcf} = I_{gcf}\sqrt{2}\frac{3}{\pi} = 8863.7 \text{ A}$$

It is seen that the analytical fault-current magnitude is in good agreement with the simulations in Figure D.17
6. Observe peak currents on Id and Iq immediately after the fault in Figure D.18. By calculation $I = \sqrt{(Id^2 + Iq^2)}$ it can be concluded that peak magnitude is around 1.8, which is acceptable.

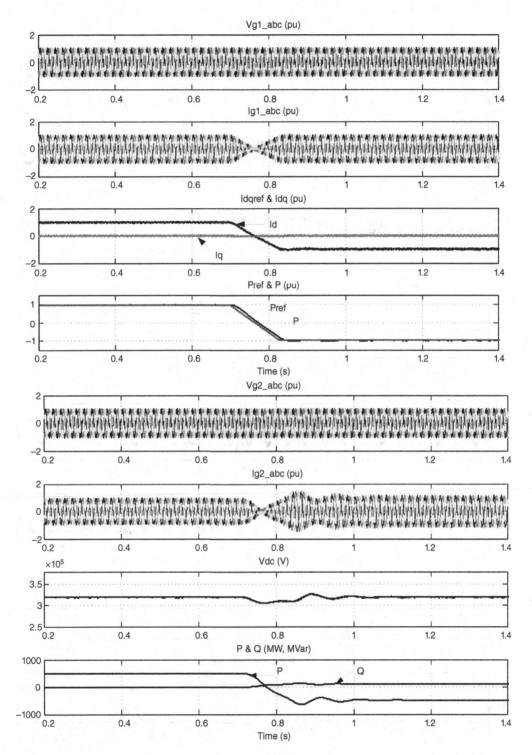

Figure D.15 Full power reversal in Example D.8 (top graph station 1 and lower graph station 2).

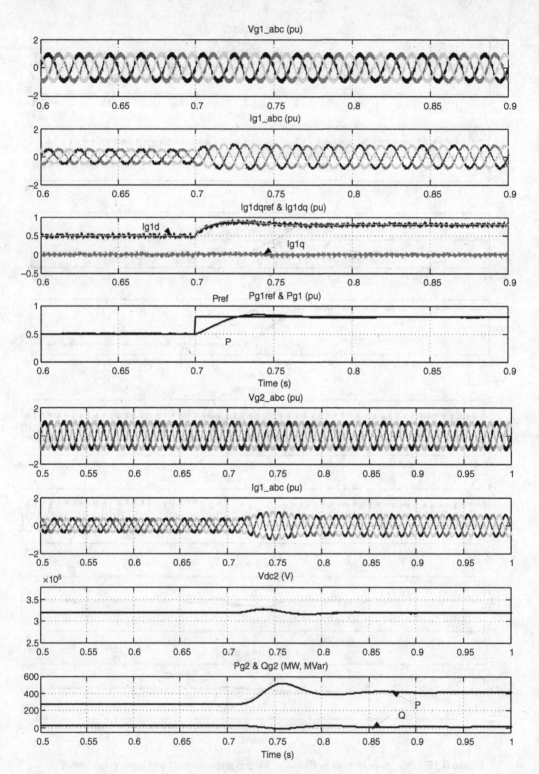

Figure D.16 Power step (0.5pu to 0.8pu) in Example D.8 (top graph station 1 and lower graph station 2).

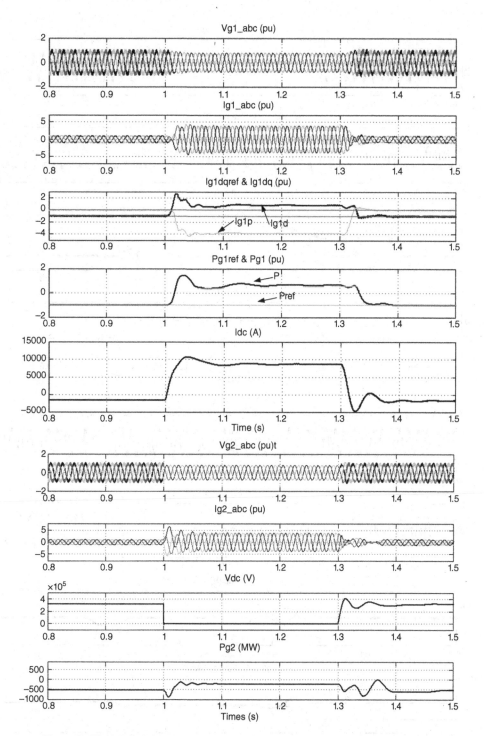

Figure D.17   DC fault in Example D.8 (top graph station 1 and lower graph station 2).

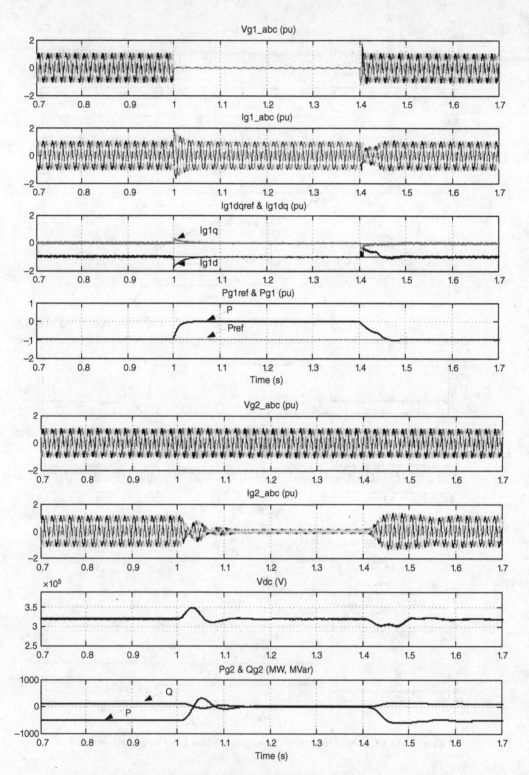

Figure D.18 AC fault in Example D.8 (top graph station 1 and lower graph station 2).

Index

High-Voltage Direct-Current Transmission: Converters, Systems and DC Grids, First Edition.
Dragan Jovcic and Khaled Ahmed.
© 2015 John Wiley & Sons, Ltd. Published 2015 by John Wiley & Sons, Ltd.

Printed in the United States
By Bookmasters